Taxonomic Tapestries

The Threads of Evolutionary, Behavioural and Conservation Research

Taxonomic Tapestries

The Threads of Evolutionary, Behavioural and Conservation Research

Edited by Alison M Behie and Marc F Oxenham

Chapters written in honour of Professor Colin P Groves

Australian National University

PRESS

ANU PRESS

Published by ANU Press
The Australian National University
Acton ACT 2601, Australia
Email: anupress@anu.edu.au
This title is also available online at http://press.anu.edu.au

National Library of Australia Cataloguing-in-Publication entry

Title:	Taxonomic tapestries : the threads of evolutionary, behavioural and conservation research / Alison M Behie and Marc F Oxenham, editors.
ISBN:	9781925022360 (paperback) 9781925022377 (ebook)
Subjects:	Biology--Classification. Biology--Philosophy. Human ecology--Research. Coexistence of species--Research. Evolution (Biology)--Research. Taxonomists.
Other Creators/Contributors:	Behie, Alison M., editor. Oxenham, Marc F., editor.
Dewey Number:	578.012

Cover design and layout by ANU Press

Cover photograph courtesy of Hajarimanitra Rambeloarivony

Contents

PART I

PART II

PART III

List of Contributors

Peter Andrews Natural History Museum, Cromwell Road, London, UK SW7 5BD

Debbie Argue School of Archaeology and Anthropology, The Australian National University, Canberra, Australia, 2601

Robert Attenborough School of Archaeology and Anthropology, The Australian National University, Canberra, Australia, 2601; Division of Biological Anthropology, Department of Archaeology & Anthropology, University of Cambridge, Pembroke Street, Cambridge, UK, CB2 3DZ

Alison M Behie School of Archaeology and Anthropology, The Australian National University, Canberra, Australia, 2601

David Bulbeck Department of Archaeology and Natural History, School of Culture, History and Language, College of Asia and the Pacific, The Australian National University, Canberra, Australia, 2601

Chris Clarkson Department of Archaeology, School of Social Sciences, The University of Queensland, Brisbane, St Lucia, QLD, Australia, 4072

Juliet Clutton-Brock Natural History Museum, Cromwell Road, London, UK, SW7 5BD

Arthur Durband Department of Sociology, Anthropology and Social Work, Kansas State University, 204 Waters Hall, Manhattan, KS, 66502

Natasha Fijn School of Archaeology and Anthropology, The Australian National University, Canberra, Australia, 2601

Peter Hiscock Department of Archaeology, School of Philosophical and Historical Inquiry | Faculty of Arts and Social Sciences, The University of Sydney, Sydney, Australia, 2006

Richard J Johnson University of Colorado, Denver Aurora, USA, CO 80045

David Lambert Griffith School of Environment, Nathan campus, Griffith University, 170 Kessels Road Nathan, QLD, Australia, 4111

Angela Meder Berggorilla & Regenwald Direkthilfe c/o Rolf Brunner Lerchenstr. 545473 Muelheim, Germany

Erik Meijaard People and Nature Consulting International, Country Woods house no. 306, JL WR Supratman, Pondok-Ranji-Rengas, Ciputat, Jakarta 15412, Indonesia

Guy G Musser Division of Vertebrate Zoology (Mammalogy), American Museum of Natural History, New York, NY 10024

John Oates Department of Anthropology, Hunter College, City University of New York, 695 Park Avenue, New York, NY 10065

Marc F Oxenham School of Archaeology and Anthropology, The Australian National University, Canberra, Australia, 2601

Mary SM Pavelka Department of Anthropology, The University of Calgary, 2500 University Drive NW, Calgary, Alberta, T2N 1N4

Benjamin Rawson Vietnam Programme, Fauna & Flora International, 340, Nghi Tam, Hanoi, Vietnam

Kees Rookmaaker Department of Biological Sciences, National University of Singapore, 14 Science Drive 4 Singapore 117543; Rhino Resource Centre

Travis S Steffens Department of Anthropology, University of Toronto, 19 Russell Street, Toronto, Ontario, M5S 2S2

Nelson Ting 308 Condon Hall, Department of Anthropology, 1218 University of Oregon, Eugene, OR 97403

Ulrich Welsch Ludwig Maximilians University Munich, Institute for Anatomy and Cell Biology Schillerstr. 4280336 München Germany

Michael C Westaway Griffith School of Environment, Nathan campus, Griffith University, 170 Kessels Road, Nathan, QLD, Australia, 4111

Tracy M Wyman Department of Anthropology, The University of Calgary, 2500 University Drive NW, Calgary, Alberta, T2N 1N4

List of Figures and Tables

Figures

Tables

PART I

1. The Groves effect: 50 years of influence on behaviour, evolution and conservation research

Alison M Behie and Marc F Oxenham

This volume explores the complexity, diversity and interwoven nature of taxonomic pursuits primarily within the context of explorations of humans and related species, although it also delves into more distantly related species to show how taxonomy has impacted fields outside of human research. Essentially we are interested in showcasing recent research into that somewhat unique species we call humankind through the theoretical and conceptual approaches afforded by the discipline of biological anthropology. Structurally, our approach to understanding human uniqueness is tripartite in focusing on: (1) the evolution of the human species, (2) the behaviour of primates and other species, and (3) how humans affect the distribution and abundance of other species through anthropogenic impact. In this manner we weave together these three key areas of bio-anthropological endeavour and scrutinise how changes in taxonomic theory and methodology, including our fluctuating understanding of speciation, have recrafted the way in which we view animal behaviour, human evolution and conservation studies.

Taxonomy forms perhaps the most fundamental structural principle of arguably all biological knowledge and research. Indeed, taxonomy is the epistemological cornerstone of the biological sciences. In this context it is somewhat astonishing to note that within the last 20 years significant gaps in taxonomic knowledge have appeared, ostensibly due to a dearth of adequately trained taxonomists in the current generation of scholars. This lacuna, referred to as the taxonomic impediment, is in our view exacerbated by a recent over reliance on 'geno-hype' (Holtzmann, 1999), which refers to our scientific love affair with genetic-based approaches, at the expense of traditional taxonomic principles. Taxonomy, however, is more than its constituent parts, with DNA but one piece of the taxonomic fabric.

While the invention of improved and non-invasive ways to collect DNA has resulted in its resurgence in the field of taxonomy, this reliance on genetics to define species is nothing new. The Biological Species Concept defines species as 'groups of actually or potentially interbreeding natural populations, which are reproductively isolated from other such groups' (Mayr, 1940: 256). This concept, which relies heavily on interbreeding, or lack thereof, to define species, has been a mainstream way to view and define species for decades. Despite its common acceptance, there are controversies surrounding it, many of which were brought to the forefront by Professor Colin Groves. One major criticism is that this concept can only be applied to species with overlapping ranges, meaning we cannot differentiate species that do not have the opportunity to attempt breeding due to non-overlapping ranges. In addition, our ability to analyse the genetics of many wild populations has allowed us to see that in some groups hybrid animals may actually be more numerous than the parent species, suggesting that reproductive isolation may not in fact separate species (Groves, 2012).

Up to 24 species concepts have been proposed over the years, but in the interest of brevity, just a few major players will be described here. The Ecological Species Concept proposes that species are groups of animals 'evolving separately from others with its own evolutionary role and tendencies' (Simpson, 1963: 153), which by anyone's definition is a vague concept that would be nearly impossible to apply in practice. The Genetic Species Concept defines species based on the amount of genetic variation both within and between species, assuming more variation between than within species. While simple and logical, it is nearly impossible to determine a cut off point for determining too much (multiple species) or too little (single species) genetic variation. The Phylogenetic Species Concept provides a more encompassing view of species in that they are defined by their possession of unique features and/or characteristics, either primitive or derived, that separate them from other groups (Nixon and Wheeler, 1990). It does not ignore genetics, but does not require different species be genetically incompatible, while also taking into account other features such as morphological similarities (Kimbel and Martin, 1993). It is this concept that has been strongly supported by Colin Groves throughout his career.

To date, debate over the definition of species continues, creating uncertainty for any researcher using species as their basic taxonomic unit of study. While this is an important issue, no volume has yet attempted to examine how changing views on the nature and processes of taxonomy have shaped modern research agendas and interpretations in the key areas of biological anthropology. It is, however, an issue with profound implications and one may ask: Does it matter if different scientists take different approaches to defining species? Will our understanding of evolution and behaviour, or the way in which we attempt to conserve species, really be impacted by the taxonomic classification system we choose to use? Important questions such as these are what this volume explores by using new data to interpret how a fluid and ever changing understanding of taxonomy has led to diverse and disparate research outcomes. The chapters in this volume will explore the significant impacts that these changes have had by investigating both historical perspectives relating to taxonomy and employing new data to uncover how the taxonomic impediment has influenced research processes and outcomes, with particular emphasis on humans and our close primate relatives.

Colin Groves

As one of the world's leading taxonomists, this volume pays tribute to Colin Groves. Professor of Biological Anthropology at The Australian National University, where he has worked for the past 40 years, Colin's influence has spread through multiple disciplines including, but certainly not limited to, animal morphology, animal behaviour, human evolution and conservation. Colin sees himself as a taxonomist above all else, and his dedication to taxonomic pursuits stems directly from his true love of animals, which can perhaps be traced back to his childhood when his grandfather bought him a book on animals. This fascination grew in his teen years, which would see him heading to the London Natural History Museum and requesting access to their bone collections. When it was time to attend university, it is no surprise Colin wanted to study zoology, but as his father had a preference for him to undertake a degree in linguistics, he settled on anthropology as a compromise between the two. And the rest, as they say, is history.

Immediately following his undergraduate years, Colin entered the PhD program at the University of London under the tutelage of the esteemed John Napier, who was at the time regarded as the leader in primate taxonomy. His doctoral research, completed in 1966, involved a large-scale survey of gorilla skulls. Somewhat serendipitously Colin's work led to a meeting with Dian Fossey, who invited him to Karisoke to see the population of mountain gorillas she is now so famous for studying. Colin's work with gorillas resulted in numerous high impact publications at a very young age, vastly expanding our knowledge of gorilla ecology, biology and morphology. The influence these early years had on Colin can still be seen today in his passion for and involvement in gorilla conservation.

Colin's doctoral research success not only fuelled his future academic pursuits, such as his two-year postdoctoral fellowship at Berkeley and his fixed-term appointment at Cambridge, but also beat a path to his wife of more than 40 years, Phyll. As Colin had studied with John Napier, who worked at the Royal University Hospital at the University of London, it was not surprising that when Colin fell ill in the fall of 1973 he was admitted to that hospital with a request to be put in a private ward due to the fact he was suspected of having tuberculosis, which luckily he did not. He did, however, get placed in a ward where the Ward Sister was one Phyll Dance. She would watch Colin taking the tea cart around, not realising it may have been because he wanted a cuppa himself, and playing chess with the elderly patients. Their connection rapidly grew into a whirlwind romance leading to Phyll packing her bags and moving to Australia with Colin just a short four months later when, in January 1973, Colin took up a position at The Australian National University. This is where he has been ever since, supervising scores of Honours, Masters, Doctoral and other graduate students in addition to collaborating with myriad scientists from around the globe, resulting in close to 200 peer-reviewed publications. Lest we forget the opportunities to travel: it also gave him the opportunity to continue to conduct field work expeditions in places such as Tanzania, Indonesia, Rwanda, Kenya and Iran.

Colin's field work adventures have included trips to museums and national parks around the globe, allowing him to study hundreds of species of animal, including every species of rhinoceros, which is still one of his greatest personal achievements. This love of the rhinoceros may have its genesis in the work he did as an undergraduate, which resulted in his first publication, entitled 'On the Rhinoceroses of South-East Asia'. This connection to rhinoceroses remains strong, which would be apparent to anyone who entered the Groves' home and noticed the vast collection of rhinoceros paraphernalia crowding the shelves and walls. When Phyll first met Colin and realised she may too have to participate in these trips, she asked him 'Do you get chased by wild animals?' to which Colin replied 'No. Animals don't chase people.' An answer he may now regret considering they have been chased by lions, a herd of banteng, as well as by a rhinoceros in Ujung Kulon. We might add that this was a rhinoceros they had been cautioned not to get too close to, but Colin with his ever inquisitive nature began to follow, not realising it would soon be following them.

Despite these close calls, Colin has always been a true conservationist. In this respect he was well ahead of his time in understanding the impact of humans on the environment. Phyll can recall when she first took Colin to her home town shortly after they met, he refused a plastic bag from a shop keeper. Although she was embarrassed by this at the time, it just goes to show that he has always been ahead of the game in his devotion to conservation – a devotion that quickly becomes apparent to anyone who has ever heard Colin speak about the plight of wild animals and the need to conserve them. He is avidly involved with conservation organisations, and right here in Canberra he often speaks at fundraising events where he is never shy about voicing his opinion on controversial topics such as the boycotting of products containing palm oil due to the impact palm oil plantations have on Asian wildlife. As taxonomy is the basis for conservation and as 'threats to the natural world and its biodiversity are ubiquitous and accelerating, it affects conservation strategies'. After all, how can we know what to conserve if we don't know what species are out there? Links such as these, tying taxonomy to practical and real research outputs, are what have set Colin apart.

Colin has also devoted much of his career to refining mammalian taxonomy, which has resulted in his naming more than 40 taxa, including species of rats, civets, possums and, most famously, the human ancestor, *Homo ergaster*, which was undertaken with colleague Vratislav Mazák in 1975. This is likely the thing he is most remembered for, but his contributions to the field of human evolution certainly do not end there. As a renowned skeptic and a 30-year member of the Australian Skeptics, Colin is always pushing the boundaries and looking for new ways to understand the world and challenge those around him. This has led to him arguably making more contributions to mammalian taxonomy than any other modern scientist, including the addition of his two influential books *Primate Taxonomy* and *Ungulate Taxonomy*, which are now used as landmark taxonomy guides. More recently, Colin has also been a major player in the recent debate surrounding *Homo floresiensis* and its relationship to modern humans. Never shy to turn away from a debate, he gets great pleasure in educating young people about many things including the truth behind creationism, which is one of the only things he seems to take offence to.

Colin has undoubtedly influenced, directly or indirectly, thousands of people far and wide, which embodies what he truly is: a teacher. Having students in his lab or speaking in schools and getting young people excited about behaviour, morphology, evolution or conservation, or any other topic of interest, makes Colin happy. He is an educator on every level, whether working as a supervisor, mentor or even simply by taking the time to write numerous letters to newspapers to discuss topics of interest. There is no doubt his many undergraduate students enjoy hearing him lecture, his graduate students appreciate his support, supervision and insight, and the staff he has mentored welcome his helpful nature and encyclopaedic knowledge of the discipline.

Despite everything Colin has obviously achieved, what might make him the most endearing to those who know him is his humility. Just recently he was made an honorary member of the American Society of Mammalogists, something which has been bestowed on less than 100 people. He was also the 2014 recipient of an award from the Margot Marsh Biodiveristy Fund for Excellence in Contribution to Primate Conservation. Both awards were a surprise, and to hear Colin tell the story, he didn't even realise that the great person being described as they began the ceremonies was himself. This is not the first time that we have heard Colin say such things. His gentle spirit and true passion for what he does leave no room for ego or arrogance, and all of us who have had the pleasure to work with or even briefly chat with Colin Groves are the better for it.

Volume structure

This volume is broken down into four main sections, prefaced by this introduction as Part I. Part II, Chapters 2 through 7, explores the influence of changing taxonomic and speciation mechanisms on studies of behaviour and morphology. Chapter 2 describes a new species of murid rodent endemic to the island of Sulawesi, *Lenomys grovesi*, a genus that up until now has been thought to be monotypic. Chapters 3 to 5 focus on primatology, with Chapter 3 providing new insights into the evolution of gibbons through consideration of the morphology of their last common ancestor. Chapter 4 considers how natural disasters may play a role in the speciation of New World monkeys, while Chapter 5 summarises, for the first time, the influence of the primatologist Adolf Remane on primatological studies. Chapter 6 turns to the hominid lineage by using new ways of thinking about lithic technology to weigh in on current debates about hominid and Neanderthal cognition. Finally, Chapter 7 explores how changes to the taxonomy of anopheline (malarial) mosquitoes have contributed to our understanding of human malaria transmission.

Part III shifts focus to studies of evolution, starting with Chapter 8, which revisits the contributions of Lamarck's ideas on the nature of species, re-evaluating his contributions to the field of biological evolution. Chapter 9 investigates the changing nature of taxonomy from Aristotle to the current day, with a focus on what this has meant to studies of domestication. The remainder of this section focuses on human evolution, starting with Chapter 10, followed by reviews of how relationships between diet, farming and cooking techniques with reductions in tooth size and corresponding cranial changes can affect evolution in the Holocene human skeletal record. Chapter 11 uses cladistic analyses to test existing hypotheses regarding the phylogenetic relationships of Ceprano, Daka, Kabwe and Bodo in the Early and Middle Pleistocene. Chapter 12 focuses on important new insights that studies of ancient DNA (aDNA) can contribute to our rewriting of the human evolutionary narrative in Sunda and Sahul.

Part IV moves from studying species themselves, to studies that consider how to conserve them. Chapter 13 describes how Mongolians have cooperated with western conservation organisations to enable the takhi to be released back on the Mongolian steppe from a captive existence in zoos and reserves. Chapter 14 describes the history of systematic research of the small group of currently existing species of rhinoceros to better understand rhinoceros conservation needs and plans. Chapters 15 through 17 consider primate conservation. Chapter 15 examines how the ever changing taxonomy of red colobus monkeys has impacted conservation efforts in Africa, while Chapter 16 evaluates the claim that use of the phylogenetic species concept undermines conservation efforts by focusing on two groups of Southeast Asian mammals, pigs (Suidae) and gibbons (Hylobatidae). Chapter 17 tries to make sense of the complexity of gorilla conservation in light of ever changing species designations as well as threats. Finally, Chapter 18 synthesises our taxonomic tapestry.

References

Groves CP. 2012. Species concepts in primates. *Am J Primatol* 74:687–691.

Holtzman N. 1996. Are Genetic Tests Adequately Regulated? *Science* 286:409.

Kimbel WH, Martin LB, editors. 1993. *Species, species concepts, and primate evolution*. New York: Plenum Press.

Mayr E. 1940. Speciation Phenomena in Birds. *The American Naturalist* 74:249–278.

Nixon KC, Wheeler QD. 1990. An amplification of the phylogenetic species concept. *Cladistics* 6:211–223.

Simpson GG. 1961. *Principles of animal taxonomy*. Columbia University Press. p. 247.

PART II

2. Characterisation of the endemic Sulawesi *Lenomys meyeri* (Muridae, Murinae) and the description of a new species of *Lenomys*

Guy G Musser

Introduction

In 1969, DJ Mulvaney sent me a batch of subfossils excavated by him and his colleagues from caves and rock-shelters in the Makassar region of Sulawesi (Mulvaney and Soejono, 1970). Among them are samples from a cave (Leang Burung 1) containing examples of *Lenomys meyeri*, a large-bodied rat endemic to Sulawesi, and a single right dentary with an intact molar row that is from a smaller-bodied *Lenomys*, which represents a new species. The small *Lenomys* along with specimens of *L. meyeri* were excavated from a level with a radiocarbon date of 2820 ± 210 BP, and other examples of *L. meyeri* were found in an underlying stratum dated at 3420 ± 400 BP (Mulvaney and Soejono, 1970: 171).

Here I name and describe the new *Lenomys* within the context of first characterising *L. meyeri* by briefly summarising external, cranial, dental, and other traits, sketching its geographic and elevational distributions based upon the small available samples, looking into geographic variation of cranial and dental morphometrics, and providing notes on ecology. The phylogenetic relationships of *Lenomys* to other murids will not be explored here.

Materials and methods

Specimens examined (mostly museum study skins and their associated skulls) are stored in the following institutions: the American Museum of Natural History, New York (AMNH); Natural History Museum (formerly British Museum of Natural History), London (BMNH); Museum Zoologicum Bogoriense, Cibinong, Java (MZB; now the Indonesian National Museum of Natural History); Naturhistorisches Museum Basel, Switzerland (NMB); Nationaal Museum of

Natural History Naturalis (formerly the Rijksmuseum van Natuurlijke Historie), Leiden (RMNH); and National Museum of Natural History, Smithsonian Institution, Washington, DC (USNM).

External measurements were either taken by me or transcribed from collector's notations on skin labels: total length; length of tail (LT); length of hind foot, including claw (LHF); length of ear, from notch to crown (LE); and body weight or mass (W). Length of head and body (LHB) was derived by subtracting tail length from total length.

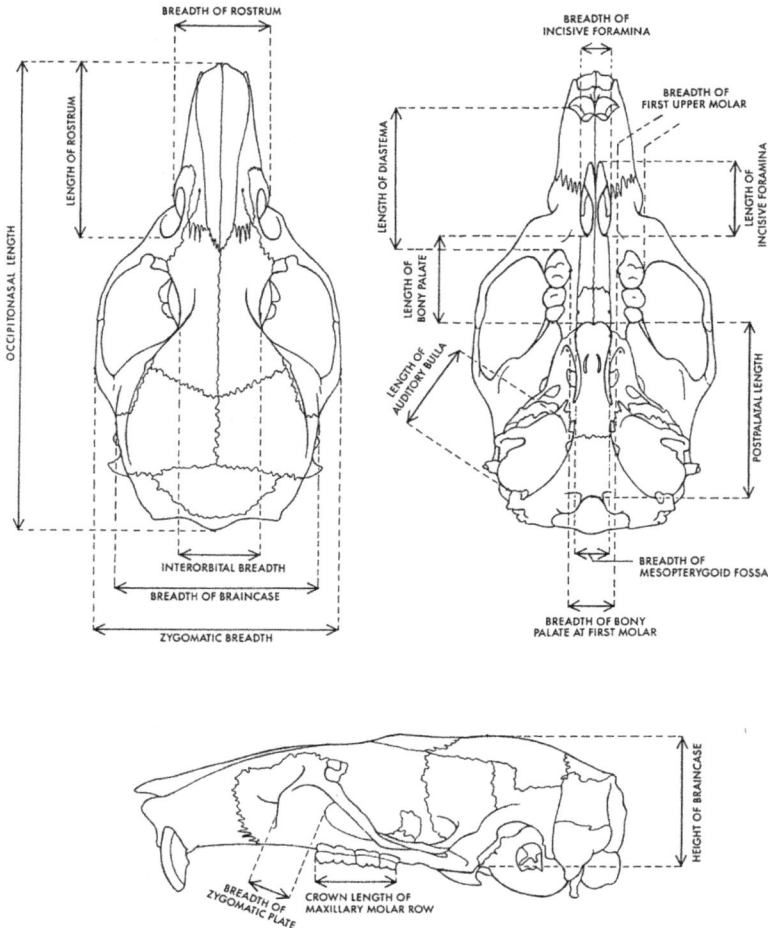

Figure 2.1: An adult *Bunomys chrysocomus* skull illustrating limits of cranial and dental measurements employed. See text for additional definitions.

Source: Drawing by Patricia Wynne.

Using dial calipers graduated to tenths of a millimetre, I measured the following cranial and dental dimensions (illustrated in Figure 2.1):

ONL = occipitonasal length (greatest length of skull; distance from tip of nasals to posterior margin of occiput)

ZB = zygomatic breadth (greatest breadth across zygomatic arches)

IB = interorbital breadth (least distance across the frontal bones between the orbital fossae)

LR = length of rostrum (from tip of nasal bones to posterior margin of zygomatic notch)

BR = breadth of rostrum (greatest breadth across rostrum, including bony nasolacrimal capsules)

BBC = breadth of braincase (measured from just above the squamosal root of each zygomatic arch)

HBC = height of braincase (from top of braincase to ventral surface of basisphenoid)

BZP = breadth of zygomatic plate (distance between anterior and posterior edges of zygomatic plate)

LD = length of diastema (distance from posterior alveolar margins of upper incisors to anterior alveolar margin of M1)

PPL = postpalatal length (distance from posterior margin of palatal bridge to posterior edge of basioccipital – ventral lip of foramen magnum)

LBP = length of bony palate (distance from posterior edge of incisive foramina to posterior margin of bony palate)

BBP = breadth of bony palate at M1 (least distance between lingual alveolar margins of first molars)

LIF = length of incisive foramina (distance from anterior to posterior margins)

BIF = breadth across incisive foramina (greatest distance across both foramina)

BMF = breadth of mesopterygoid fossa (distance from one edge of mesopterygoid fossa to the other)

LB = length of ectotympanic (auditory) bulla (greatest length of bullar capsule, excluding the bony eustachian tube)

15

CLM1-3 = crown length of maxillary molar row (from anterior enamel face of M1 to posterior enamel face of M3)

alm1-3 = alveolar length of mandibular molar row (from anterior alveolar rim of m1 to posterior alveolar rim of m3)

clm1-3 = crown length of mandibular molar row (from anterior enamel face of m1 to posterior enamel face of m3)

BM1 = breadth of first maxillary (upper) molar (taken across widest part of molar)

bm1, bm2, bm3 = breadths of first, second and third mandibular (lower) molars (taken across widest part of molar).

M = maxillary (upper) molars

M = mandibular (lower) molars

Cranial and dental measurements were obtained only from adults: animals clothed in adult pelage with occlusal surfaces of molars expressing the range from slight (young adults) to moderate (adults) to well worn where cusp patterns are nearly obliterated (old adults). Sexes were not separated in any of the statistical analyses because of the small number of specimens in each geographic sample. Furthermore, weak sexual dimorphism in cranial and dental variables generally characterises nongeographic sexual variation among muroid rodents (Musser, 2014; Musser and Durden, 2014).

Standard univariate descriptive statistics (mean, standard deviation, and observed range) were calculated for samples containing modern and fossil examples of *L. meyeri* and the new species. Principal-component analyses were computed using original cranial and dental measurements transformed to natural logarithms. Principal components were extracted from a variance-covariance matrix; loadings (correlations) of the variables are given as Pearson product-moment correlation coefficients of the extracted principal components. Probability levels denoting significance of the correlations are unadjusted. The statistical packages in SYSTAT 11 for Windows, Version 11 (2005), were used for all analytical procedures.

Gazetteer and specimens

Figure 2.2a: Collection localities for modern and subfossil samples of *Lenomys meyeri*. Numbers key to localities described in the gazetteer. The map in Figure 2.2b contains collection localities 8 and 9. Locality 12 designates Leang Burung 1, the cave where subfossil dentaries of *L. meyeri* and *L. grovesi* n. sp. were excavated (see gazetteer); the other subfossils come from caves and rock-shelters at locality 14. The Tempe Depression mentioned in text is at about the level of Parepare (locality 11).

Notes: Sulawesi consists of a central region from which four arms or peninsulas radiate: the northern peninsula, which ends in a northeastern jog; the eastern peninsula; the southeastern peninsula; the southwestern peninsula. I use these informal labels when describing the distribution of *Lenomys*, and refer to the central portion as Sulawesi's core. I also use the Indonesian names for mountain (Gunung), stream or river (Sungai), and lake (Danau).

Source: Drawing by Patricia Wynne.

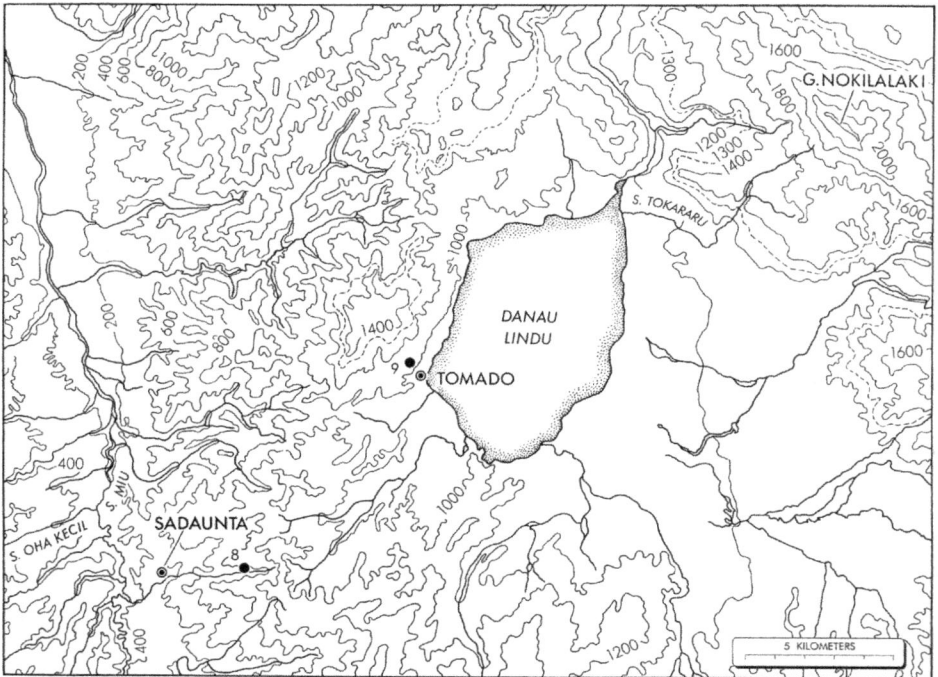

Figure 2.2b: Collection localities for modern and subfossil samples of
***Lenomys meyeri*. Numbers key to localities described in the gazetteer. The**
map in Figure 2.2b contains collection localities 8 and 9; dashed contour
line at 1300 m marks approximate boundary between tropical lowland
evergreen and tropical lower montane rainforests.

Notes: Sulawesi consists of a central region from which four arms or peninsulas radiate: the northern peninsula, which ends in a northeastern jog; the eastern peninsula; the southeastern peninsula; the southwestern peninsula. I use these informal labels when describing the distribution of *Lenomys*, and refer to the central portion as Sulawesi's core. I also use the Indonesian names for mountain (Gunung), stream or river (Sungai), and lake (Danau).

Source: Drawing by Patricia Wynne.

Listed below are the localities at which the 27 modern and 20 subfossil specimens of *Lenomys meyeri* I examined were collected. The number preceding each place keys to the same numbered locality on the maps in Figures 2.2a and 2.2b. Cartographic and gazetteer sources for latitudes, longitudes, and spellings are referenced in Musser and others (2010).

1. *Manado*, 01°30′N, 124°50′E, coastal plain near sea level: RMNH 21233.

2. *Rurukan*, 01°21′N, 124°52′E, 3500 ft (1067 m): BMNH 97.1.2.19.

3. *Gunung Masarang*, 01°19′N, 124°51′E, 3800 ft (1159 m): BMNH 97.1.2.20.

4. *Tomohon*, 01°19'N, 124°49'E, 700–800 m BMNH 99.10.1.9; NMB 1208, 3326, 3327, 1110/4759.

5. *Langoon*, 01°09'N, 124°50'E, 700–800 m: RMNH 18302 (holotype of *Mus meyeri*).

6. *Amurang*, 01°11'N, 124°35'E, coastal plain near sea level: BMNH 21.2.9.4; MZB 384, 5810.

7. *Bumbulan*, 00°29'N, 122°04'E, coastal plain near sea level: AMNH 153011.

8. *Valley of Sungai Miu, Sungai Sadaunta*, 01°23'S, 119°58'E, 3000 ft (915 m): AMNH 226813, 226814.

9. *Valley of Danau Lindu, forest near Tomado* (a village on western shore of Danau Lindu), 01°19'S, 120°03'E, 1000 m: AMNH 224317.

10. *Gimpu*, 01°36'S, 120°02'E, 400 m: USNM 219712 (holotype of *Lenomys longicaudus*).

11. *Parepare*, 04°01'S, 119°38'E, coastal plain near sea level: RMNH 18303. This is the cranium discussed, figured, and identified as *Mus callitrichus* by Jentink (1890: pl. 10, Figures 4–6, Plate 120).

12. *Leang Burung 1*, a cave near the village of Pakalu, 'about 2 km north of the main road from Maros, 10 km to the west' (Mulvaney and Soejono, 1970: 169), less than 100 m: AMNH 265022-265028 (subfossils). The dentary of *Lenomys grovesi* n. sp. was also collected here (Table 2.8).

13. *Wawokaraeng, Gunung Lompobatang*, an extinct volcanic mountain near the tip of the southwestern peninsula, 05°20'S, 119°55'E: *2200 m*, AMNH 101125, 101127, 101128 (holotype of *Lenomys meyeri lampo*), 101129; *2300 m*, AMNH 101126; *2400 m*, ANMH 101124.

14. *Panganreang Tudea*, a rock shelter at the southern tip of the southwestern peninsula (see description and map in Mulvaney and Soejono, 1970: 166): subfossil Specimen Numbers 2–10 documented by Musser (1984). *Batu Ejaja*, a cave 'a few hundred meters distant' from Panganreang Tudea (Mulvaney and Soejono, 1970: 166), 275 m: subfossil Specimen Number 1 documented by Musser (1984). *Batu Edaja 2*, a rock shelter near Batu Edjaja (see description Mulvaney and Soejono, 1970): AMNH 265029-265031 (subfossils).

The following specimens were examined but their provenances were not mapped:

A. Minahassa, Warumbungan: RMNH 2797.

B. RMNH 18304; a skull listed as specimen 'c' under '*Mus giganteus*' in Jentink's 'Catalogue Ostéologique des Mammifères', published in 1887 (p. 210); 'Java' is listed as provenance, but the skull is *Lenomys meyeri*.

Lenomys (Thomas, 1898)

The mounted skin and partial skull shown in Figures 2.3 and 2.4 are the basis for Jentink's (1879: 12–13) short and undiagnostic description of *Mus meyeri*. Later, Thomas (1898: 409) recognised the distinctive attributes of *meyeri*, proposed the genus *Lenomys* to contain it, and provided a diagnosis that applies to the genus and to *L. meyeri*. The characterisation of *L. meyeri* follows.

Lenomys meyeri (Jentink, 1879)

Mus meyeri Jentink, 1879: 12.

Lenomys longicaudus Miller and Hollister, 1921b: 5.

Lenomys meyeri lampo Tate and Archbold, 1935b: 5.

Holotype and type locality

The holotype is RMNH 18302; the skin and skull of an adult (sex not determined) obtained in September 1875 by SCIW van Musschenbroek, and is listed as specimen 'a' in Jentink's (1888: 65) 'Catalogue Systématique des Mammifères'. External, cranial, and dental measurements, along with other relevant data, are listed in Table 2.2.

The skin is mounted in a pose meant to simulate the living rat (Figure 2.3); the fur is discoloured, a straw-brown. The cranium is incomplete: pieces of nasals and right zygomatic arch, and ventral and occipital portions of the braincase are missing (Figure 2.4). Except for worn angular processes, the mandible is intact. All incisors and molars are present and undamaged.

The type locality is Langoon (01°09'N, 124°50'E), 700–800 m (locality 4 in gazetteer and Figure 2.2a), northeastern tip of the northern peninsula, Propinsi Sulawesi Utara, Indonesia. Jentink (1879: 13) noted the 'habitat' to be 'Celebes, Menado', but information attached to the holotype indicates it came from 'Menado, Langowan' (now spelled Langoon).

Figure 2.3: Holotype of *Mus meyeri* (RMNH 18302; listed as specimen '*a*' in Jentink's [1888: 65] 'Catalogue Systématique des Mammifères'). Measurements are listed in Table 2.2.

Source: Photograph by Peter Goldberg.

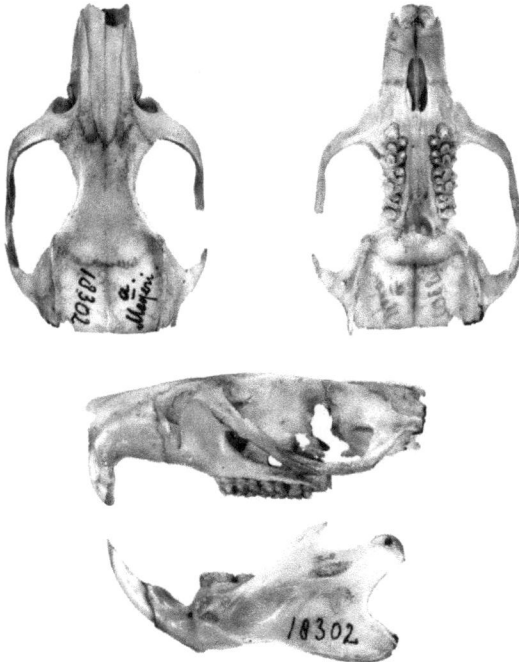

Figure 2.4: Skull of the holotype of *Mus meyeri* (RMNH 18302. See Table 2.2 where measurements are listed. X2

Source: Photograph by Peter Goldberg.

Geographic and elevational distributions

Because provenances of specimens are few and scattered, the actual range of *L. meyeri* is unknown; most samples are from the northeastern tip of the northern peninsula and the southern end of the southwestern peninsula (Figure 2.2a). While broad swaths of Sulawesi, including all of the eastern and southeastern peninsulae as well as most of the central portion, are without records of *L. meyeri*, I suspect its range extends over most of Sulawesi wherever suitable forest habitats remain.

Modern specimens from north of the southwestern peninsula have been collected from the coastal plain near sea level to 1,159 m, an elevational range in which tropical lowland evergreen rainforests dominate, or at least did in the past (I use Whitmore's [1984] terminology for forest formations). Maryanto and others (2009: 47) record a specimen of *'Lenomys meyeri'* from about 2,100 m in 'cloud forest' in Lore Lindu National Park, which is in the northern section of central Sulawesi. I had worked in the same region but obtained *L. meyeri* only at lower elevations in tropical lowland evergreen rainforest, and *Eropeplus canus*, which externally closely resembles *Lenomys*, only high in mountain forest; I suspect their specimen is an example of *E. canus*.

South of the Tempe Depression (the swampy and lake-filled lowlands bisecting the peninsula at its northern margin at about the latitude of Parepare), examples of *L. meyeri* are represented by subfossils excavated from caves in lowlands (100–275 m) and modern samples obtained in montane forest at 2,200–2,400 m on Gunung Lompobatang – the only place where the species is accurately recorded from montane forest habitats.

Table 2.1: Descriptive statistics for measurements of lengths of head and body, tail, hind foot, and ear, and for weight, derived from modern samples of *Lenomys meyeri*.

Variable	Northeast	Northcentral	Core	Southwest Peninsula
LHB	281.5 ± 13.63 (270–301) 4	256	253.3 ± 16.80 (235–268) 3	265.7 ± 11.34 (245–275) 6
LT	277.4 ± 2.51 (275–280) 5	242	270.7 ± 10.07 (260–280) 3	277.3 ± 19.74 (240–298) 6
LT/LHB (%)	99	95	107	104
LHF	47.2 ± 0.84 (46–48) 5	47	47.7 ± 1.53 (46–49) 3	48.2 ± 1.72 (46–50) 6
LE	-	28	27.0 ± 1.41 (26–28) 2	26.3 ± 1.51 (24–28) 6
W	-	-	322.5 ± 3.54 (320–325) 2	-

Notes: Measurements are in millimetres, weight in grams. Mean ± 1 SD, observed range (in parentheses), and size of sample are listed. Mean values were used to compute LT/LHB. Variable abbreviations are defined in the text. Specimens measured are listed below:

Northeast – Amurang: BMNH 21.2.9.4, MZB 5810. Gunung Masarang: BMNH 97.1.2.20. Rurukan: BMNH 97.1.2.19. Langowan: RMNH 18302 (holotype of *Mus meyeri*). All were measured by collectors other than me.

Northcentral – Bumbulan: AMNH 153011 (measured by JJ Menden).

Core – Sadaunta: AMNH 226813. Tomado: AMNH 224317 (these two were measured by me). Gimpu: USNM 219712 (holotype of *Lenomys longicaudus*; measured by HC Raven).

Southwest Peninsula – Gunung Lompobatang: AMNH 101124-27, 101128 (holotype of *Lenomys meyeri lampo*), 101129 (all measured by G Heinrich).

Source: Author's data.

Table 2.2: Age, sex, number of teats, and external, cranial, and dental measurements for holotypes associated with *Lenomys meyeri*.

	Mus meyeri RMNH 18302	*Lenomys longicaudus* USNM 219712	*Lenomys meyeri lampo* AMNH 101128
Age	Adult	Adult	Adult
Sex	?	Female	Female
LHB	275	235	275
LT	280	260	276
LT/LHB (%)	102	111	100
LWS	128	82	145
LWS/LT (%)	46	32	53
TSR/CM	7	7	7
LHF	46	48	47
LE	-	-	25
LDF	20–25	25–30	25–30
Teats	?	2 inguinal pairs	2 inguinal pairs
ONL	-	56.2	57.2
ZB	27.5	27.1	28.2
IB	7.5	7.3	7.0
LR	-	15.7	17.4
BR	10.1	10.3	9.6
BBC	-	19.3	18.7
HBC	-	14.4	13.5
BZP	6.5	5.6	6.1
LD	13.1	13.9	16.0
PPL	-	19.2	19.0
LBP	14.4	14.4	14.8
BBP	2.6	2.8	3.1
BMF	3.5	3.3	3.0
LIF	7.3	6.9	9.1
BIF	2.8	2.8	3.0
LB	-	7.8	8.8
CLM1-3	12.1	10.8	11.0
BM1	3.5	3.2	3.3

Notes: Measurements are in millimetres. Unless otherwise indicated, I copied from skin tags the values for external measurements; I measured the cranial and dental dimensions. Variable acronyms are defined in the text.

For the holotype of *meyeri*, I measured lengths of head and body, tail, and white tail segment on the mounted skin. In the original description of RMNH 18302, Jentink listed 290 mm and 270 mm, respectively, for lengths of head and body and tail. For the holotype of *longicaudus*, I measured lengths of tail, white tail segment, and hind foot on the stuffed museum study skin; 280 mm for the tail and 45 mm for the hind foot are recorded on the skin label.

Number of scale rings per centimetre on the tail (TSR/CM) was counted about one-third the distance from its base.

To measure lengths of overfur and guard hairs on the dorsum (LDF), I placed a ruler at a right angle to the skin surface on the back near the rump and recorded the approximate mark where ends of the bunched hairs rested.

Source: Author's data.

Description

Adult *Lenomys meyeri* are physically large with a short and wide head, stocky body, small ears, and moderately long tail (LHB = 235–301 mm, LT = 240–298 mm, LHF = 46–50 mm, LE = 24–28 mm, W = 320–325 mm; Figure 2.5, Table 2.1). Fur covering upperparts is dense, woolly, soft, and long (length of underfur = 15–30 mm, depending on elevation of provenance), with guard hairs extending 5–10 mm beyond the underfur layer; overall colour ranges from grayish brown to dark brownish gray and is without patterning except for brownish black areas at the base of the mystacial vibrissae. The dark dorsal coat extends onto the basal 50–55 mm of the tail. The small ears are dark gray or brownish gray and finally haired on the margins. Ventral coat is short (10–15 mm) and grayish white (hairs with gray bases and unpigmented tips) or whitish gray. Tail is bicoloured, scantily haired, and ranges from being slightly shorter than length of head and body to slightly longer (LT/LHB = 95–107%); scale hairs are moderately large (Table 2.2), each scale bearing three short hairs; basal portion of tail is dark brownish gray, distal 32–63% is white (Figure 6). Front and hind feet are short and broad; all dorsal surfaces (tarsal, metatarsal, and digital) are dark brown, claws are stout and unpigmented. Palmar and plantar surfaces are naked, unpigmented and adorned with large fleshy pads (three interdigitals, thenar and hypothenar on palmar surface; four interdigitals, thenar and hypothenar on plantar surfaces).

Females have two pairs of inguinal teats.

The cranium is large and stocky (Figure 2.7). Characteristic features are its short rostrum; stout zygomatic arches that bow widely out from sides of the rostrum and braincase; strong supraorbital ridges that transform posteriorly into postorbital shelves that extend posteriorly as pronounced temporal ridges all the way to the occiput where they fuse with the lamboidal ridges; trapezoid-shaped interparietal; wide zygomatic plate; large premaxillary foramen, moderately short and wide incisive foramina; long, deeply scored bony palate; wide mesopterygoid region with moderately long and wide sphenopalatine vacuities; large ectotympanic bulla; and no alisphenoid strut.

Figure 2.5: Young adult female *Lenomys meyeri* (AMNH 224317) from tropical lowland evergreen rainforest near Tomado, 1000 m. Collected May 16, 1974.

Source: Photograph by Author.

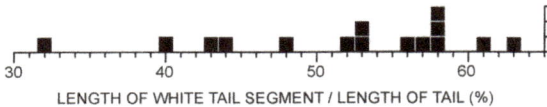

LENGTH OF WHITE TAIL SEGMENT / LENGTH OF TAIL (%)

Figure 2.6: Length of distal white tail segment relative to total length of tail in the sample of *Lenomys meyeri*. I measured the distal white segment (in mm) on freshly caught rats in the field and on dry museum preparations.

Source: Drawing by Patricia Wynne.

A carotid arterial circulation that is derived for muroid rodents in general but primitive for members of subfamily Murinae (character-state 2 of Carleton, 1980; pattern 2 described by Voss, 1988; diagrammed for *Oligoryzomys* by Carleton and Musser, 1989), is present in *L. meyeri* as indicated by the presence of a large stapedial foramen in the petromastoid fissure, a deep groove extending from the middle lacerate foramen to the foramen ovale on the ventral posterolateral surface of each pterygoid plate, and no sphenofrontal foramen or squamosal-alisphenoid groove. Each robust dentary has a pronounced coronoid process and stout condyloid and angular processes; the ramus anterior to the molar row is short and thick.

Incisors are broad and robust, enamel layers are orange in the uppers and pale orange in the lowers; uppers emerge from the rostrum at either a right angle or curve slightly back.

Maxillary (upper) and mandibular (lower) molars are robust, moderately high-crowned, with sharply defined large and high angular cusps, and only slight overlapping occurs between molars in each row (Figure 2.9). Typically, M1 is anchored by five roots, M2 by four, and M3 by three. The same pattern of roots is found on the lower molars, with the exception of AMNH 265022, in which m4 has five roots and m3 has four.

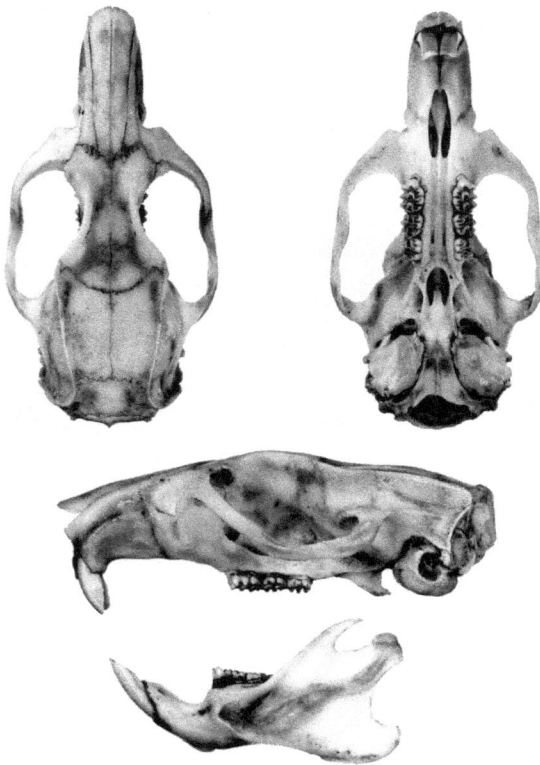

Figure 2.7: Cranium and left dentary of *Lenomys meyeri lampo* (AMNH 101128, the holotype), an adult female collected at 2200 m on Gunung Lompobatang. X2.

Source: Photograph by Peter Goldberg.

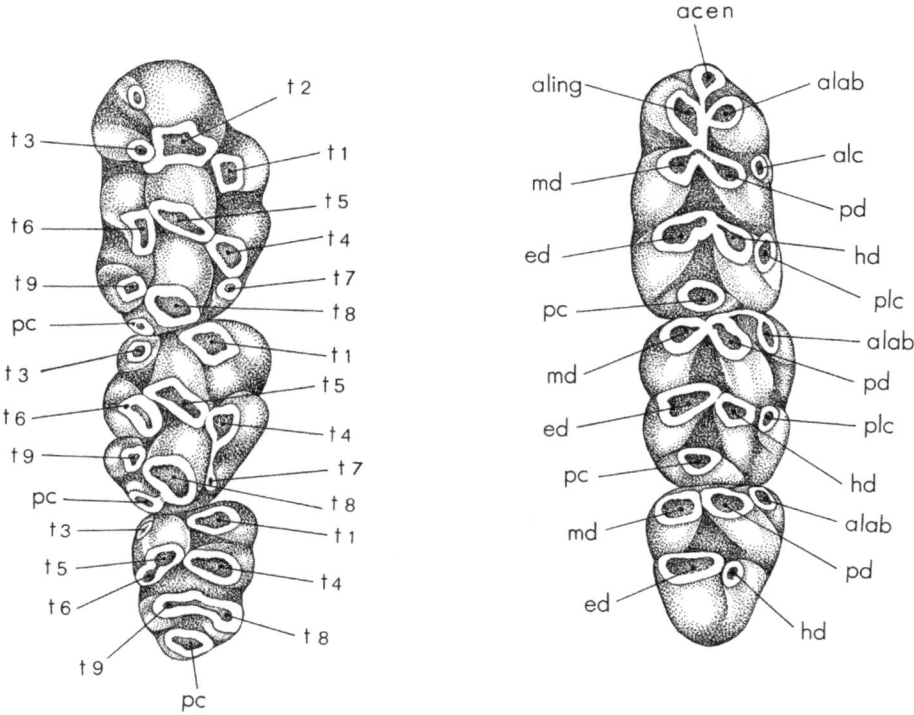

Figure 2.8: Nomenclature of dental structures using right upper and lower molars of *Lenothrix canus*. Maxillary molars: Cusps are numbered according to Miller's (1912) scheme and are referred to in the text with the prefix 't'; pc, posterior cingulum. Mandibular molars: The anterocentral cusp (acen), anterolabial cusp (alab), and anterolingual cusp (aling) form the anteroconid; an anterior labial cusplet (alc) is present on the first molar and posterior labial cusplets (plc) occur on all three teeth; primary cusp rows are formed by the protoconid (pd) and metaconid (md), and the hypoconid (hd) and entoconid (ed); a posterior cingulum (pc) sits at the back of first and second molars (adapted from van de Weerd, 1976: 44).

Source: Drawing by Patricia Wynne.

Occlusal cusp patterns seen in *Lenomys* are unique among endemic Sulawesi murids. Distinctive features of the maxillary molars, in addition to the tall and discrete cusps, are a prominent cusp t7 on all molars (indistinct on M3 in some specimens because it has partially fused with the adjacent cusp t8); a prominent cusp t3 on M2 and M3; a posterior cingulum that is a moderately large and discrete cusp on all molars of many specimens, but an enamel ridge or bump on other specimens; prominent crests (enamel and dentine) extending backwards from cusps t1 and t3 on M1, from cusp t1 on M2, from cusps t4 and t6 on M1 and M2 (stephanodonty; see Misonne, 1969: 55). In some specimens, the crests extending from cusps t4 and t6 merge with the anterior faces of cusps t7 and t9 to form a circle consisting of cusps and crests on M1 and M2.

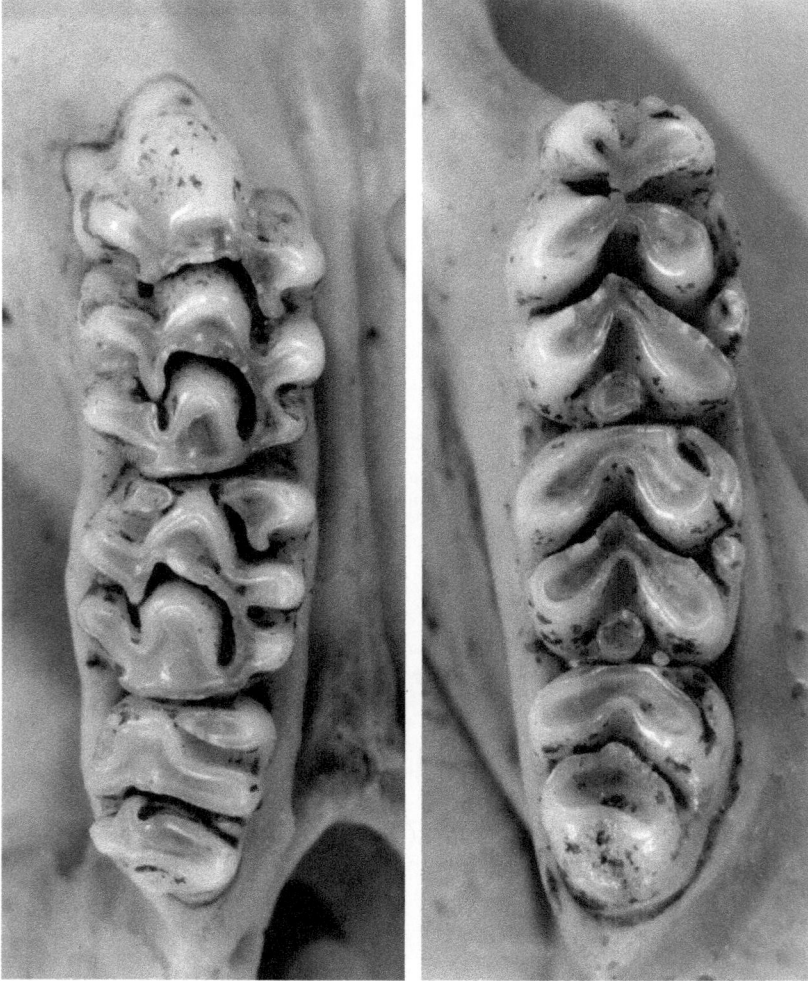

Figure 2.9: Occlusal views of right maxillary (left image) and mandibular (right image) molar rows of *Lenomys meyeri* (AMNH 101127) from Gunung Lompobatang. X10. See Figure 2.8 for cusp labels.

Source: Photograph by Peter Goldberg.

Occlusal cusp patterns of mandibular molars are equally distinctive (Figures 2.9, 2.15). In many specimens, a mid-sagittal enamel crest connects the anteroconid and the second cusp row, and that row is also connected with the posterior cusp row by a mid-sagittal crest in some specimens. In all specimens, the anteroconid is formed by two large and discrete cusps, and a small anterocentral cusp is present on some molars (Figure 2.15, centre) but absent from other specimens (Figure 2.15, right). Additional elaborations are anterior and posterior cusplets on m1, an anterolabial cusp and posterior labial cusplet on m2, an anterolabial cusp on m3, and prominent posterior cingula on m1 and m2.

In addition to the cranial and dental traits summarised here, there is information covering spermatozoan morphology. The sperm head is symmetrical without hooks, and the spermatozoan tail is attached to the midbasal region of the head (Breed and Musser, 1991). Among the Sulawesian species of murines surveyed by Breed and Musser, only *Eropeplus canus* and *Taeromys celebensis* share similar morphologies.

Geographic variation

If there is more than one species in *Lenomys*, it is not clearly evident in the modern material at hand, which consists of small samples (1–6 specimens) from widely separated localities (Figure 2.2). Variation in fur colour (upper parts range from grayish brown to dark brownish gray, underparts whitish gray to grayish white, dorsal fur thick and woolly to thinner and harsher) forms no apparent geographic pattern. Cranial and dental measurements suggest a pattern that must be substantiated by measurements from larger samples. The montane sample from Gunung Lompobatang averages a longer and wider skull (ONL and ZB), zygomatic plate (BZP) and bony palate (BBP); longer diastema (LD) and incisive foramina (LIF); and larger bullae (LB) than other geographic samples (Table 2.3). The variation is summarised by the distribution of specimen scores projected on first and second principal components in Figure 2.10. Size influences spread of the scores along the first axis (largest to the right), with loadings of the variables noted above being most forceful ($r = 0.62$–0.88) – the three scores for the Gunung Lompobatang series are farthest to the right. Long incisive foramina and large bullae were among the traits used by Tate and Archbold (1935: 5) to distinguish *L. m. lampo*. The cranium from Parepare (at the southern margin of the Tempe Depression) also has long incisive foramina (8.7 mm according to Hooijer, 1950: 76), thus falling within the range of the Lompobatang sample. Scatter of scores for the other specimens likely reflect individual and age variation within adults. For example, the points for specimens from the northeast (filled circle) are scattered along the first component and all are adults. Aside from the possible separation of the Gunung Lompobatang sample, no other apparent geographic pattern emerges from the analysis; samples, however, are small. A principal-component analysis (not illustrated) employing only dental variables, which allowed me to use the subfossil material, revealed no significant geographic pattern in dental measurements, and no separation of the Lompobatang toothrows.

Table 2.3: Descriptive statistics for cranial and dental measurements derived from modern samples of *Lenomys meyeri*.

Variable	Northeast $N = 5$	Northcentral $N = 1$	Core $N = 2$	Southwest Peninsula $N = 3$
ONL	53.8 ± 1.70 (52.3–56.4)	54.8	54.7 ± 2.12 (53.2–56.2)	55.4 ± 2.02 (53.2–57.2)
ZB	27.0 ± 1.06 (25.8–28.4)	27.2	26.6 ± 0.71 (26.1–27.1)	27.5 ± 0.64 (27.0–28.2)
IB	8.0 ± 0.21 (7.8–8.3)	8.0	7.1 ± 0.28 (6.9–7.3)	7.2 ± 0.29 (7.0–7.5)
LR	15.6 ± 0.38 (15.2–16.2)	16.1	15.4 ± 0.42 (15.1–15.7)	15.9 ± 1.50 (14.4–17.4)
BR	10.1 ± 0.30 (9.7–10.4)	10.5	10.2 ± 0.14 (10.1–10.3)	9.6 ± 0.25 (9.4–9.9)
BBC	19.3 ± 0.18 (19.1–19.5)	18.9	18.9 ± 0.64 (18.4–19.3)	19.0 ± 0.35 (18.7–19.4)
HBC	13.4 ± 0.45 (12.8–14.0)	13.5	14.1 ± 0.49 (13.7–14.4)	13.7 ± 0.25 (13.5–14.0)
BZP	5.9 ± 0.30 (5.5–6.3)	6.5	5.8 ± 0.28 (5.6–6.0)	6.2 ± 0.36 (5.9–6.6)
LD	13.8 ± 0.84 (13.0–15.1)	15.1	13.7 ± 0.35 (13.4–13.9)	14.9 ± 1.15 (13.7–16.0)
PPL	19.8 ± 0.35 (19.4–20.2)	19.1	19.5 ± 0.42 (19.2–19.8)	19.7 ± 1.02 (19.0–20.9)
LBP	13.6 ± 0.38 (13.1–14.0)	13.0	13.3 ± 1.56 (12.2–14.4)	14.3 ± 0.44 (14.0–14.8)
BBP	2.7 ± 0.29 (2.3–3.1)	2.8	2.8 ± 0.00	3.0 ± 0.31 (2.7–3.3)
BMF	3.1 ± 0.17 (3.0–3.4)	3.5	3.3 ± 0.07 (3.2–3.3)	3.0 ± 0.00
LIF	7.6 ± 0.31 (7.1–7.9)	8.5	7.4 ± 0.71 (6.9–7.9)	8.9 ± 0.47 (8.4–9.3)
BIF	3.1 ± 0.16 (2.9–3.3)	3.2	3.2 ± 0.49 (2.8–3.5)	3.1 ± 0.07 (3.0–3.3)
LB	8.3 ± 0.39 (7.9–8.7)	8.1	7.8 ± 0.07 (7.7–7.8)	8.9 ± 0.21 (8.7–9.1)
CLM1–3	11.8 ± 0.34 (11.3–12.2)	10.5	11.1 ± 0.42 (10.8–11.4)	11.5 ± 0.56 (11.0–12.1)
BM1	3.5 ± 0.23 (3.3–3.8)	3.2	3.5 ± 0.35 (3.2–3.7)	3.4 ± 0.23 (3.3–3.7)

Notes: Measurements are in millimetres. Mean ± 1 SD and observed range (in parentheses) are listed. Specimens measured are indicated below.

Northeast – Gunung Masarang: BMNH 97.1.2.20. Tomohon: NMB 3327, 4759. Amurang: BMNH 21.2.9.4; MZB 5810.

Northcentral – AMNH 153011.

Core – Tomado: AMNH 224317. Gimpu: USNM 219712 (holotype of *Lenomys longicaudus*).

Southwest Peninsula – Gunung Lompobatang: AMNH 101125, 101126, 101128 (holotype of *Lenomys meyeri lampo*).

Source: Author's data.

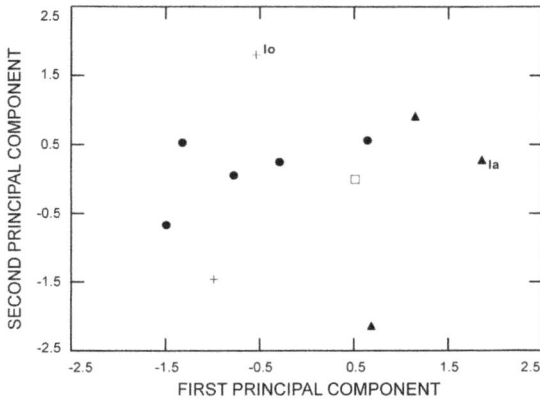

Figure 2.10: Specimen scores representing modern examples of *Lenomys meyeri* projected on first and second principal components extracted from principal-components analysis of 16 cranial and two dental log-transformed variables. Symbols: filled circle = northern peninsula east of Gorontalo region (*N* = 5; Amurang, Gunung Masarang, Tomohon); empty square = Bumbulan, west of Gorontalo region (*N* = 1); cross = central region (*N* = 2; Gimpu, Tomado); filled triangle = southwest peninsula (*N* = 3; Gunung Lompobatang). Abbreviations identify scores for holotypes: la = *Lenomys meyeri lampo*; lo = *Lenomys longicaudus*. See Table 2.4 for correlations of variables and percent variance.

Source: Author's data.

Table 2.4: Results of principal-components analysis comparing modern specimens of *Lenomys meyeri*.

Variable	Correlations	
	PC1	PC2
ONL	0.70(C)	0.23
ZB	0.71(C)	0.40
IB	−0.43	0.35
LR	0.45	−0.08
BR	−0.26	0.09
BBC	−0.14	0.21
HBC	0.19	0.23
BZP	0.62(C)	−0.40
LD	0.88(A)	−0.03
PPL	−0.11	−0.66(C)
LBP	0.56	0.43
BBP	0.76(B)	0.40
BMF	−0.43	0.22
LIF	0.82(B)	−0.52
BIF	−0.23	−0.85(A)
LB	0.64(C)	−0.19
CLM1−3	−0.16	−0.43
BM1	−0.44	−0.63(C)
Eigenvalue	0.022	0.012
% Variance	38.4	19.9

Notes: Correlations (loadings) of log-transformed values for 16 cranial and 2 dental variables are based on 11 specimens; see Figure 2.10.

(A) P ≤ 0.001; (B) P ≤ 0.01; (C)P ≤ 0.05.

Source: Author's data.

The specimen from Gimpu is the holotype of *L. longicaudus*, which Miller and Hollister (1921: 95) thought was distinctive because its tail is longer than length of head and body, but they had no comparative material at hand, and among the specimens I examined tail length ranges from slightly shorter to slightly longer than length of head and body (Table 2.1). The dental traits they highlighted in their diagnosis are not diagnostic and fall within the range of variation seen in larger samples of *L. meyeri*.

Ecological notes

Observations provided here come from my capture of a young adult male from Tomado (AMNH 224317) and an adult female from Sungai Sadaunta (AMNH 226813) in lowland tropical evergreen rainforest (see Table 2.5 for details).

Figure 2.11: Hillside primary forest along Sungai Sadaunta in area near where *Lenomys meyeri* was caught. Photographed in 1976.

Source: Photograph by Author.

Figure 2.12: Ground cover over terrace where burrow of *Lenomys meyeri* was located. Burrow system was in a stable area near edge of a gently sloping terrace above and back from a stream. Dry leaves and broken branches covered the ground beneath a moderately dense undergrowth of shrubs, palm and tree seedlings, ferns, and gingers. Surrounding forest was tall second-growth regenerating over clearing made by a fallen canopy tree. Primary forest covered slopes back of this disturbed area; see Table 2.5. Photographed in 1976.

Source: Photograph by Author.

Most ecological information comes from the rat collected on Sungai Sadaunta, which I kept captive for about a month. It eagerly accepted and consumed katydids, grasshoppers, and moths (usually eating everything except legs and wings); large and small fruit from several species of figs (*Ficus* spp.); small and young tender green plants, tips of fern fronds, palm heart (mostly the thin embryonic leaves); and bait (peanut butter, raisins, oatmeal, and bacon ground into a mash). Tougher leaves of sapling or understorey trees were rejected as were several other kinds of non-fig fruits. It also required a constant source of drinking water.

Table 2.5: Summary of microhabitats at trapping sites and other relevant information for two specimens of *Lenomys meyeri* collected in Central Sulawesi, 1974–1975.

Locality, AMNH and (ASE-field) numbers	Elevation (m)	Date	Trap site and other information
Tomado 224317 (1612)	1000	May 1974	Caught during the night on decaying, moss and vine-covered trunk bridging narrow stream about 2 m above water level. On one bank of stream is tall primary forest that has been slightly disturbed by removal of a tree for lumber, some rattan and occasionally a sugar palm. The opposite bank is scrubby, consisting of a few coffee trees next to scrub and low second-growth forest, which abuts a grassy meadow; ground cover here is mostly grass, shrubs, and ferns. Minimum and maximum ambient temperatures (°F) during a 30-day period were 65.0 (56–67) and 80.9 (70–94), respectively.
Sungai Sadaunta 226813 (4401a)	915	April 1976	Dug out of burrow near edge of a gently sloping terrace about 1.5 m above a stream and 2.5 m back from it. The area was stable, the soil firmly held in place by roots, and the ground covered with dry leaves and litter from rotting small branches and rattan. Undergrowth was knee-to-waist high consisting of shrubs, palm and tree seedlings, ferns, gingers, and rattan rosettes. Forest shading this spot was tall second-growth festooned with woody vines in the understorey and represented regenerating growth over an area where a large canopy tree fell, marked now only by the decaying base of its trunk on the terrace edge near the burrow and the remainder of the trunk lying on the terrace partially covered by undergrowth. Tall primary forest surrounded this disturbed spot. Minimum and maximum ambient temperatures (°F) during a 4-month period was 66.9 (62–70) and 75.2 (70–86), respectively. See the habitat images in Figures 2.11 and 2.12.

Notes: Descriptions of the trapping sites are summarised from my field journals (in Mammalogy archives at AMNH). *ASE* designates Archbold Sulawesi Expedition field numbers. Both sites were in lowland tropical evergreen rainforest (as defined by Whitmore, 1984).

Source: Author's data.

Both the Tomado and Sadaunta rats consumed their own faeces (coprophagous), either picking the pellets from the cage floor or taking them directly from the anus.

The Sadaunta animal gave birth to one young: the pup began uncoordinated movements and ears opened on day 17, pup actively wandered around cage and climbed on supports by day 21 at which time the eyes opened and although still nursing the pup nibbled on bait and palm heart.

The Tomado and Sadaunta animals slept in a leaf nest during the day and were active only at night. They could also move along branches; their short and wide feet with huge fleshy pads and curved claws are fit for climbing – this species nests in ground burrows and likely forages during the night on substrates above ground in the understorey.

A diagram of the burrow and other details are provided in Figure 2.13. During the night the burrow entrance was open but in the morning had been sealed by a plug of dry leaves.

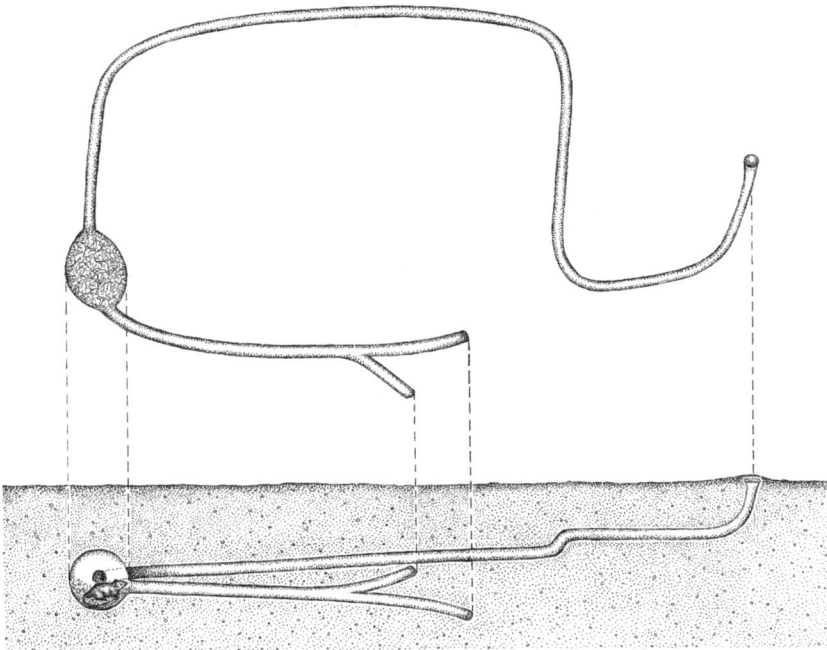

Figure 2.13: Burrow system of *Lenomys meyeri* excavated on a stream terrace adjacent to Sungai Sadaunta, 915 m (3 April 1976). Tunnel system was 5.6 m long (90 cm at its widest and 248 cm at its longest, tunnels were 7–10 cm in diameter) with a single entrance and two blind passages; entire structure lay about 45 cm below ground surface. The oval chamber was 30 cm long and 20 cm in diameter and partially filled with compacted dry leaves forming a shallow bowl-shaped nest. An adult female (AMNH 226813) was found in the burrow.

Source: Drawing by Patricia Wynne.

Subfossils

Subfossil specimens of *L. meyeri* are identified and documented in Table 2.6.

Table 2.6: Subfossil fragments of *Lenomys meyeri* from the southwest peninsula of Sulawesi.

Cave and specimen	Age	Description
BATU EJAJA		
Specimen 1	adult	Most of left dentary, molar row intact, incisor missing.
AMNH 272963	adult	Small fragment of right pelvic girdle.
BATU EJAJA 2		
AMNH 265029	adult	Fragment of left dentary containing m2 and portion of incisor.
AMNH 265030	adult	Incomplete right upper incisor.
AMNH 265031	adult	Distal tip of right upper incisor.
Specimen 2	adult	Most of right dentary, molar row intact, incisor missing.
Specimen 3	adult	Piece of right dentary, m1 and m2 present, incisor intact.
Specimen 4	adult	Fragment of right dentary, basal piece of incisor, no molars.
Specimen 5	adult	Piece of left dentary, m1 and m2 present, incisor intact.
Specimen 6	adult	Fragment of left dentary, molar row intact, incisor missing.
Specimen 7	adult	Fragment of left dentary, molar row intact, incisor missing.
Specimen 8	adult	Piece of right dentary, m1 and m2 present, incisor intact.
Specimen 9	adult	Fragment of right dentary, m1 and basal half of incisor present.
Specimen 10	adult	Piece of left dentary, all molars present, basal piece of incisor.
LEANG BURUNG 1		
AMNH 265022-A	adult	Most of left dentary, molar row and incisor intact (Figure 2.14).
AMNH 265023-A	adult	Anterior fragment of left dentary, molars and incisor missing.
AMNH 265028-B	young	Fragment of left dentary, most of incisor present, molars gone.
AMNH 265024-B	adult	Most of right lower incisor.
AMNH 265025-A	adult	Distal piece of right upper incisor.
AMNH 265026-A	adult	Distal fragment of right upper incisor.
AMNH 265027-A	adult	Nearly complete left pelvic girdle.
AMNH 272961	adult	Intact left femur.
AMNH 272962	adult	Distal portion of right tibia.

Note: Identities of Specimens 1–10 from Batu Ejaja 2 were discussed more fully elsewhere (Musser, 1984).

Source: Author's data.

Lenomys grovesi, new species

Holotype and type locality

The holotype and only example of the species is AMNH 265021, a subfossil right dentary from an adult collected on 26 July 1969 by members of the Australian-Indonesian Archaeological Expedition to Sulawesi under the leadership of DJ Mulvaney and RP Soejono. The dentary is mostly undamaged except for the missing tip of the coronoid process; all molars are intact; root of the incisor is embedded within the dentary, the rest of the incisor is missing (Figures 2.14 and 2.15). Dental measurements are listed in Table 2.7.

The type locality is Leang Burung 1, a cave near the village of Pakalu, 'about 2 km north of the main road from Maros, 10 km to the west' (Mulvaney and Soejono, 1970: 169 and map on p. 170; the site is also mapped by Bulbeck [2004: 130]), less than 100 m (locality 12 in Figure 2.2a), in the southeastern portion of the southwestern peninsula, Propinsi Sulawesi Tengah, Indonesia.

Mulvaney and Soejono (1970) identified the cave as 'Leang Burung' but it has been designated 'Leang Burung 1' in subsequent literature to distinguish it from the nearby 'Leang Burung 2' (Glover, 1981; Bulbeck, 2004). The subfossil was excavated from 'Trench A'. Mulvaney and Soejono (1970: 170) wrote that

> Most of Trench A cut through recent limestone rubble that reached to bedrock. Just inside the line of overhang, however, bedrock dipped vertically, and excavation showed that beneath the rubble lay a zone of disturbed occupation material, and underneath, undisturbed deposit. The depth reached in our test trench was 4 m, at which depth excavation became impossible because of massive fallen rocks.

Mulvaney and Soejono (1970: 171) indicated that charcoal from a depth of about 270 cm in Trench A was dated at 2820 ± 210 (2,360–3,460) years BP, and D. Bulbeck (in a letter to me, 1997; 2004: 136) indicated that 3,000–2,000 years BP are the outside limits for Trench A; the time frame is late Holocene.

Table 2.7: Measurements of mandibular molars from subfossil and modern samples of *Lenomys meyeri* and the subfossil specimen of *Lenomys grovesi*.

Species, locality and specimen	clm1-3	alm1-3	bm1	bm2	bm3
L. meyeri					
SUBFOSSIL					
Southwest Peninsula					
Leang Burung 1					
AMNH 265022	11.5	11.9	3.0	3.1	2.7
Batu Ejaja					
AMNH 265029	-	-	-	3.1	-Batu Ejaja
-Batu Ejaja Panganreang Tudea					
Specimens 1–10 (Musser, 1984: 69)	11.2 ± 0.47 (10.7–11.9) 5	11.7 ± 0.82 (10.9–13.1) 8	3.1 ± 0.11 (3.0–3.3) 8	3.1 ± 0.08 (3.0–3.2) 8	2.8 ± 0.14 (2.6–2.9) 5
MODERN					
Northeast	11.2 ± 0.34 (10.7–11.6) 7	11.6 ± 0.51 (10.9–12.2) 5	3.2 ± 0.17 (3.0–3.4) 7	3.2 ± 0.22 (2.9–3.5) 7	2.8 ± 0.16 (2.6–3.0) 7
Northcentral	10.0	9.9	2.9	2.9	2.7
Core					
AMNH 224317	11.0	11.4	3.3	3.3	3.0
USNM 219712	10.4	10.5	3.0	3.1	2.7
Southwest Peninsula	10.7 ± 0.38 (10.3–11.2) 5	11.3 ± 0.43 (10.7–11.8) 5	3.1 ± 0.18 (2.9–3.3) 5	3.1 ± 0.16 (2.9–3.3) 5	2.8 ± 0.16 (2.6–3.0) 5
Lenomys grovesi					
SUBFOSSIL					
Southwest Peninsula					
Leang Burung I					
AMNH 265021	9.3	9.6	2.7	2.6	2.3

Notes: Measurements are in millimeters. Mean ± 1 SD, observed range (in parentheses), and size of sample is listed. Modern specimens measured are identified below:

Northeast – Gunung Masarang: BMNH 97.1.2.20. Tomohon: NMB 3326, 3327, 4759. Langoon: RMNH 18302, holotype of *Mus meyeri*. Amurang: (BMNH 21.2.9.4). 'Minahassa': RMNH 21233.

Northcentral – Bumbulan: AMNH 153011.

Core – Tomado and Gimpu (holotype of *Lenomys longicaudus*).

Southwest peninsula – Gunung Lompobatang: AMNH 101124-28 (includes holotype of *Lenomys meyeri lampo*).

Source: Author's data.

Diagnosis

In morphology of the dentary and molars, and the occlusal patterns formed by molar cusps, *Lenomys grovesi* is a diminutive version of the larger-bodied *L. meyeri*, with the added distinction that m3 is relatively narrower as contrasted with the dental proportions in *L. meyeri*.

Etymology

The patronym honours Colin Groves, a professional colleague and friend who over the years has consistently impressed me with his unique intellect and the breadth of his scholarship – his impact on the study of mammalian diversity will influence the nature of research long after the rest of us disappear into that place where the winds begin and the sky reflects infinity.

Geographic distribution and habitat

Leang Burung 1 is located on the western coastal plain at about 100 m in the Maros karst region at the southern end of the southwestern peninsula of Sulawesi (see the map in Simons and Bulbeck, 2004: 168). Forest cover on the coastal plain during middle and late Holocene times likely consisted of tropical lowland evergreen or semi-evergreen rainforest formations. Based upon analyses of the vertebrate faunal composition excavated from caves and rock shelters, Simons and Bulbeck (2004: 178) speculate that the environment in this region '… consisted of a mosaic of copses of primary forest, expanses of secondary forest, and possibly more open habitats between about 7000 and 2000 BP.' I strongly suspect *L. grovesi* to be restricted to forested terrain; the mosaic landscape suggested by Simons and Bulbeck would support populations of *L. grovesi* and the other murids represented by subfossils excavated at Leang Burung 1 (Table 2.8), and from other caves in the Maros region (Table 2.9). During my trapping in the central part of Sulawesi, I found *Lenomys meyeri*, *Paruromys dominator*, *Maxomys hellwaldii*, and *Rattus hoffmanni*, which are represented by subfossils from Leang Burung 1 (Table 2.8), to inhabit primary forest or secondary growth that had nearly assumed the structure of old-growth formations.

Table 2.8: Subfossil representatives of murids excavated from Leang Burung 1.

Species	Trench A (2820 ± 210 BP)	Trench B (3420 ± 400 BP)
Lenomys grovesi	AMNH 265021	
Lenomys meyeri	AMNH 265022, 265023, 265025-265027	AMNH 265024, 265028
Paruromys dominator	AMNH 265002-265004	AMNH 265005, 265006, 269960, 269962
Maxomys hellwaldii	—	AMNH 265016
Rattus bontanus	AMNH 265032	AMNH 269957
Rattus hoffmanni	AMNH 265032	—
Rattus tanizumi	—	AMNH 265040

Notes: Radiocarbon dates are from Mulvaney and Soejono (1970); also see Figure 2 in Bulbeck (2004: 133).

Of the seven species listed, all are endemic to Sulawesi except for *Rattus tanizumi*, which is a member of the geographically broadly distributed *Rattus rattus* complex (Aplin et al., 2011), and represents an unintended anthropogenic introduction to the endemic Sulawesi murid fauna.

Results from analysis of external traits and morphometric data from skull and dental measurements in population samples (Musser, unpublished data) coupled with analyses of DNA sequences (Achmadi et al., 2013) indicate that *Maxomys hellwaldii* as listed here will likely prove to be a complex of species, reminiscent of the diversity uncovered in populations of *Maxomys surifer* and *M. whiteheadi* from mainland Indochina and islands on the Sunda Shelf, each previously considered as a single species (Gorog et al., 2004).

Source: Author's data.

Whether *L. grovesi* is extinct or still occurs on the southwestern peninsula is unknown. Most of the lowland forest was long ago removed and the region converted to agriculture (Fraser and Henson, 1996; Whitten et al., 1987), which comprises much of the present landscape south of the Tempe Depression (see map 8d in Mackinnon, 1997); however, remnant tracts of lowland forest are present, especially in karst areas (Whitten et al., 1987: 102; Froehlich and Supriatna, 1996). Unfortunately, these patches have never been adequately, if it all, surveyed for endemic species of murids. Most samples of modern specimens come from the flanks of the volcano Gunung Lompobatang in either montane or mid elevational forests, which have yielded *Lenomys meyeri*, *Paruromys dominator*, *Maxomys hellwaldii* and *M. musschenbroekii*, *Rattus hoffmanni* and *R bontanus*, but not *L. grovesi* (Musser, ms).

Table 2.9: Murid species represented by middle and late Holocene subfossils excavated from caves and rock shelters on the southern end of the southwestern peninsula of Sulawesi.

Species	Batu Ejaja	Bola Batu	Panganreang Tudea	Panisi Ta'batu	Leang Burung 1	Ulu Leang 1	PattaE	Leang Karassa
Lenomys grovesi					+			
Lenomys meyeri	+	+	+		+			
Paruromys dominator	+	+	+		+	+		+
Taeromys celebensis			+					
Taeromys punicans	+	+			.		+	
Maxomys hellwaldii	+				+			
Maxomys musschenbroekii		+						
Bunomys chrysocomus						+		
Bunomys andrewsi	+					+		
Rattus bontanus	+	+	+	+	+	+		+
Rattus hoffmanni	+				+			
Rattus tanezumi				+	+	+		

Notes: The sites are mapped and described in Mulvaney and Soejono (1970), Bulbeck (2004), and Simons and Bulbeck (2004).

My identifications of the samples are documented in Musser (1984, in press), Simons and Bulbeck (2004), and in unpublished manuscripts either being prepared by me or submitted for publication. Some of the names in the table differ from those recorded in the publications. The specimens of *Bunomys andrewsi* were originally identified by me as *B. heinrichi* (listed in tables by Simons and Bulbeck, 2004), which is the name available for the population of *B. andrewsi* occurring on the southwestern peninsula. The *Rattus bontanus* have been documented as either *R. xanthurus foramineus* (Musser, 1984) or *R. foramineus* (Simons and Bulbeck, 2004), but *foramineus* is a synonym of *R. bontanus*, a southwestern peninsular endemic (Musser and Carleton, ms). Examples determined as *Rattus tanezumi* have either been referred to as *R. rattus* (Musser, 1984) or *R. tanezumi* (Simons and Bulbeck, 2004); see note to Table 2.8.

Results from analysis of external traits and morphometric data from skull and dental measurements in population samples (Musser, unpublished data) coupled with analyses of DNA sequences (Achmadi et al., 2013) indicate that *Maxomys hellwaldii* and *M. musschenbroekii* as listed here will likely prove to be a complex of species, reminiscent of the diversity uncovered in populations of *Maxomys surifer* and *M. whiteheadi* from mainland Indochina and islands on the Sunda Shelf, each previously considered as one species (Gorog et al., 2004).

Source: Author's data.

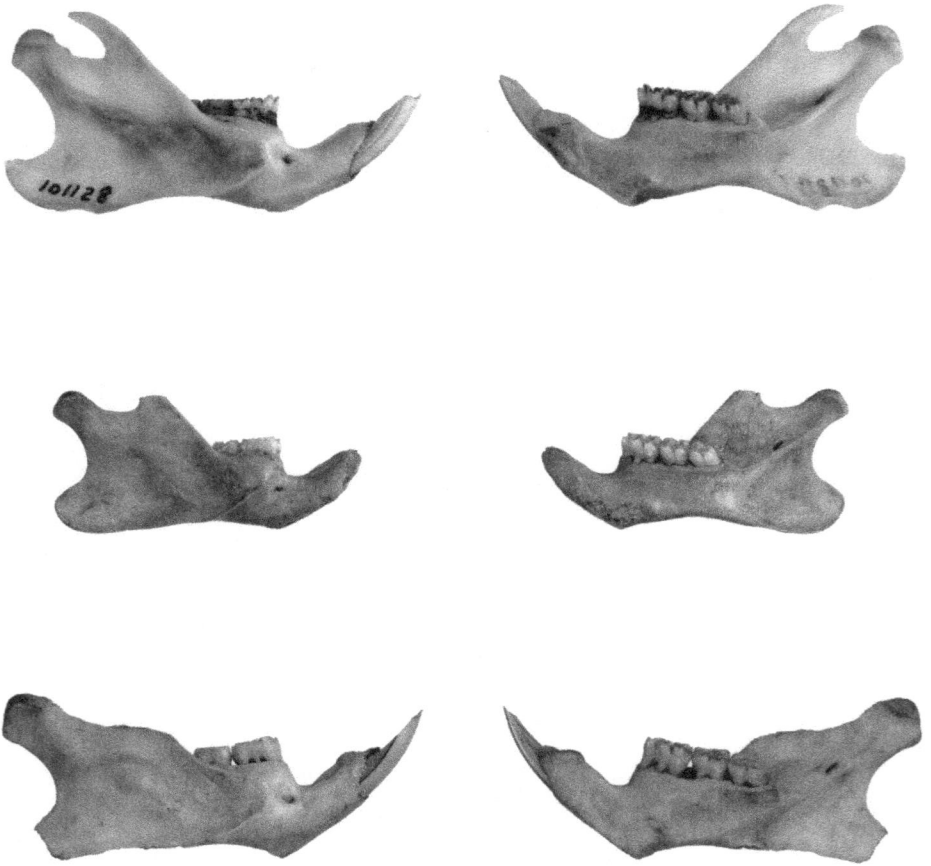

Figure 2.14: Dentaries from *Lenomys* obtained in the southwest peninsula. Upper row: right dentary of modern *Lenomys meyeri* (AMNH 101128, holotype of *L. m. lampo*) from Gunung Lompobatang, 2,200 m. Middle row: right dentary of subfossil *L. grovesi* (AMNH 265021) from Leang Burung 1 on coastal plain. Lower row: left dentary (image was flipped for comparison with the two other specimens) of subfossil *L. meyeri* (AMNH 265022) from Leang Burung 1. X2.

Source: Photograph by Peter Goldberg.

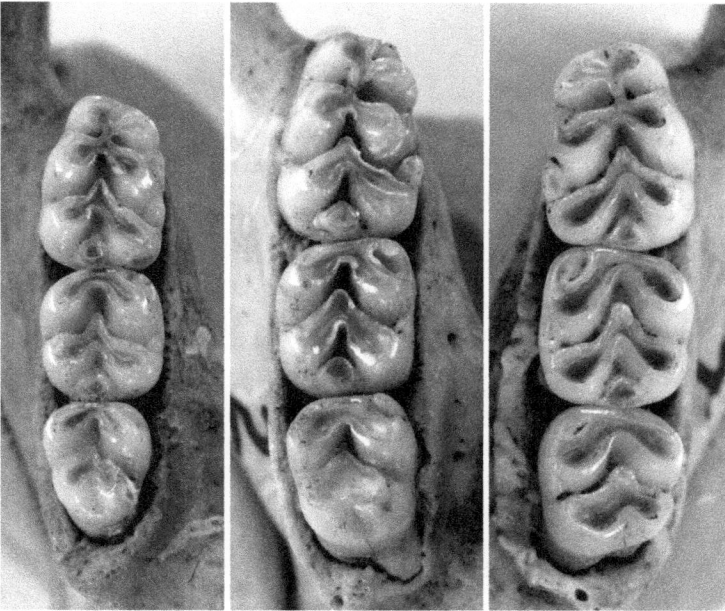

Figure 2.15: Occlusal views of subfossil mandibular molar rows from two species of *Lenomys*. Left, *L. grovesi* (AMNH 265021; clm1–3 = 9.3 mm; right toothrow) from Leang Burung I; Middle, *L. meyeri*, (Specimen 2; clm1–3 = 11.3 mm; right toothrow) from Panganreang Tudea; Right, *L. meyeri*, (Specimen 1; clm1–3 = 10.7 mm; left toothrow) from Batu Ejaja.

Source: Photograph by Peter Goldberg.

Description and comparisons

Lenomys grovesi is a miniature version of *L. meyeri*, assuming size of dentary and molars reflect overall physical size of the animals; no other examples of *L. meyeri* at hand, either subfossil or modern, matches the smaller dimensions characteristic of *L. grovesi*. Configuration of the dentary in *L. grovesi* conforms to that described previously for *L. meyeri*, but is simply much smaller (Figure 2.14). So are the molars, a difference evident visually (Figure 2.15), by measurements (Table 2.7), and quantitatively by the distribution of specimen scores for samples of *L. grovesi* and *L. meyeri* projected on first and second principal components extracted from principal-components analysis (Figure 2.16). Along the first component, covariation among all variables isolates the score for *L. grovesi*, reflecting its smaller molars and shorter toothrow, and the greater size of these dental elements in *L. meyeri* ($r = 0.84$–0.94; Table 2.10). Position of the score for *L. grovesi* along the second axis indicates its relatively narrower m3 ($r = -0.53$) compared with the proportion in *L. meyeri*.

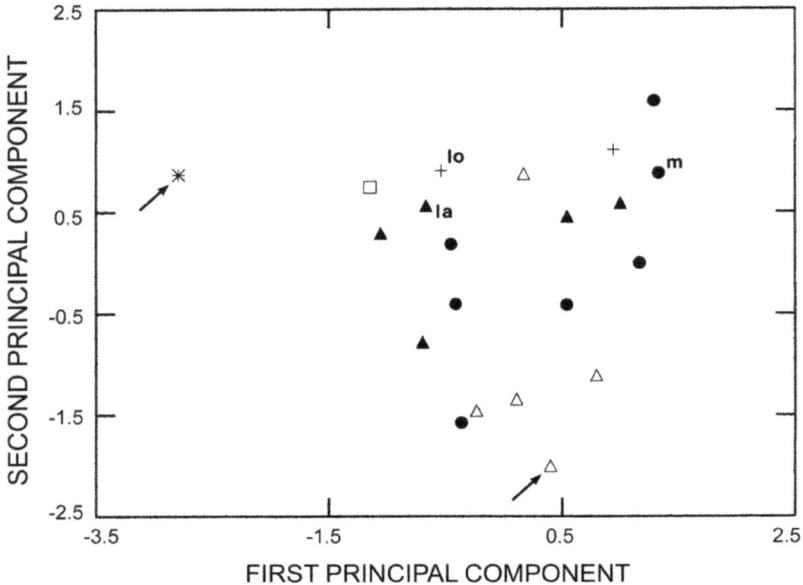

Figure 2.16: Scores representing specimens of modern and subfossil *Lenomys meyeri* and subfossil *L. grovesi* projected on first and second principal components extracted from principal-components analysis of four log-transformed dental variables. Symbols for *Lenomys meyeri*: filled circle = northern peninsula east of Gorontalo region (N = 7; Amurang, Gunung Masarang, Tomohon, Langoon); empty square = Bumbulan, west of Gorontalo region (N = 1); cross = central region (N = 2; Gimpu, Tomado); filled triangle = southwest peninsula (N = 5; Gunung Lompobatang); open triangle = southwest peninsula (N = 5; subfossils from Batu Ejaja, Panganreang Tudea, and Leang Burung 1). Symbol for *L. grovesi*: asterisk = southwest peninsula (N = 1; Leang Burung 1). Abbreviations identify scores for holotypes: la = *Lenomys meyeri lampo*; lo = *Lenomys longicaudus*; m = *Mus meyeri*. Arrows point to scores for subfossils from Leang Burung 1, the holotype of *L. grovesi* (asterisk) and AMNH 265022 representing *L. meyeri* (empty triangle). See Table 2.10 for correlations of variables and per cent variance.

Source: Authors' data.

Table 2.10: Results of principal-components analysis comparing modern and subfossil samples of *Lenomys meyeri* with the subfossil specimen of *Lenomys grovesi*.

Variable	Correlations	
	PC 1	PC 2
clm1-3	0.94[A]	0.19
bm1	0.94[A]	0.23
bm2	0.94[A]	0.02
bm3	0.84[A]	−0.53[B]
Eigenvalue	0.013	0.001
% Variance	83.8	8.8

Note: (A) $P \leq 0.001$; (B) $P \leq 0.01$. Correlations (loadings) of four dental log-transformed variables are based on 21 specimens; see Figure 2.16.

Source: Author's data.

Except for size, other traits are shared by the two species. In both, m1 is typically anchored by five roots, m2 by four, and m3 by three. Occlusal patterns formed by molar cusps are closely similar in both (Figure 2.15): major cusps are large and sharply defined; the anteroconid and cusp row behind it are joined by a mid-sagittal ridge; an anterocentral cusp is absent in *L. grovesi*, but is either present or absent in samples of *L. meyeri*; anterior and posterior labial cusplets adorn m1, an anterolabial cusp is on m2 and m3, and a posterior labial cusplet sits on m2 but is absent from m3.

Three examples of *L. meyeri* are represented by molar and dental elements also excavated from level A in Leang Burung 1, as well as upper incisor fragments and a pelvic girdle (Table 2.8).

Conclusion: Hypotheses

Lenomys meyeri

That the samples of *L. meyeri* analysed here represent one widely geographically spread species is a hypothesis that has to be tested by analyses of more specimens from a broader coverage of Sulawesi than is now available, and the use of DNA sequences as well as qualitative anatomical and morphometric data. The most likely candidate for separation is the population inhabiting montane forest habitats on Gunung Lompobatang at the south end of the southwestern peninsula (locality 13 in Figure 2.2a). Presently, I assume the montane animals and those from the adjacent lowlands are members of the same population;

dental measurements support this view, but thorough documentation remains elusive without intact modern specimens from the lowlands south of the Tempe Depression.

Lenomys grovesi

That the single small dentary and associated molars I designate *L. grovesi* represent a species and not just a genetic aberration in a population of *L. meyeri* seems a reasonable hypothesis set against the background of size variation in mandibular and dental elements documented for available samples of that larger-bodied species. It is also reasonable to postulate that *L. grovesi* is another murid endemic to the southwestern peninsula south of the Tempe Depression; *Bunomys coelestis*, *Rattus mollicomulus*, and *Rattus bontanus* are other members of that endemic group.

Future search in subfossil samples from the southwestern peninsula may undercover additional specimens of *L. grovesi*. Clason (1976: 66), for example, listed 65 fragments identified only as 'rodent' that were excavated from Ulu Leang I; this material bears critical reexamination.

Colin Groves

'Guy, this is Colin Groves,' explained Paul Ryan as he brought a thin, bespectacled Englishman sporting a Beatles-style haircut over to where I was standing, which happened to be next to a row of specimen cabinets extending the length of the rodent range in the Mammal Division of the National Museum of Natural History in Washington, DC. Colin and I shook hands and he expounded in his understated way on his research interests; I then countered with my reasons for rummaging through the trays of Asian rodents. That was in the late 1960s. Colin was then focusing on primates and I mistakenly pigeonholed him as a primatologist. Between then and now and through many visits with one another at museums in the United States and England, over dinners of Indian delicacies and standard British fare, and in one of my forest camps in Sulawesi, I came to understand the intellectual acuity of this man and the depth and range of Colin's interest and expertise in all mammal groups, their taxonomy, biogeography, and evolutionary history. Yes, he was deeply interested in primates, including fossil humans, but he also published on other groups, extending from monotremes and marsupials to rodents and ungulates. Perissodactyls and artiodactyls, however, were his first love. While sharing tea with John Hill at the British Museum, John reminisced about the time a young Colin Groves appeared in the Mammal Section to show him a thick manuscript revising ungulate taxonomy, asking about the next steps to publication, absolutely dumbfounding all the curators.

And he never quit gathering data for his ungulate research. I fondly remember one of his visits to the American Museum when he merrily disappeared with his huge set of wooden calipers into the depths of the rhinoceros collection to emerge later with measurements of not just some but all the specimens – his eyes sparkled merrily.

Applying the honorary patronym to the small-bodied *Lenomys* endemic to Sulawesi also serves as a tribute to Colin for his numerous contributions to the science of mammalogy, for his friendship, for his glittering intelligence, and dedication to his profession. I would like to think that my contributions are much better than they would have been without Colin's influence.

Acknowledgements

My fieldwork in Sulawesi was supported by the former Celebes Fund of the American Museum of Natural History and Archbold Expeditions, Inc.; I was sponsored in Indonesia by the Lembaga Ilmu Pengetahuan Indonesia (LIPI) and the Museum Zoologicum Bogoriense (MZB), and also assisted by members of United States Navy Medical Research Unit No. 2 (NAMRU-2) in Jakarta. The drawings in Figures 2.1, 2.2, 2.6, 2.8, and 2.13 are the work of Patricia Wynne; photographic prints of the skulls, molar rows, dentaries, and habitats were produced by Peter Goldberg. Curators and supporting staff attached to the various institutions I visited provided access to specimens and archival data. Without the assistance of artist, photographer, and museum staffs, I could not have accomplished my research goals.

References

Achmadi AS, Esselstyn JA, Rowe KC, Maryanto I, Abdullah MT. 2013. Phylogeny, diversity, and biogeography of Southeast Asian spiny rats (*Maxomys*). *J Mammal* 94(6):1412–1423.

Aplin KP, Suzuki H, Chinen AA, Chesser RT, Ten Have J, Donnellan SC, Austin J, Frost A, Gonzalez JP, Herbreteau Y, Catzeflis F, Soubrier J, Fang Y-P, Robins J, Matisoo-Smith E, Bastos ADS, Maryanto I, Sinaga MH, Denys C, Van Den Bussche RA, Conroy C, Rowe K, Cooper A. 2011. Multiple geographic origins of commensalism and complex dispersal history of black rats. *PLoS ONE* 6(11):e26357, doi:10.1371/journal.pone.0026357.

Breed WG, Musser GG. 1991. Sulawesi and Philippine rodents (Muridae): A survey of spermatozoa morphology and its significance for phylogenetic inference. *Am Mus Novit* 3003:1–15.

Bulbeck D. 2004. Divided in space, united in time: The Holocene prehistory of South Sulawesi. In: Keates SG, Pasveer JM, editors. Quaternary research in Indonesia. *Mod Quat Re* 18:129–166.

Carleton MD. 1980. *Phylogenetic relationships in Neotomine-Peromyscine rodents (Muroidea) and a reappraisal of the dichotomy within New World Cricetinae.* Miscellaneous Publications Museum of Zoology, University of Michigan 157:1–146.

Carleton MD, Musser GG. 1989. Systematic studies of oryzomyine rodents (Muridae, Sigmodontinae): A synopsis of *Microryzomys*. *Bull Am Mus Nat Hist* 191:1–83.

Clason AT. 1976. A preliminary note about the animal remains from the Leang I Cave, South Sulawesi, Indonesia. *Mod Quat Re* 2:53–67.

Fraser BJ, Henson SM. 1996. *Survai jenis-jenis burung endemik di Gunung Lompobattang, Sulawesi Selatan* [Survey of endemic bird species on Gunung Lompobattang, south Sulawesi]. Bogor: PHPA/BirdLife International-Indonesia Programme, Technical Memorandum No. 12.

Froehlich JW, Supriatna J. 1996. Secondary intergradation between *Macaca maurus* and *M. tonkeanus* in south Sulawesi, and the species status of *M. togeanus*. In: Fa JE and Lindburg DG, editors. *Evolution and ecology of macaque societies*, Cambridge: Cambridge University Press. pp. 43–70.

Glover IC. 1981. Leang Burung 2: An Upper Palaeolithic rock shelter in South Sulawesi, Indonesia. *Mod Quat Re* 6:1–38.

Gorog AJ, Sinanga MH, Engstrom MD. 2004. Vicariance or dispersal? Historical biogeography of three Sunda shelf murine rodents (*Maxomys surifer, Leopoldamys sabanus* and *Maxomys whiteheadi*). *Biol J Linn Soc Lond* 81:91–109.

Hooijer DA. 1950. Man and other mammals from Toalian sites in South-western Celebes. Koninklije Nederlandsche Akademie van Wetenschappen, Verhandelingen Afdeling. *Natuurkunde (Tweede Sectie)* 46(2):1–164.

Jentink FA. 1879. On various species of *Mus*, collected by S. C. I. van Musschenbroek Esq. in Celebes. *Notes of the Royal Zoological Museum of the Netherlands*, Leyden 1(note 2):7–13.

Jentink FA. 1887. *Catalogue ostéologique des Mammifères. Muséum d'Histoire Naturelle des Pays-Bas*, Tome 9. Leide: E.J. Brill. pp. 1–360.

Jentink FA. 1888. *Catalogue systématique des Mammifères. Museum d'Histoire Naturelle des Pays-Bas*, Tome 12. Leide: E.J. Brill. pp. 1–280.

Jentink FA. 1890. Mammalia from the Malay Archipelago. II. Rodentia, Insectivora, Chiroptera. In: Weber M, editor. *Zoologische Ergebnisse einer Reise in Niederländisch Ost-Indien*. Leiden: E.J. Brill. Erster Band. pp. 115–130, pls VIII–XI.

MacKinnon J, editor. 1997. *Protected areas systems review of the Indo-Malayan Realm* (prepared on behalf of the World Bank by the Asian Bureau for Conservation in collaboration with the World Conservation Monitoring Centre). Canterbury: Asian Bureau for Conservation Limited.

Maryanto I, Prijono S, Yani M. 2009. Distribution of rats at Lore Lindu National Park, Central Sulawesi, Indonesia. *J Trop Biol Conserv* 5:43–52.

Miller GS, Jr. 1912. *Catalogue of the mammals of Western Europe (Europe exclusive of Russia) in the collection of the British Museum*. London: British Museum (Natural History).

Miller GS Jr, Hollister N. 1921. Descriptions of sixteen new murine rodents from Celebes. *Proc Biol Soc Wash* 34:93–104.

Misonne X. 1969. African and Indo-Australian Muridae: Evolutionary trends. *Ann Musée Roy Afr Centr Ser 8 Sci Zool* 172:1–219.

Mulvaney J and Soejono RP. 1970. The Australian-Indonesian expodition to Suwalesi. *AP* 13:163–177.

Musser GG. 1984. Identities of subfossil rats from caves in southwestern Sulawesi. *Mod Quat Re* 8:61–94.

Musser GG. 2014. A systematic review of Sulawesi *Bunomys* (Muridae, Murinae) with the description of two new species. *Bull Am Mus Nat Hist* 392:1–313.

Musser GG, Durden LA. 2014. Morphological and geographic definitions of the sulawesian shrew rats Echiothrix leucura and E. centrosa (muridae, murinae), and description of a new species of sucking louse (Phthiraptera: Anoplura). *Bull Am Mus Nat Hist* 871:1–87.

Musser GG, Durden LA, Holden ME, Light JE. 2010. Systematic review of endemic Sulawesi squirrels (Rodentia, Sciuridae), with descriptions of new species of associated sucking lice (Insecta, Anoplura), and phylogenetic and zoogeographic assessments of sciurid lice. *Bull Am Mus Nat Hist* 339:1–260.

Simons A, Bulbeck D. 2004. Late quaternary faunal successions in South Suwalesi, Indonesia. *Mod Quat Re* 18:167–190.

Tate GHH, Archbold R. 1935. Results of the Archbold Expeditions. No. 3. Twelve apparently new forms of Muridae (other than *Rattus*) from the Indo-Australian region. *Am Mus Novit* 803:1–9.

Thomas O. 1898. VIII. On the mammals collected by Mr. John Whitehead during his recent expedition to the Philippines with field notes by the collector. *Trans Zool Soc Lond* 14:377–414.

Van de Weerd A. 1976. Rodent faunas of the Mio-Pliocene continental sediments of the Teruel-Alhambra region, Spain. *Utrecht Micropaleontol Bull*, Special Publication 2:1–218.

Voss RS. 1988. Systematics and ecology of ichthyomyine rodents (Muroidea): Patterns of morphological evolution in a small adaptive radiation. *Bull Am Mus Nat Hist* 188:259–493.

Whitmore TC. 1984. *Tropical rainforests of the Far East*. Second edition. Oxford: Clarendon Press.

Whitten AJ, Mustafa M, Henderson GS. 1987. *The ecology of Sulawesi*. Yogyakarta: Gadjah Mada University Press.

3. Gibbons and hominoid ancestry

Peter Andrews and Richard J Johnson

Introduction

Gibbons form a monophyletic group that differs from other hominoid primates, both in behaviour and anatomy. They are found exclusively in eastern Asia, and although evidence from DNA suggests their lineages diverged close to the time of hominoid origins (Goodman et al., 1998), the species array seen today in Asia did not speciate until closer to 6 Mya (Hayashi et al., 1995; Groves, 2001). Their taxonomy has been clarified by the work of Groves (2001), but their evolutionary history is still poorly understood. There is next to no fossil evidence to show when and where the gibbon lineage emerged and speciated, but comparisons with the great apes provide evidence of their shared common ancestor. Gibbon locomotor morphology, which formed the basis of our shared work with Colin Groves (Andrews and Groves, 1976), is unique in the animal world, but attempts to link gibbon anatomy with that of the great apes has generated much confusion not only in ape evolution but also in human evolution. The question 'were human ancestors brachiators?' provides a good instance of this, and it seriously retarded evolutionary interpretations of human evolution during the middle part of the twentieth century (and is still with us today). We will review this evidence here, followed by the fossil evidence for gibbon evolution, little as it is, and attempt to reconstruct the gibbon common ancestor. Three evolutionary scenarios will be presented based on these two lines of evidence, with a third based on the evolutionary significance of the loss of the uricase gene during gibbon evolution (Johnson and Andrews, 2010).

Hylobatid taxonomy and morphology

Gibbons (family Hylobatidae) share some characters with the great apes, including the relatively large size and the configuration of the brain; the morphology of the teeth; long clavicle; the orientation and dorsal positioning of the scapula; the cranial orientation and shape of the head of the humerus; the free rotatory movements of the radioulnar joints; the mobility of the wrist and hand , in particular the meniscus development of the wrist; the shortened caudal and lengthened sacral regions of the vertebral column; the expanded ilium; the loss of the tail; the shape of the thorax; the presence of a vermiform appendix;

and the disposition of the abdominal viscera (Napier, 1960, 1963; Lewis, 1971, 1989; Groves, 1972, 1986; Preuschoft et al., 1984). They have many adaptations for below branch suspensory locomotion, or brachiation, for example their elongated arms and the automatic hook formed by their hands when they extend their arms – they literally cannot extend their fingers when their arms are extended, an excellent device for hanging securely on to branches. They may move bipedally on larger branches of trees (Avis, 1962), and their legs are relatively long compared with the size of their bodies. They are strictly arboreal, living in tropical rainforest, with a diet consisting mainly of fruit, and they are 5 to 12 kilograms in body weight. In all of these morphological characters, gibbons appear to be derived relative to monkeys and other primates.

In many characters, gibbons are also derived relative to the earliest known fossil apes, so that they may have branched off from the other apes soon after the appearance of this fossil group. The earliest fossil that can be shown to share hominoid synapomorphies is *Proconsul heseloni* (Ward et al., 1991, 1993), which is from 18 Ma deposits on Rusinga Island, Kenya, but other species of the genus extend back in time to 20 to 22 Ma. There is also a fossil monkey from similar aged deposits, and recently a monkey-like tooth has been described from the Rukwa rixft in southern Tanzania dated to 25 Ma (Stevens et al., 2013). From the same site at Rukwa is a fossil tooth row showing a remarkable degree of similarity to *Rangwapithecus gordoni* (Stevens et al., 2013), but it should be pointed out that there are no characters that establish either this species or the new *Rukwapithecus fleaglei* as members of the hominoid lineage. Their elongated molars and adaptations for more folivorous diets (Kay, 1977) are shared with monkeys rather than apes, and it may be that *R. gordoni* and *R. fleaglei* should be distinguished as a separate family, together with 'nyanzapithecines' and separate from hominoids (Harrison, 2002). Be that as it may, there is also evidence for gibbon divergence prior to the loss of the uricase gene (see below), a loss common to all living hylobatids, before 13.1 Ma (Keebaugh and Thomas, 2010) or 9.8 Ma (Oda et al., 2002).

The family is divided into four genera and as many as 14 to 18 species (Groves, 2001; Brandon-Jones et al., 2004; Thinh et al., 2010). Genus *Hylobates* (named as subgenera in Groves, 2001) is the most widespread with seven species; *Bunopithecus hoolock* and *Symphalangus syndactylus* are monospecific genera, and *Nomascus* has six species. They live exclusively in tropical and subtropical forests of eastern Asia where they have a unique form of locomotion, brachiation, which is common to all 14 species (Napier, 1963; Avis, 1962; Lewis, 1971). At the time when Colin and lead author, PA, wrote their paper on gibbon locomotion, there was much discussion about brachiation, what it is and how common it is in other primates. Much of this was the result of anatomical studies of the primate shoulder and forearm (Napier, 1963; Ashton and Oxnard, 1963, 1964), of the skeleton (Schultz, 1973) and of the hand (Lewis, 1971), and following in this tradition we undertook a series of anatomical dissections of the gibbon shoulder and forearm. Colin and PA found that:

> The most striking thing, perhaps, about the musculature of the gibbon is the prevalence of interlinked muscle systems; indeed *Hylobates lar* may be crudely characterised as a mass of muscle chains. A long chain runs from pectoralis major via biceps brachii to flexor digitorum sublimis, and this is reinforced by a chain from latissimus dorsi via dorsoepitrochlearis to biceps. Further cleidodeltoideus (in the *lar* group only) inserts into pectoralis major. Separate from this multified chain is a second linking the caudal head of subscapularis with the deep fibres of teres major. Functionally a muscle chain acts to transmit the contraction of one muscle to the action of a second. (Andrews and Groves, 1976: 207)

There are variations within the hylobatids, as we and others have pointed out (references in Andrews and Groves, 1976), but these are minor compared with the species of great apes and humans. Even the spider monkey, which comes closest to gibbons in its form of locomotion and in its specialisations for below-branch locomotion, lacks the specialised interlinked muscle systems so characteristic of gibbons. They are absent in the great apes, and we concluded that characters for brachiation, and for suspensory locomotion in general, must be tied to the hylobatids, and the absence of these characters in great apes suggests a non-suspensory evolutionary history. Characters such as the broad thorax, which are sometimes put forward as evidence of suspensory function in the great apes, together with associated characters of elongated clavicle and position of the scapula, are rather allometrically associated with increase in body size within primates, and the broad thorax in hylobatids is an exception to this allometric gradient (Andrews and Groves, 1976). This is consistent with fossil evidence (see below), which shows that for the first eight million years of the known fossil record of apes, the thorax was narrow and deep like that of monkeys. The same argument applies to the 'long' arms in gibbons and great apes; in the latter, their length in the African apes is on the same allometric gradient as that of monkeys (and humans), and while orangutan arms are slightly longer than expected for their body size, it is only the gibbons that have arm lengths outside the allometric gradient (Biegert and Maurer, 1972; Aiello, 1981; Jungers, 1984). We concluded by saying:

> Where differences in morphology occur between gibbons and Great Apes, we conclude that, potentially, the condition seen in gibbons is that adaptive for brachiation. In many cases the functional interpretation significantly relates the condition to brachiation, but in some cases the features can be interpreted as adaptive for upright posture of mobility of forelimb, both necessary but not exclusive attribute of brachiation. It is

in these features that Great Apes resemble gibbons, and it is concluded that they share a common feeding adaptation, despite their very different habitats, involving feeding in a stationary upright posture by reaching all round with the mobile forelimbs. (Andrews and Groves 1976: 213)

References to and justifications of these conclusions are set out in full in our long 1976 article and cannot be repeated here. The message we wish to convey here is that the gibbons formed a monophyletic group marked by unique behaviour and morphology that are shared to a great extent by all hylobatid species. It has also become apparent from molecular studies that gibbon speciation was both recent and sudden, either as a vicariant event or a rapid radiation at the end of the Miocene period (Israfil et al., 2011). We will now look to the fossil record to see if any of the features present today in living hylobatids may be seen in any known fossil ape.

Fossil evidence

It is a remarkable thing that despite their origin in the Miocene, hylobatids are not known in the fossil record earlier than the Middle Pleistocene. There are isolated teeth from Middle Pleistocene deposits in China and Indonesia, and Matthews and Granger (1923) described *Bunopithecus sericus* from Szechuan, which is indistinguishable from hoolock gibbons. Delson (1977) described a number of isolated teeth, but without attributing them to species, and Hooijer (1960) described numerous siamang teeth from Middle to Late Pleistocene sites in Indonesia. All are indistinguishable from living species of hylobatid and tell us nothing about the evolution of the lineage.

There are several reasons that might explain why the fossil record is so poor. Gibbons did not differentiate until relatively late in hominoid evolution, about the same time as the African ape and human clade, but while there is a good fossil record for early humans there is almost none for the great apes. DNA evidence shows that gibbons split off from the other apes and humans well before the orangutan divergence, but it may be that early species of fossil gibbon were extremely rare, and the sparse fossil record of fossil apes has so far failed to recover any. It is possible, even likely, that early gibbons did not look anything like modern gibbons, the characteristics of which almost certainly appeared late in their evolution, and it may be that some fossil gibbons are already known, but since they are not recognisable as gibbons they are not generally accepted as such. Finally, it might be also that gibbons were restricted to dense tropical forest, unlike the majority of fossil apes (see below), and the rarity of fossil sites representing forest habitats means that no fossil gibbons have been recovered.

It is also the case with the great apes that few fossil apes can definitively be assigned to any of the great ape lineages. There is a fossil orangutan skeleton from middle Pleistocene deposits in Vietnam (Bacon and Long, 2001), fragmentary remains of chimpanzees in Africa, also from the Middle Pleistocene (McBrearty and Jablonsky, 2005; Pickford and Senut, 2005), and no fossil gorillas are known at all. It is the human lineage that is by far the best represented in the fossil record.

Morphology of fossil apes

We will first briefly review the evidence for morphological variation in fossil apes. Fossil apes span the last 20 million years, restricted initially to Africa in the early Miocene and spreading into Europe and Asia during the middle Miocene.

The earliest known apes are the proconsulids from the early Miocene, and they are characterised by the following key morphologies (described in Table 3.1).

Table 3.1: Morphological features of proconsulids related to their form of locomotion.

Torsion of the humeral head
Mobile shoulder joint
Mobile elbow joint
Narrow chest
Long curved back
Relatively long thumb
Opposable thumb, non-rotatory
Non-weight-bearing wrist and hand
Relatively short hand
Narrow gripping foot
Powerful flexor muscles for gripping branches
Molars had low degrees of shearing

Source: Compiled from sources on proconsulid morphology. All sources in reference list.

The conclusion to be drawn from the morphology of proconsulids is that they were arboreal climbers, moving on the tops of branches, but they were not habitual leapers rather moved slowly and powerfully in the trees. They were mainly fruit eaters with body sizes varying from 9–11 kg to 63–83 kg, from siamang size to larger than chimpanzees, and some degree of terrestrial activity is indicated, particularly for the larger species, which were as big as chimpanzees or even bigger (Le Gros Clark and Leakey, 1951; Napier and Davis, 1959; Napier, 1960; Andrews, 1978, 1992; Walker and Pickford, 1983; Walker et al., 1983,

1993; Rose, 1983, 1984; Beard et al., 1986; Gebo et al., 1988; Rafferty, 1988; Lewis, 1989; Walker and Teaford, 1988, 1989; Ward, 1993; Begun et al., 1994; Teaford, 1994; Rafferty et al., 1995; Ward et al., 1995; Harrison, 2002; Gebo et al., 2009). Primates at the upper end of the size range and living in woodland (non-forest) environments, must have been partly terrestrial, as in chimpanzees and gorillas today, for they were too large to move easily between arboreal pathways. Mike Rose has made a particularly telling point when he said: When I look at the postcranial bones from the Miocene apes, I get a fairly consistent pattern from many species, but it is nothing like what we see in modern apes. Maybe we should consider the ones that survived as the bizarre ones.

Dendropithecus macinnesi, an early Miocene ape from the same sites and levels as the proconsulids, was described originally by Le Gros Clark and Thomas (1951), who showed its similarities to hylobatids based on the gracile limb bones, which were taken to indicate suspensory locomotion in trees. Limb proportions were not gibbon-like, however, but more similar to those of spider monkeys. Andrews and Simons (1977) agreed with this interpretation, placing the new genus *Dendropithecus* in Hylobatidae, but in many respects it was shown that the morphology of the limb bones was more like that of colobine monkeys, the most arboreal of Old World monkeys.

There is also increasing evidence that early Miocene apes did not, for the most part, live in tropical forests but are mostly found associated with woodland habitats. The proconsulids (and *Dendropithecus*) at Rusinga Island are associated with a flora, which preserved a rich plant assemblage (Collinson et al., 2009) dominated by deciduous woodland tree species. There were very few twigs with thorns such as are found on more arid adapted species of *Acacia* or *Balanites*, and there was no evidence of forest trees. Broadleaved deciduous woodland is therefore indicated by the Rusinga flora. Evidence of large forest trees is known from Mfwangano Island, and the faunas from a few levels at Songhor and Koru suggest localised forest as well (Collinson et al., 2009).

Early in the middle Miocene, apes emigrated from Africa, initially in small numbers but later in the middle Miocene in greater numbers. Three groups are known at this stage, afropithecines, kenyapithecines and griphopithecines, and they share the following characters: relatively broad upper central incisors; lower crowned and relatively robust canines; enlarged premolars that are relatively long; molars with thick enamel, low dentine relief; long curved back; forelimbs adapted for both climbing and terrestrial locomotion; stiffer lower back (than proconsulids) analogous but not homologous to the condition in the living great apes; hand proportions indicate both arboreal and terrestrial locomotion; the foot was adapted for powerful grasping.

Body sizes were within the range seen in the proconsulid species, estimated at 35 to 55 kg, and environments were mainly woodland or even open woodland in Africa and subtropical woodlands in Europe (Harrison, 2002). Clearly, primates of this size, living in relatively open canopy woodlands, would have had to spend part of their time on the ground. The thick enamel of their teeth suggest a harder, coarser fruit and nut diet compared with proconsulids (Tekkaya, 1974; Alpagut et al., 1990, 1996; Teaford, 1988, 1991; Harrison, 1992; McCrossin and Benefit, 1997; McCrossin et al., 1998; Nakatsukasa et al., 1998, 2007; Begun and Güleç, 1998; King et al., 1999; Ishida et al., 1999, 2004; Ward, 1993; Ward et al., 1995; Kelley et al., 2000, 2002, 2008; Kelley, 2002, 2008; Ungar, 2007; Ersoy et al., 2008; Nakatsukasa, 2008).

During the second half of the middle Miocene and extending into the late Miocene, there was a greater proliferation of fossil apes in Europe and Asia. There are also a few genera and species known in Africa. The taxonomic status of subfamily Dryopithecinae has passed through several stages in its history, from the time when it included almost all known fossil apes, after the 1965 revision by Elwyn Simons and David Pilbeam (Simons and Pilbeam, 1965), to the later part of the twentieth century when almost everything except *Dryopithecus* itself had been removed (Begun, 2002). This situation will certainly change in the future, with some of the species and genera perhaps being combined and new ones found. Their characters are as follows: skulls with lightly built crania with relatively prominent brow ridges; variable prognathism from low to high; strong angle between face and skull (klinorhynchy) in *Hispanopithecus laietanus;* reduced maxillary sinus; broad triangular nose; broad palate; high zygomatic root; primitive teeth in *D. fontani,* molars with broad basins between cusps elongated molars and premolars in *Pierolapithecus;* teeth with thick enamel in *Anoiapithecus* and *Pierolapithecus;* reduced M3 in the three earlier species but not in *H. laietanus* and *R. hungaricus;* orthograde (upright) posture; broad chest region; long clavicle; scapula shifted on to back; stiff lumbar region; mobile elbow joint, stable at full extension; mobile wrist; long slender hand phalanges (short and less curved in some); femur head above greater trochanter; femur neck steeply angled.

Not all these characters are known for all species, but where they are known for two or more species the characters are consistent, with the conclusion that upright posture, and/or suspensory locomotion had evolved in some species of dryopithecines, particularly in *Hispanopithecus laietanus* (Crusafont-Pairo and Hurzeler, 1961; Pilbeam and Simons, 1971; Kretzoi, 1975; Morbeck, 1983; Begun et al., 1990, 2003; Moyà-Solà et al., 1993, 2004, 2009a, 2009b; Kordos, 1991; Begun and Kordos, 1993; Moyà-Solà S. and Köhler, 1993, 1995, 1996, 1997; Kordos and Begun 1997; Ungar and Kay 1995; Kordos and Begun, 1997; Kordos and Begun, 1997; Begun, 2002, 2009; Ungar, 2005; Alba et al., 2010; Begun et al.,

2012). Some of the characters supposedly indicating suspensory locomotion are absent in gibbons, the most suspensory of the apes, for example the stiff lower back. Similarly, the combination of mobility and stability in the elbow joint is seen as far back in time as the early Miocene in *Proconsul heseloni*, which had no suspensory adaptations. Crompton et al. (2008) points out that the adaptations of the trunk are related to upright posture, not necessarily to suspensory activity in trees, although it may be a pre-adaptation to the specialised brachiation in living gibbons. The large and elongated hand and cranial orientation of the head of the humerus are the two major adaptations that can be related to overhead suspension during locomotion, and these would certainly be pre-adaptations to gibbon-style locomotion.

The smaller dryopithecine species (15 to 35 kg, Begun, 2002) have been found associated with subtropical to warm temperate swamp forests, with mesophytic broadleaved trees and deciduous conifers. These forests are deciduous and have open canopy, but there may have been a lower canopy of evergreen sclerophyllous bushes such as palms and laurels (Kretzoi et al., 1974; Axelrod, 1975; Myers and Ewel, 1990; Kovar-Eder et al., 1996; Kordos and Begun, 2002; Andrews and Cameron, 2010; Merceron et al., 2007; Marmi et al., 2012). There is no indication that these species of fossil ape were terrestrial.

More generally, all five dryopithecine species share some characters with the living great apes, and can be grouped with them in Hominidae. Some characters are shared only with the African great apes and others only with the orangutan. None are shared exclusively with gibbons, although the characters of the shoulder joint and hand are most similar to those of gibbons and orangutans, and it does not seem possible to link dryopithecines with one or other of the extant apes. The mosaic nature of evolutionary change depicted by the dryopithecines suggests that many of the cranial and dental similarities shared by the great apes evolved independently and should not be expected to be present in their common ancestor.

Mosaic evolution is also shown by the morphology of *Oreopithecus bambolii*, a late Miocene ape from Italy. It also had long arms, a broad thorax, short trunk, mobile hindlimbs, and powerful grasping hands and feet, and, like *Hispanopithecus*, it was adapted for forelimb suspension and arboreal climbing (Harrison, 1991; Harrison and Rook, 1997), but its skull and dental morphology show it to be different from the dryopithecines and probably an aberrant side branch of ape evolution.

Late Miocene fossil apes are less well known. A few fragmentary specimens have been recovered in Africa, but they are best known in Asian deposits, from Pakistan and India to Southeast Asia. Their associated habitats appear to be subtropical to tropical woodland (Badgley, 1984, 1989). Some of the species extend back into the middle Miocene, for example *Sivapithecus sivalensis*, and they are similar functionally to middle Miocene European apes, with relatively robust jaws and thick-enamelled teeth. Some have similarities of the skull with the orangutan, but the few postcrania show no suspensory adaptations and indicate a strong element of terrestriality in their locomotion (Pilbeam, 1982, 1996, 2004; Pilbeam et al., 1990; Rose, 1984, 1986, 1988, 1989, 1994, 1997). *Laccopithecus robustus* from late Miocene deposits in China is an ape similar to hylobatids in its skull and dental formation, but a single proximal phalanx is long and curved, like that of *Hispanopithecus* and gibbons (Wu and Pan, 1984; Meldrum and Pan, 1988; Begun, 2002).

In summary, the spectrum of fossil apes as known at present appears to have little or no bearing on the evolution of gibbons and nor, for that matter, to the great apes and humans. In 2006, Terry Harrison and PA tried to define the common ancestor between apes and humans by looking at the full extent of this spectrum, and we found that the last common ancestor of apes was probably not great-ape-like at all (Andrews and Harrison, 2006), and we suggest here that, similarly, the ancestral gibbons for most of their evolutionary history also did not look anything like recent gibbons.

Divergence date of gibbons

In the absence of fossil evidence, it has been proposed that the divergence of genera within the hylobatid clade was about 6 Ma. This is based in large part on the similarities in sequence diversity in mtDNA within the hylobatid clade and the African ape and human clade (Hayashi et al., 1995). Hylobatids are accepted as the outgroup to Hominidae (Hylobatidae(Hominidae(Ponginae(Ho mininae)))), and so, clearly, their separation from hominids must have predated the earliest divergence within Hominidae, that of the orangutan, and this gives a minimum age for the emergence of hylobatids. It has already been observed that the divergence of hylobatids from the great ape and human clade after that of the proconsulids does not itself give a maximum age of divergence. A minimum date is also provided by the uricase mutation mentioned above (and see below) of 13.1 to 9.8 Ma (Keebaugh and Thomas, 2010; Oda et al., 2002).

There are many unresolved issues in determining divergence ages. For a start, the calibration point from which molecular phylogenies are generated from DNA trees is that of the separation between monkeys and apes, and this is commonly taken to be about 30 Ma (Raaum et al., 2005; Locke et al., 2011; Disotell, 2013). However, as Disotell points out, the earliest fossil apes or monkeys date to about 20 Ma, and there is no real basis for assuming a date much earlier than this. It is on the basis of a 30 Ma split between monkeys and apes that the orangutan lineage is thought to have diverged at 12 to 13 Ma, but if the monkey/ape split was closer to 20 Ma, the orangutan divergence would have been closer to 9 or 10 Ma and the separation of the gibbon lineage from other apes not much earlier than that.

The other main problem with the divergence of the orangutan lineage is that two species of fossil ape could be implicated. One is the Indian ape *Sivapithecus sivalensis*, the earliest record for which is about 12 Ma, and which gives a minimum age for the origin of the orangutan lineage (Pilbeam, 1996; Pilbeam et al., 1990). However, the postcranial skeleton of this fossil ape is nothing like that of the orangutan (Rose, 1984, 1986, 1989), whereas the skeleton and some aspects of the skull of *Hispanopithecus laietanus* from nine-million-year-old deposits in Spain have many similarities with the orangutan (Moyà-Solà and Köhler, 1996; Moyà-Solà et al., 2004, 2009a, 2009b). There is little likelihood, however, that *S. indicus* and *H. laietanus* are closely related, and it is clear that one or the other is converging on the orangutan, but which one? The fossils provide a range of dates of 9 to 12 Ma, and all we can say at present is that the gibbons branched off earlier.

Hylobatid common ancestor

To truly reflect evolutionary history, phylogenies must be based on characters inherited from recent ancestors, that is, homologous characters, and convergent, or non-homologous, homoplasies must be discarded. Further to this, the characters shared between two species are only significant in evolutionary terms if they were uniquely shared with their common ancestor, so that they are both homologous and derived relative to other species. Homology can sometimes be clearly evident, as for instance in the loss of the tail in all apes, but they can be difficult to distinguish from primitive retentions from an anthropoid ancestor. As new data are introduced into the analysis, the potential for error is compounded once the individual traits are combined into an ancestral morphotype. With this in mind, it is important to view ancestral morphotypes as approximations with relatively low resolution, rather than precise and accurate formulations of the ancestral condition. This is particularly important in stem forms where the proportion of potentially phylogenetically meaningful characters is small in relation to the number of primitive features, and the level of resolution may, therefore, exceed the capability to confidently differentiate their preserved anatomy from the ancestral morphotypes. If this is the case, there is a serious danger that the outcome of phylogenetic analyses might be influenced or skewed by the introduction of a few characters of uncertain or dubious utility.

From a theoretical perspective it may be argued that any common ancestor is essentially unknowable, because the characters by which it may be linked with its descendant species are not yet present. Closely related species are certain to share many characters as well as having developed different characters after their separation, but by the time that any one of these characters are present in a putative ancestor, it is no longer the common ancestor but belongs to one or other of the descendent lineages. For example, one of the most visible morphologies distinguishing living apes from all other primates is the loss of the tail, and any fossil primate lacking a tail, such as *Proconsul*, cannot be the common ancestor between monkeys and apes but must already be considered an ape. On the other hand, the common ancestor of apes and monkeys could have had a tail without being placed on the line leading to monkeys, for all other primates have tails, and this is the ancestral condition for all primates, primitively retained by monkeys. It is likely, therefore, that the common ancestor of apes and monkeys had a tail, but this does not help to identify it, just as lack of tail would disqualify it as the common ancestor.

Similarly with the gibbons: they share a whole suite of characters relating to their suspensory locomotion and orthograde posture, and this shows them to be a monophyletic group. Absence of some or even most of these characters would not exclude any fossil ape from being ancestral to gibbons *before* they began to speciate, and several possible scenarios for gibbon ancestry can be suggested and potentially tested against future fossil evidence. One such scenario is that the gibbon lineage will show the progressive acquisition of suspensory characters. In this scenario, it is suggested that early gibbon ancestors retained mainly primitive catarrhine characters, with a gradually developing suite of suspensory and orthograde adaptations, most of which were almost certainly acquired independently of the great apes and humans. Two examples illustrate this. *Dendropithecus macinnesi* from 18 Ma in Africa had clear suspensory adaptations, but they were analogous with those of colobine or ateline monkeys, and they cannot be identified with any certainty as being homologous with gibbons. Evidence is lacking to show if the fossil species was orthograde or not. Similarly, the late Miocene *Laccopithecus robustus* from China combined primitive skull morphology with a single phalanx that also showed tantalising evidence of suspensory locomotion. Either or both could be ancestral to gibbons, but it is also the case that there were many other small-sized catarrhine primates in the Miocene, any one of which could have given rise to the gibbons.

An alternative scenario is that the gibbon lineage arose out of one of the lineages already known to have well developed suspensory adaptations, for example from *Hispanopithecus laietanus*. The adaptations of the upper arm and shoulder, the greatly elongated hand and the cranial orientation of the head of the femur could all be precursors to the highly specialised suspensory adaptations in living gibbons. Reduction in size, with increasingly gracile skulls, is not a major evolutionary step. Shea (2013) has shown that many of the apparent differences in the skull between great apes and gibbons is due to their size differences, characters such as palate depth, naso-alveolar clivus morphology, nasal aperture size and shape and the height of the zygomatic root. It is apparent that reduction in body weight by about 50 per cent of a fossil ape such as *Hispanopithecus laietanus* would leave its skull close to the morphology of gibbon skulls. The apparent adaptation of this dryopithecine, together with that of *Rudapithecus hungaricus*, to below branch suspensory locomotion in warm temperate to subtropical swamp forests in southern Europe (Merceron et al., 2007; Andrews and Cameron, 2010; Marmi et al., 2012) could have led to increasing specialisation to life in tropical rainforests.

A third scenario can be based on evidence of the mutation leading to the shutting down of the uricase gene. Uricase is an enzyme that breaks down uric acid, and exists in all mammals except the hominoids. There is evidence that the uricase enzyme progressively lost its activity in hominoids during the early Miocene due to mutations in its promoter region (Oda et al., 2002). However, complete silencing of the uricase gene occurred separately in the great ape–human clade (in codon 33 of exon 2) and in the hylobatids (in codon 18 of exon 2) during the mid-Miocene. The timing of the uricase mutations is not known with certainty, but it has been calculated in the great ape–human clade to have occurred either 12.9 Ma (Keebaugh and Thomas, 2010), 15.4 Ma (Oda et al., 2002), or between 15.7 and 20 Ma (Eric Gaucher, pers. comm.), respectively. The silencing mutation in the hylobatid lineage has been estimated to occur at 13.1 Ma (Keebaugh and Thomas, 2010) or 9.8 Ma (Oda et al., 2002) based on proposed separations of the hominoids and Old World monkeys at 23 Ma and of the *Catarrhini* and hominoids at 35 Ma, respectively. The presence of a monkey-like tooth in 25 Ma deposits of the Nsungwe Formation (Stevens et al., 2013) suggests an earlier divergence of monkeys and apes. These data are also consistent with a separation of hylobatids from the great ape–human lineage between 15 and 20 Ma.

It has been hypothesised that the loss of the uricase gene had a positive adaptive function in the great ape and human clade (Johnson and Andrews, 2010). For fruit eaters such as gibbons, this might appear to be a serious matter, for fructose, the primary sugar present in fruit, stimulates fat synthesis and accumulation due in part to an increase in intracellular and serum uric acid that occurs as a consequence of its unique metabolism (Lanaspa et al., 2012). The ability of fructose to stimulate fat accumulation in the liver is enhanced when uricase is inhibited (Tapia et al., 2013). The uric acid generated by fructose also has an important role in driving the elevations in serum triglycerides, induction of insulin resistance, and elevations in blood pressure in response to fructose (Nakagawa et al., 2006). Therefore, the loss of uricase may have enhanced the ability of gibbons and other ancestral apes to increase their fat stores from the ingestion of ripe fruits rich in fructose that could aid survival through the adverse conditions in seasonal habitats (Johnson and Andrews, 2010). In this scenario, it may be that the late Miocene gibbon ancestors passed through a traumatic phase in increasingly hostile environments before ending up in the rich tropical rainforests of Asia.

Whichever of these scenarios proves to be correct, or any other not so far proposed, it is evident from molecular studies that the gibbon radiation was recent and rapid, either as a vicariant event or extremely rapid radiation (Thinh et al., 2010; Israfil et al., 2011; Disotell, 2013). The trigger for this was probably the rise and fall of sea levels, combined with expansion and contraction of the Southeast Asian rainforests during the glaciations (Geissmann, 1995; Brandon-Jones, 1998). This diversification, however, probably took place during the last two to three million years, and it does not answer the question of what prompted the emergence of the highly specialised gibbon adaptations.

Acknowledgements

We are grateful to the organisers and editors of the book to honour the many contributions made by Colin Groves. Working with Colin has always been enjoyable and stimulating, both in the field on the Tana River project, and in the laboratory, when we did our work on gibbon morphology. PA benefited greatly from the association and owes much to Colin during a crucial stage of his career.

References

Aiello LC. 1981. The allometry of primate body proportions. In: Stringer CB, editor. *Aspects of human evolution*. London: Taylor & Francis. pp. 331–358.

Alba DM, Fortuny J, Moyà-Solà S. 2010. Enamel thickness in the middle Miocene great apes *Anoiapithecus, Pierolapithecus* and *Dryopithecus*. *Proc R Soc Lond* 277:2237–2245.

Alpagut B, Andrews P, Martin L. 1990. New hominoid specimens from the middle Miocene site at Paşalar, Turkey. *J Hum Evol* 19:397–422.

Alpagut B, Andrews P, Fortelius M, Kappelman J, Temizsoy I, Celebi H, Lindsay W. 1996. A new specimen of *Ankarapithecus meteai* from the Sinap Formation of central Anatolia. *Nature* 382:349–351.

Andrews P. 1978. A revision of the Miocene Hominoidea of East Africa. *Bull Br Mus Nat Hist (Geol)* 30:85–224.

Andrews P. 1992. Evolution and environment in the Hominoidea. *Nature* 360:641–646.

Andrews P, Cameron D. 2010. Rudabànya: Taphonomic analysis of a fossil hominid site from Hungary. *Palaeogeog, Palaeoclimat, Palaeoecol* 297:311–329.

Andrews P, Groves CP. 1976. Gibbons and brachiation. In: Rumbaugh D, editor. *Gibbon and Siamang*, Vol. 4. Basel: Karger. pp. 167–218.

Andrews P, Harrison T. 2005. The last common ancestor of apes and humans. In: Lieberman DE, Smith RJ, Kelley J, editors. *Interpreting the past: Essays on human, primate, and mammal evolution in honor of David Pilbeam*. Boston: Brill Academic Publishers. pp. 103–121.

Andrews P, Simons EL. 1977. A new African gibbon-like genus *Dendropithecus* (Hominoidea, Primates) with distinctive postcranial adaptations: Its significance to origin of Hylobatidae. *Folia Primatol* 28:161–169.

Ashton EH, Oxnard CE. 1963. The musculature of the primate shoulder. *Trans Zool Soc Lond* 29:553–650.

Ashton EH, Oxnard CE. 1964. Locomotor patterns in primates. *Proc Zool Soc Lond* 142:1–28.

Avis V. 1962. Brachiation: The crucial issue for man's ancestry. *Southwestern J Anthrop* 18:119–148.

Axelrod DI. 1975. Evolution and biogeography of the Madrean-Tethyan sclerophyll vegetation. *Ann Miss Bot Garden* 62:280–334.

Bacon A-M, Long VT. 2001. The first discovery of a complete skeleton of a fossil orang-utan in a cave of the Hoa Binh Province, Vietnam. *J Hum Evol* 41:227–241.

Badgley C. 1984. The palaeoenvironment of South Asian Miocene hominoids, In: White RD, editor. *The evolution of East Asian environments*. Hong Kong: Centre for Asian Studies. pp. 796–811.

Badgley C. 1989. Community analysis of Siwalik mammals from Pakistan. *J Vert Paleont* 9:11A.

Beard KC, Teaford MF, Walker A. 1986. New wrist bones of *Proconsul africanus* and *P. nyanzae* from Rusinga Island, Kenya. *Folia Primatol* 47:97–118.

Begun DR. 2002. European hominoids. In: Hartig WC, editor. *The primate fossil record*. Cambridge: Cambridge University Press. pp. 339–368.

Begun DR. 2009. Dryopithecins, Darwin, de Bonis and the European origin of the African apes and human clade. *Geodiversitas* 31:789–816.

Begun DR, Geraads D, Guleç E. 2003. The Çandır hominoid locality: Implications for the timing and pattern of hominoid dispersal events. *Courier Forsch Senckenberg* 240:251–265.

Begun DR, Güleç E. 1998. Restoration of the type and palate of *Ankarapithecusmeteai*: Taxonomic and phylogenetic implications. *Am J Phys Anthrop* 105:279–314.

Begun DR, Kordos L. 1993. Revision of *Dryopithecus brancoi* Schlosser, 1910, based on the fossil hominid material from Rudabanya. *J Hum Evol* 25:271–286.

Begun DR, Moyá-Sola S, Köhler M. 1990. New Miocene hominoid specimens from Can Llobateres (Valles Penedes, Spain) and their geological and paleoecological context. *J Hum Evol* 9:255–268.

Begun DR, Nargolwalla MC, Kordos L. 2012. European Miocene hominids and the origin of the African ape and human clade. *Evol Anthrop* 21:10–23.

Begun DR, Teaford MF, Walker A. 1994. Comparative and functional anatomy of Proconsul phalanges from the Kaswanga primate site, Rusinga Island, Kenya. *J Hum Evol* 26:89–165.

Biegert J, Maurer R. 1972. Rumpfskelettlange, Allometrien und Korperproportionen bei catarrhinen Primaten. *Folia Primatol* 17:142–156.

Brandon-Jones D. 1998. Pre-glacial Bornean primate impoverishment and Wallace's line. In: Hall R, Holloway JD, editors. *Biogeography and geological evolution of SE Asia*. Leiden: Blackhuys Publishers. pp. 393–403.

Brandon-Jones D, Eudey AA, Geissmann T, Groves CP, Melnick DJ, Morales JC, Shekelle M, Stewart C-B. 2004. Asian primate classification. *Int J Primatol* 25:97–164.

Collinson ME, Andrews P, Bamford M. 2009. Taphonomy of the early Miocene flora, Hiwegi Formation, Rusinga Island, Kenya. *J Hum Evol* 57:149–162.

Crompton RH, Vereeke EE, Kalb JE. 2008. Locomotion and posture from the common hominoid ancestor to fully modern hominins, with special reference to the common Panin/Hominin ancestor. *J Anat* 212:501–543.

Crusafont-Pairó M, Hürzeler J. 1961. Les pongidés fossilles d'Espagne. *Comptes Rendus bebdomadaires d l'Academie des Sciences Paris* 254:582–584.

Delson E. 1977. Vertebrate paleontology, especially of non-human primates from China. In: Howells WW, Tsuchitani PJ, editors. *Paleoanthropology in the People's Republic of China*. Washington: National Academy of Sciences, pp. 40–65.

Disotell TR. 2013. Genetic perspectives on ape and human evolution. In: Begun DR, editor. *A companion to paleoanthropology*. Oxford: Wiley-Blackwell. pp. 291–305.

Ersoy A, Kelley J, Andrews P, Alpagut B. 2008. Hominoid phalanges from the middle Miocene site of Paşalar, Turkey. *J Hum Evol* 54:518–529.

Gebo DL, Beard KC, Teaford MF, Walker A, Larson SG, Jungers WL, Fleagle JG. 1988. A hominoid proximal humerus from the early Miocene of Rusinga Island, Kenya. *J Hum Evol* 17:393–401.

Gebo DL, Malit NR, Nengo IO. 2009. New proconsuloid postcranials from the early Miocene of Kenya. *Primates* 50:311–319.

Geissmann T. 1995. Gibbon systematics and species identification. *Zoo News* 42:467–501.

Goodman M, Porter CA, Czelnusiak J, Page SL, Schneider H, Shoshani J, Gunnell G, Groves CP. 1998. Towards a phylogenetic classification of primates based on DNA evidence complemented by fossil evidence. *Mol Phyl Evol* 9:585–598.

Groves CP. 1972. Systematics and phylogeny of gibbons. In: Rumbaugh D, editor. *Gibbon and Siamang*, Vol. 1. Basel: Karger. pp. 1–89.

Groves CP. 2001. *Primate taxonomy*. Washington, DC: The Smithsonian Institution.

Harrison T. 1991. The implications of *Oreopithecus* for the origins of bipedalism. In: Coppens Y, Senut B, editors. *Origine(s) de la Bipédie Chez les Hominidés*. Paris: Cahiers de Paléoanthropologie, CNRS. pp. 235–244.

Harrison T. 1992. A reassessment of the taxonomic and phylogenetic affinities of the fossil catarrhines from Fort Ternan, Kenya. *Primates* 33:501–522.

Harrison T. 2002. Late Oligocene to middle Miocene catarrhines from Afro-Arabia. In: Hartwig WC, editor. *The primate fossil record*. Cambridge: Cambridge University Press. pp. 311–338.

Harrison T, Rook L. 1997. Enigmatic anthropoid or misunderstood ape? The phylogenetic status of *Oreopithecus bambolii* reconsidered. In: Begun, DR, Ward CV, Rose MD, editors. *Function, phylogeny, and fossils: Miocene hominoid evolution and adaptations*. New York: Plenum Press. pp. 327–362.

Hayashi S, Hayasaka K, Takenaka O, Horai S. 1995. Molecular phylogeny of gibbons inferred from mitochondrial DNA sequences: Preliminary report. *J Mol Evol* 41:359–365.

Hooijer DA. 1960. Quaternary gibbons from the Malay archipelago. *Zool Verh Mus Leiden* 46:1–42.

Ishida H, Kunimatsu Y, Nakatsukasa M, Nakano Y. 1999. New hominoid genus from the middle Miocene of Nachola. *Anthropological Science* 107:189–191.

Ishida H, Kunimatsu Y, Takano T, Nakano Y, Nakatsukasa M. 2004. *Nacholapithecus* skeleton from the Middle Miocene of Kenya. *J Hum Evol* 46:69–103.

Israfil H, Zehr SM, Mootnick AR, Ruvolo M, Steiper ME. 2011. Unresolved molecular phylogenies of gibbons and siamangs (Family Hylobatidae) based on mitochondrial, Y-linked and X-linked loci indicate a rapid Miocene radiation or sudden vicariance event. *Mol Phyl Evol* 58:447–455.

Johnson RJ, Andrews P. 2010. Fructose, uricase, and the back to Africa hypothesis. *Evol Anthrop* 19:250–257.

Jungers WL. 1984. Aspects of size and scaling in primate biology with special reference to the locomotor skeleton. *Yrbk Phys Anthrop* 27:73–97.

Keebaugh AC, Thomas JW. 2010. The evolutionary fate of the genes encoding the purine catabolic enzymes in hominoids, birds, and reptiles. *Mol Biol Evol* 27:1359–1369.

Kay, RF. 1977. Diet of early Miocene African hominoids. *Nature* 268:628–630.

Kelley J. 2002. The hominoid radiation in Asia. In: Hartwig WC, editor. *The primate fossil record*. Cambridge: Cambridge University Press. pp. 339–368.

Kelley J. 2008. Identification of a single birth cohort in *Kenyapithecus kizili* and the nature of sympatry between *K. kizili* and *Griphopithecus alpani* at Paşalar. *J Hum Evol* 54:530–537.

Kelley J, Andrews P, Alpagut B. 2008. A new hominoid species from the middle Miocene site of Paşalar, Turkey. *J Hum Evol* 54:455–479.

Kelley J, Ward S, Brown B, Hill A, Downs W. 2000. Middle Miocene hominoid origins: Response. *Science* 287:2375a.

Kelley J, Ward S, Brown, B, Hill A, Duren DL. 2002. Dental remains of *Equatorius africanus* from Kipsaramon, Tugen Hills, Baringo District, Kenya. *J Hum Evol* 42:39–62.

King T, Aiello L, Andrews P. 1999. Dental microwear of *Griphopithecus alpani*. *J Hum Evol* 36:3–31.

Kordos L. 1991. Le *Rudapithecus hungaricus* de Rudabànya (Hongrie). *L'Anthropologie* 95:343–362.

Kordos L, Begun DR. 1997. A new reconstruction of RUD 77, a partial cranium of *Dryopithecus brancoi* from Rudabánya, Hungary. *Am J Phys Anthrop* 103:277–294.

Kordos L, Begun DR. 2002. Rudabánya: A late Miocene subtropical swamp deposit with evidence of the origin of the African apes and humans. *Evol Anthrop* 11:45–57.

Kovar-Eder J, Kvacek Zastawniak E, Givulescu R, Hably L, Mihajlovic D, Teslenko J, Walther H. 1996. Floristic trends in the vegetation of the Paratethys surrounding areas during Neogene times. In: Bernor RL, Fahlbusch V, Mitmann H-W, editors. *The evolution of Western Eurasian Neogene mammal faunas.* New York: Columbia University Press. pp. 395–413.

Kretzoi M. 1975. New ramapithecines and *Pliopithecus* from the lower Pliocene of Rudabànya in northeastern Hungary. *Nature* 257:578–581.

Kretzoi M, Krolopp E, Lorincz H, Palfalvy I. 1974. A Rudabanyai Alsopannonai prehominidas lelohely floraja, faunaja es retegtani helyzete. *M All Foldtani Intezet Evi Jelentezi* 1974:365–394.

Lanaspa MA, Sanchez-Lozada LG, Choi YJ et al. 2012. Uric acid induces hepatic steatosis by generation of mitochondrial oxidative stress: Potential role in fructose-dependent and -independent fatty liver. *J Biol Chem* 287:40732–44.

Le Gros Clark WE, Leakey LSB. 1951. The *Miocene Hominoidea of East Africa.* London: British Museum (Natural History).

Le Gros Clark WE, Thomas DP. 1951. *Associated jaws and limb bones of* Limnopithecus macinnesi. London: British Museum (Natural History).

Lewis OJ. 1971. Brachiation and the early evolution of the Hominoidea. *Nature* 203:577–579.

Lewis OJ. 1989. *Functional morphology of the evolving hand and foot.* Oxford: Clarendon Press.

Locke DP, Hillier LW, Warren WC, Worley KC, Nazareth LV, Muzny DM, Yang S-P, Wang Z, Chinwall AT, Minx P. 2011. Comparative and demographic analysis of orang utan genomes. *Nature* 469:529–533.

McBrearty S, Jablonsky NG. 2005. First fossil chimpanzee. *Nature* 437:105–108.

McCrossin ML, Benefit BR. 1997. On the relationships and adaptations of *Kenyapithecus*, a large-bodied hominoid from the middle Miocene of eastern Africa. In: Begun DR, Ward CV, Rose MD, editors. *Function, phylogeny and fossils: Miocene hominoid evolution and adaptations.* New York: Plenum Press. pp. 241–267.

McCrossin ML, Benefit BR, Gitau SN, Palmer AK, Blue KT. 1998. Fossil evidence for the origin of terrestriality among Old World higher primates. In: Strasser E, Fleagle J, Rosenberger A, McHenry H, editors. *Primate locomotion: Recent advances.* New York: Plenum Press. pp. 353–396.

Marmi J, Casanovas-Vilar I, Robles JM, Moyá-Sola S, Alba DM. 2012. The paleoenvironment of *Hispanopithecus laietanus* as revealed by paleobotanical evidence from the late Miocene of Can Llobateres 1 (Catalonia, Spain). *J Hum Evol* 62:412–423.

Matthews WD, Granger W. 1923. New fossil mammals from the Pliocene of Szechuan, China. *Bull Amer Mus Nat Hist* 48:563–598.

Meldrum DJ, Pan Y. 1988. Manual proximal phalanx of *Laccopithecus robustus* from the latest Miocene site of Lufeng. *J Hum Evol* 17:719–731.

Merceron G, Schulz E, Kordos L Kaiser TM. 2007. Paleoenvironment of *Dryopithecus brancoi* at Rudabánya, Hungary: Evidence from dental meso- and micro-wear analyses of large vegetarian mammals. *J Hum Evol* 53:331–349

Morbeck ME. 1983. Miocene hominoid discoveries from Rudabánya: Implications from the postcranial skeleton. In: Ciochon RL, Corruccini RS, editors, *New interpretations of ape and human ancestry.* New York: Plenum Press. pp. 369–404.

Moyà-Solà S, Alba DM, Almecija S, Casanovas-Vilar I, Köhler M, Esteban-Trivigno S de, Robles JM, Galindo J, Fortuny J. 2009a. A unique middle Miocene European hominoid and the origins of the great ape and human clade. *PNAS* 106:1–6.

Moyà-Solà S, Köhler M. 1993. Recent discoveries of *Dryopithecus* shed new light on evolution of great apes. *Nature* 365:543–545.

Moyà-Solà S, Köhler M. 1995. New partial cranium of *Dryopithecus* Lartet, 1863 (Hominoidea, Primates) from the upper Miocene of Can Llobateres, Barcelona, Spain. *J Hum Evol* 29:101–139.

Moyà-Solà S, Köhler M. 1996. The first *Dryopithecus* skeleton: Origins of great ape locomotion. *Nature* 379:156–159.

Moyà-Solà S, Köhler M. 1997. The phylogenetic relationships of *Oreopithecus bambolii* Gervais, 1872. *C R Acad Sciences* 324:141–148.

Moyà-Solà S, Köhler M, Alba DM, Casanovas-Vilar I, Galindo J. 2004. *Pierolapithecus catalaunicus*, a new middle Miocene great ape from Spain. *Science* 306:1339–1344.

Moyà-Solà S, Köhler M, Alba DM, Casanovas-Vilar I, Galindo J, Robles JM, Cabrera L, Garces M, Almecija S, Beamud E. 2009b. First partial face and upper dentition of the middle Miocene hominoid *Dryopithecus fontani* from Abocador de Can Mata (Valles-Penedes Basin, Catalonia, NE Spain): Taxonomic and phylogenetic implications. *Am J Phys Anthrop* 139:126–145.

Myers RL, Ewel JJ. 1990. *Ecosystems of Florida*. Gainesville: University of Central Florida Press.

Nakagawa T, Hu H, Zharikov S et al. 2006. A causal role for uric acid in fructose-induced metabolic syndrome. *Am J Physiol Renal Physiol* 290:F625–631.

Nakatsukasa M. 2008. Comparative study of Moroto vertebral specimens. *J Hum Evol* 55:581–588.

Nakatsukasa M, Kunimatsu Y, Nakano Y, Ishida H. 2007. Vertebral morphology of *Nacholapithecus kerioi* based on KNM-BG 35250. *J Hum Evol* 52:347–369.

Nakatsukasa M, Yamanaka A, Kunimatsu Y, Shimizu D, Ishida H. 1998. A newly discovered *Kenyapithecus* skeleton and its implications for the evolution of positional behaviour in Miocene East African hominoids. *J Hum Evol* 34:657–664.

Napier JR. 1960. Studies of the hands of living primates. *Proc Zool Soc Lond* 134:647–657.

Napier JR. 1963. Brachiation and brachiators. *Symp Zool Soc Lond* 10:183–195.

Napier JR, Davis PR. 1959. *The fore-limb skeleton and associated remains of* Proconsul africanus. London: British Museum (Natural History).

Oda M, Satta Y, Takenaka O, Takahata N. 2002. Loss of urate oxidase activity in hominoids and its evolutionary implications. *Mol Biol Evol* 19(5):640–653.

Pickford M, Senut B. 2005. Hominoid teeth with chimpanzee- and gorilla-like features from the Miocene of Kenya: Implications for the chronology of ape-human divergence and biogeography of Miocene hominoids. *Anthrop Sci* 113:95–102.

Pilbeam DR. 1982. New hominoid skull material from the Miocene of Pakistan. *Nature* 295:232–234.

Pilbeam DR. 1996. Genetic and morphological records of the Hominoidea and hominid origins: A synthesis. *Mol Phyl Evol* 5:155–168.

Pilbeam DR. 2004. The anthropoid postcranial axial skeleton: Comments on development, variation and evolution. *J Experiment Zool* 302:241–267.

Pilbeam DR, Rose MD, Barry JC, Shah MI. 1990. New *Sivapithecus* humeri from Pakistan and the relationship of *Sivapithecus* and *Pongo*. *Nature* 348:237–239.

Pilbeam DR, Simons EL. 1971. Humerus of *Dryopithecus* from Saint Gaudens, France. *Nature* 229:406–407.

Preuschoft H, Chivers DJ, Brockelman WY, Creel N. 1984. *The Lesser Apes: Evolutionary and Behavioural Biology*. Edinburgh: Edinburgh University Press.

Raaum R, Sterner KN, Noviello CM, Stewart C-B, Disotel DR. 2005. Catarrhine primate divergence dates estimated from complete mitochondrial genomes: Concordance with fossil and nuclear DNA evidence. *J Hum Evol* 48:237–257.

Rafferty KL. 1998. Structural design of the femoral neck in primates. *J Hum Evol* 34:361–383.

Rafferty K, Walker A, Ruff C, Rose M, Andrews P. 1995. Postcranial estimates of body weight in *Proconsul*, with a note on a distal tibia of P. major from Napak, Uganda. *Am J Phys Anthrop* 97:391–402.

Rose MD. 1983. Miocene hominoid postcranial morphology: Monkey-like ape-like, neither, or both? In: Ciochon RL, Corruccini RS, editors. *New interpretations of ape and human ancestry*. New York: Plenum Press. pp. 405–417.

Rose MD. 1984. Hominoid postcranial specimens from the middle Miocene Chinji Formation, Pakistan. *J Hum Evol* 13:503–516.

Rose MD. 1986. Further hominoid postcranial specimens from the late Miocene Nagri Formation of Pakistan. *J Hum Evol* 15:333–367.

Rose MD. 1988. Another look at the anthropoid elbow. *J Hum Evol* 17:193–224.

Rose MD. 1989. New postcranial specimens of catarrhines from the middle Miocene Chinji Formation, Pakistan: Descriptions and a discussion of proximal humeral functional morphology in anthropoids. *J Hum Evol* 18:131–162.

Rose MD. 1994. Quadrupedalism in some Miocene catarrhines. *J Hum Evol* 26:387–411.

Rose MD. 1997. Functional and phylogenetic features of the forelimb in Miocene hominoids. In: Begun DR, Ward CV, Rose MD, editors. *Function, phylogeny and fossils: Miocene hominoid evolution and adaptations*. New York: Plenum Press. pp. 79–100.

Schultz AH. 1973. The skeleton of the Hylobatidae and other observations on their morphology. In: Rumbaugh DM, editor. *Gibbon and Siamang*, Vol. 2. Basel: Karger. pp. 1–54.

Simons EL, Pilbeam DR. 1965. A preliminary revision of the Dryopithecinae. *Folia Primatol* 3:81–152.

Stevens, NJ, Seiffert ER, O'Connor PM, Roberts EM, Schmitz, MD, Krause C, Gorscak E, Ngasala S, Hieronymus TL, Temu J. 2013. Palaeontological evidence for an Oligocene divergence between Old World monkeys and apes. *Nature*, published online 15 April 2013.

Tapia E et al. 2013. Synergistic effect of uricase blockade plus physiological amounts of fructose-glucose on glomerular hypertension and oxidative stress in rats. *Am J Physiol Renal Physiol* 304:F727–36.

Teaford MF. 1988. A review of dental microwear and diet in modern mammals. *Scanning Microsc* 2:1149–1166.

Teaford MF. 1991. Dental microwear: What can it tell us about diet and dental function? In: Else J, Lee P, editors. *Advances in dental anthropology*. New York: Wiley-Liss Inc. pp. 341–356.

Teaford MF. 1994. Dental microwear and dental function. *Evol Anthrop* 3:17–30.

Tekkaya I. 1974. A new species of anthropoid (Primates, Mammalia) from Anatolia. *Bull Min Explor Inst*, Ankara (MTA) 83:148–165.

Thinh VN, Mootnick AR, Geissmann T, Li M, Ziegler T, Agil M, Moisson P, Nadler T, Walter L, Roos C. 2010. Mitochondrial evidence for multiple radiations in the evolutionary history of small apes. *BMC Evol Biol* 10: 74.

Ungar PS. 2005. Dental evidence for the diets of fossil primates from Rudabánya, Northeastern Hungary with comments on extant primate analogs and 'noncompetitive' sympatry. *Palaeont Ital* 90:97–112.

Ungar PS, editor. 2007. *Evolution of the human diet.* Oxford: Oxford University Press.

Ungar PS, Kay RF. 1995. The dietary adaptations of European Miocene catarrhines. *PNAS* 93:5479–5481.

Walker AC, Falk D, Smith R, Pickford M. 1983. The skull of *Proconsul africanus*: Reconstruction and cranial capacity. *Nature* 305:525–527.

Walker A, Pickford M. 1983. New postcranial fossils of *Proconsul africanus* and *Proconsul nyanzae*. In: Ciochon RL, Corruccini RS, editors. *New interpretations of ape and human ancestry*. New York: Plenum Press. pp. 325–351.

Walker A, Teaford M. 1988. The hunt for *Proconsul*. *Scient Am* 260:76–82.

Walker A, Teaford M. 1989. The Kaswanga primate site: An early Miocene hominoid site on Rusinga Island, Kenya. *J Hum Evol* 17:539–544.

Walker A, Teaford M, Martin L, Andrews P. 1993. A new species of *Proconsul* from the early Miocene of Rusinga/Mwangano Islands, Kenya. *J Hum Evol* 25:43–56.

Ward CV. 1993. Torso morphology and locomotion in *Proconsul nyanzae*. *Am J Phys Anthrop* 92:291–328.

Ward CV, Ruff CB, Walker A, Teaford MF, Rose MD, Nengo IO. 1995. Functional morphology of *Proconsul* patellas from Rusinga Island, Kenya, with implications for other Miocene-Pliocene catarrhines. *J Hum Evol* 29:1–19.

Ward CV, Walker A, Teaford MF. 1991. *Proconsul* did not have a tail. *J Hum Evol* 21:215–220.

Ward CV, Walker A, Teaford MF, Odhiambo I. 1993. Partial skeleton of *Proconsul nyanzae* from Mfangano Island, Kenya. *Am J Phys Anthrop* 90:77–111.

Wu R, Pan Y. 1984. A late Miocene gibbon-like primate from Lufeng, Yunnan Province. *Acta Anthrop Sin* 3:185–194.

4. Hurricanes and coastlines: The role of natural disasters in the speciation of howler monkeys

Alison M Behie, Travis S Steffens, Tracy M Wyman,
Mary SM Pavelka

In his highly influential book *Primate Taxonomy*, Colin Groves discusses the importance of having an accurate account of primate taxa in order to understand evolutionary relationships that exist between species. This includes understanding genetic and morphological similarities between species as well as the processes of speciation. As the most widely ranging Neotropical monkey, the evolutionary relationships of the genus *Alouatta* have been examined from behavioural, morphological and most recently genetic data. According to Groves (2001) there are nine or 10 species of *Alouatta* and up to 19 subspecies (Table 4.1). Of these species, three are found in Mesoamerica: *A. palliata*, *A. pigra* and *A. coibensis*, with the rest located in South America. However, a more recent study of the molecular genetics of these species revealed *A. coibensis* to be indistinct from *A. palliata*, leaving *A. pigra* and *A. palliata* as the two remaining species in the Mesoamerican clade of howlers (Cortes-Ortiz et al., 2003). In 2012, while at a conference in Mexico, the lead author asked Colin what his thoughts were on the potential role of severe weather in speciation events, and more specifically on the biogeographical distribution on these two closely related species. He admitted he had never given it much thought, but was intrigued by the idea. This chapter further investigates this idea, by pulling together an array of evidence for both *A. pigra* and *A. palliata* in an attempt to add another piece to the puzzle of what factors are important in defining species.

Until 1970 *A. pigra* was considered a subspecies of *A. palliata,* when they were separated into two distinct species by Smith. This distinction was based on differences in their cranial size and morphology, characteristics of the upper molar dentition as well as the colour and texture of the pelage. Horwich and Johnson (1986) later noted that testes descend much earlier in *A. pigra*, and that *A. pigra* lives in consistently smaller social groups. The taxnomonic separation of *A. pigra* and *A. palliata* was genetically confirmed by Cores-Ortiz and others (2007) who showed them to have a 5.7% difference in mitochondrial DNA.

Table 4.1: Howler species (genus *Alouatta*) as recognised by Colin Groves in *Primate Taxonomy* (2001).

Alouatta palliata group		
	Alouatta palliata	
	Alouatta pigra	
	Alouatta coibensis	
Alouatta seniculus group		
	Alouatta seniculus	
		Alouatta seniculus arctoidea
		Alouatta seniculus juara
	Alouatta macconnelli	
	Alouatta sara	
	Alouatta belzebul	
	Alouatta nigerrima	
	Alouatta guariba	
		Alouatta guariba guariba
		Alouatta guariba clamitans
Alouatta caraya group		
	Alouatta caraya	

Note: This chapter considers only the first group, which is the Central American group.

Source: After Groves (2001).

The dominant view of the colonisation of Mesoamerica by South American primate species revolves around the barrier created by the Andes, which limited the ability of taxa to move out of South America (Ford, 2006). When a land bridge formed between the two regions 3.5 million years ago, the only genera that were able to migrate were those already across this boundary in the northwest of South America. This included the genera *Alouatta*, *Ateles*, *Aotus*, *Cebus* and *Saguinus*. The divergence of the Mesoamerican *Alouatta* species into *A. pigra* and *A. palliata* could have occurred in one of two ways: colonisation by a single species that later split or speciation pre-colonisation and a first wave of *A. pigra* followed later by *A. palliata* (Smith, 1970). Cores-Ortiz and others' (2007) genetic data suggest a split of three million years ago for the two species, which coincides with the formation of the Panamanian land bridge. This is consistent with either the pre or post colonisation split. Ford (2006) supports a pre-colonisation speciation followed by four to five waves after the formation of the land bridge, however, cautions that due to the poorly resolved phylogenies of Mesoamerican primate species, the data do not support firm conclusions about this.

It has been suggested that *A. palliata* may not have been able to colonise areas where *A. pigra* are found due to the ability of *A. pigra* to live in a wider range of habitat types including swamps, mangroves and dry deciduous forests (Reid,

1997). Further, *A. pigra* is mainly found in lowland coastal regions of less than 500 m (Horwich and Johnson, 1986; Baumgarten, 2006). Coastal forests of low elevation have an increased vulnerability to storms and severe stochastic weather patterns (Ford, 2006). As part of the Northern Atlantic Cyclone Basin, the Caribbean along with Central and South America are frequently hit by hurricanes, averaging six hurricanes and two tropical storms per year (www.noaa.com). Most of these, however, pass through the northern part of the hurricane basin in the Yucatan region. This type of regular exposure to severe weather events should have profound effects on the animals living in the area and, depending on frequency of exposure, could have influenced their behaviour and demographic profile (Ford, 2006). This is certainly the case in Madagascar, where the history of regular cyclones has resulted in changes to lemur behaviour and morphology including, small group size, high degrees of energy conserving behaviours (including torpor), and a limited number of species that are dedicated frugivores (Wright, 1999).

In this chapter we explore the possible role of severe weather events in determining the biogeographical distribution of *A. pigra* and *A. palliata* through a study of the forest characteristics that predict the occurrence of *A. pigra* in Belize. We also conduct an examination of group size and evidence of energy conservation in *A. pigra* and *A. palliata* to determine if *A. pigra* show any of the behavioural features associated with living in a stochastic environment. Finally we look for evidence of different levels of environmental stochasticity faced by *A. pigra* as compared to *A. palliata* by comparing the frequency of hurricanes and other tropical storms making landfall in the regions populated by each species. While we acknowledge that Colin Groves considers *A. coibensis* a separate species (Table 4.1), in this paper we focus on *A. palliata* and *A. pigra* due to limited data available on *A. coibensis* as well as the fact that its distribution is limited to the island of Coiba, which is located outside of a hurricane belt (Rylands et al., 2006).

Materials and methods

Forest characteristics associated with *A. pigra*

We used satellite imagery, local informants, guides, and published literature to identify areas that might contain *A. pigra* populations or be suitable habitat for *A. pigra* in Belize. To determine *A. pigra* presence or absence, and relative abundance, areas that were accessible were surveyed on foot using existing trails, logging roads, new trails or by boat along rivers. When a group of *A. pigra* was spotted, location data were collected using a hand held Global Positioning System device (Garmin GPSmap 60CSx). When possible this included the exact location of the group, but when it was not possible we recorded our location and a compass bearing towards the group and a visual estimate of the distance to the group in metres. We also recorded the group size and composition, height in the canopy and group activity. When a group was heard vocalising we estimated the distance in metres and direction of the group. In previous studies (Pavelka et al., 2007), we found this overestimated the distance by 0.5 to 1.0 km, therefore we took this into account in our analysis here. All spatial data (visual and vocal contacts, confirmations of presence from local informants, and track information) were entered into ArcMap software (v.10.1). Relative abundance was calculated by dividing the number of monkeys sighted by the total distance (km) walked, in that patch.

To assess forest characteristics associated with *A. pigra*, we measured its presence or absence and relative abundance in relation to anthropogenic disturbance (road density, number of settlements, human population density, amount of edge, and presence of agriculture), natural disturbance (hurricanes), and patch characteristics (patch size, patch type, river density, and elevation). We used ArcMap GIS (v.10.1) software to measure road density, the number of settlements within a patch, human population density of the patch, amount of edge to area ratio of a patch, how many hurricanes of category 1–4 have occurred within a patch, patch size, patch type, river density, and mean area-weighted patch elevation. The data to measure the above variables were acquired from BERDS (Biodiversity and Environmental Resource Data System) for all variables except the number of hurricanes, which was acquired from the National Oceanic and Atmospheric Administration (NOAA, 2006).

Behavioural comparisons between *A. pigra* and *A. palliata*

We surveyed the available literature on *A. pigra* and *A. palliata* to determine the average and mean group size of each species as well as to determine the influence of fruit consumption on activity budgets to look for indications of energy minimising behaviour. For the behavioural studies, we only included studies that were done over greater than six months to allow for seasonal variation in fruit availability and consumption to be considered. This resulted in the inclusion of three studies for *A. pigra* and 13 for *A. palliata* (Table 4.2). Due to this limited sample size, no statistical analyses could be performed on the behavioural changes associated with seasonal fruit scarcity, however, these data were compared qualitatively.

Table 4.2: Studies included in behavioural comparisons of *Alouatta pigra* and *A. palliata*.

Behavioural response to fruit shortage	Species	Study site	Group size (# individuals)	Reference
Decrease time spent travelling/ranging	*Alouatta palliata*	Los Tuxlas, Mexico	14	Estrada, 1984
		Arroyo Liza, Mexico	6	Asensio et al., 2007
	Alouatta pigra	Monkey River	6.6	Pavelka and Knopff, 2004
		Cockscomb Basin Wildlife Sanctuary, Belize	6.5	Silver and Marsh, 2003
Group fission	*Alouatta palliata*	Agaltepec Island, Mexico	59	Asensio et al., 2007
Increase time spent feeding	*Alouatta palliata*	Los Tuxlas, Mexico	4	Sheddon-Gonzalez and Rodriguez-Luna, 2010
		La Pacifica, Costa Rica	13	Glander 1978
Increased time spent travelling/ranging	*Alouatta palliata*	La Pacifica, Costa Rica	25	Williams-Guillen, 2003
		La Pacifica, Costa Rica	15	Williams-Guillen, 2003
		La Pacifica, Costa Rica	20	Williams-Guillen, 2003
		La Pacifica, Costa Rica	13	Glander, 1978
		Agaltepec Island, Mexico	10	Rodriguez Luna et al., 2003
		Agaltepec Island, Mexico	27	Rodriguez Luna et al., 2003
		Barro Colorado Island, Panama		Milton, 1980, 1981
No change in behaviour	*Alouatta palliata*	Los Tuxlas, Mexico	7	Estrada and Coates-Estrada, 1999
		La Selva, Mexico	20	Stoner, 1996
		La Selva, Mexico	11	Stoner, 1996
		Santa Rosa, Costa Rica	40	Chapman, 1987, 1988
		Playo Escondida, Mexico	7	Asensio et al., 2007
	Alouatta pigra	Community Baboon Sanctuary, Belize	5.9	Silver et al., 1998
Increase in time spent inactive	*Alouatta pigra*	Cockscomb Basin Wildlife Sanctuary, Belize	6.5	Silver and Marsh, 2003

Source: Data compiled from behavioural and diet studies of at least nine months in duration to account for seasonality. This resulted in three studies for *A. pigra* and 13 for *A. palliata*. See reference list for full references of studies included.

Hurricane-to-coastline ratios

In order to calculate the number of hurricanes in relation to length of coastline (a hurricane to coastline ratio) for *A. pigra* and *A. palliata,* species distribution maps based on Rylands and others (2006) and hurricane tracks obtained from the National Oceanic and Atmospheric Administration (NOAA) in digital format, were plotted in ArcGIS 10.1 (ESRI, 2012). Political boundary data of all Central American countries consisted of datasets from the Digital Chart of the World (Defence Mapping Agency, 1992). These datasets have a standard 1:1,000,000 scale and were used as the base for coastline measures. For the purpose of this study we chose to use a method that would be easily repeatable yet still provide a valid measure of coastline length with respect to the potential distance that tropical storms could cross. Coastline lengths for *A. pigra* and *A. palliata* ranges were calculated using the 'detailed hull' function in the extension XTools Pro 9.2 (Data East, 2012) in ArcGIS. The detailed hull function creates a 'contour' around outer points similar to that of a standard convex hull, except that it does include some concave angles when consecutive line segments fall below a certain length – the result of this detailed hull function is a 'finer scaled convex hull'. If the actual length of coastline was used, it would overestimate length due to undulations caused by major inlets and bays along the actual coastline. These are especially apparent along the coast of Belize.

The complete dataset of hurricane tracks included hurricanes from Category 1 to 5, tropical storms, and tropical depressions of all known and recorded storms from the years 1851 to 2007. Because hurricane strength often changes once hitting land, it was not possible to calculate a reasonable comparative measure of number of hurricanes by species range area. Thus we calculated a ratio of the number of hurricanes crossing the coastline for each of the species' range. Coastline measurements were clipped to create a measure for the two species ranges' separately. Hurricane track data were filtered so that two measures could be performed: (1) all tropical depressions, tropical storms, and hurricanes Category 1 to 5; (2) hurricanes only (all categories).

Results

Forest patch characteristics of *A. pigra* in Belize

We recorded a total of 83 visual and 110 vocal contacts of *A. pigra* within Belize and collected another 284 confirmations of presence from other reliable sources. All patches were sampled in lowland habitat and a chi square test for independence (p = 0.05) found that *A. pigra* were more likely to be present in a patch classified as lowland broad-leaf moist and wet forest, and more likely to be absent from patches classified as agriculture, lowland savannah, shrubland, or wetland. The area-weighted mean elevation of all patches ranged between sea level and 222 m ASL with the highest confirmation of an *A. pigra* group at 700 m and the highest reported sighting of a group at 289 m ASL. When considering factors influencing the relative abundance of *A. pigra*, only the amount of edge was positively associated (r = 0.545; p = 0.006) accounting for 30% of the variation in relative abundance (r^2 = 0.297).

We found a significant positive relationship between the number of settlements in a patch and the presence of *A. pigra*, with more settlements in present patches (1.19) than absent patches (0.14). There was also a significant relationship between *A. pigra* presence and patch size. Patches ranged in size from 0.13 km^2 to over 1256 km^2 with the largest patch occurring in the Rio Bravo and Gallon Jug region and the smallest patches (<1.0 km^2) adjacent to the Belize River. We found present patches to be, on average, larger (117.24 km^2) than absent patches (27.29 km^2).

Behavioural comparisons of *A. pigra* and *A. palliata*

Our literature review found *A. pigra* to live in significantly smaller groups than *A. palliata* with *A. pigra* living in groups ranging from 5.9 to 9 individuals ($\bar{x}$ = 6.83, CV = 19.55) and *A. palliata* living in groups ranging from 4 to 59 individuals ($\bar{x}$ = 15.37, CV = 79.78). Group size in *A. palliata* was more variable, which may be due to the increased sample size for the comparatively well studied mantled howler or reflect group size constraints that may be present in *A. pigra* due to the stochastic nature of their environment.

When examining the relationship between fruit consumption and activity patterns we found no consistent differences within or between species in how populations adjust behaviour during periods of fruit shortage. Populations of *A. pigra* either decreased time spent travelling (from 9.52% to 5.45%), increased time spent inactive (from 69% to 77%) or did not adjust activity patterns in response to seasonal reductions in fruit intake (Behie and Pavelka, 2005; Silver et al., 1998; Pavelka and Knopff, 2004). While some groups of *A. palliata* also did not show changes in activity that correspond to fruit production, the most common response reported in the literature is an increase in either travel time (from 18.6% to 35.8% in *A. palliata* in Nicaragua, Williams-Guillen, 2003) or ranging distance (from 114.05 m to 502.88 m for a population in Mexico, Estrada, 1984), which is a response that has never been reported for a population of *A. pigra*.

Severe weather events in the ranges of *A. pigra* and *A. palliata*

The method described above resulted in a coastline length of 893 km for *A. pigra* and 1372 km for *A. palliata*. Since 1851, 118 Atlantic hurricanes, depressions or tropical storms have crossed into coastal regions populated by *A. pigra*, where only 40 have crossed into the range of *A. palliata*. There is a significantly higher storm to coastline ratio of 0.132 for *A. pigra* than the 0.0292 for *A. palliata* (X^2 =37.52, df = 1, p<0.0001) (Figure 4.1a). When considering hurricanes only, the difference is still evident. Many more hurricanes crossed the coastal regions of *A. pigra* (N = 49) compared to *A. palliata* (N = 19) (Figure 4.1b). The hurricane to coastline ratios (0.551 for *A. pigra* and 0.0138 for *A. palliata*) were significantly different (X^2 =12.36, df = 1, p=0.0004).

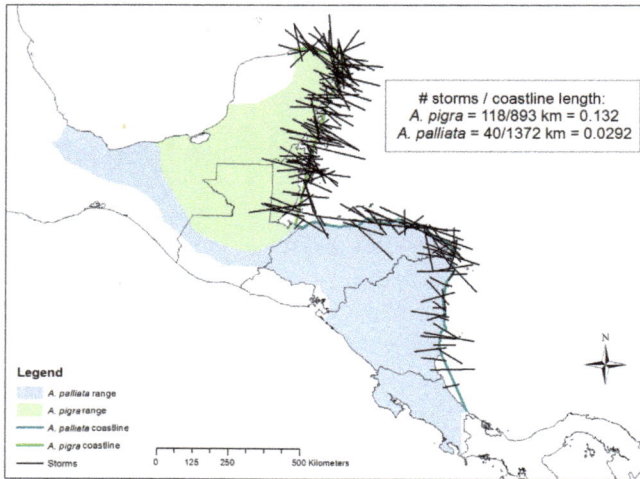

Figure 4.1a: Total number of storms that crossed through the ranges of *Alouatta pigra* and *Alouatta palliata* from 1851 to 2007.

Source: Data were obtained from the National Oceanic and Atmospheric Administration (NOAA) and plotted in ArcGIS 10.1 (ESRI, 2012). Political boundary data of all Central American countries consisted of datasets from the Digital Chart of the World (Defence Mapping Agency, 1992). These datasets have a standard 1:1,000,000 scale and were used as the base for coastline measures.

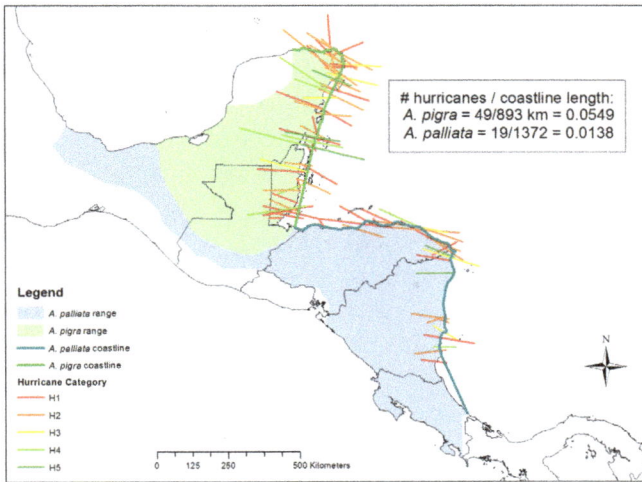

hurricanes / coastline length:
A. pigra = 49/893 km = 0.0549
A. palliata = 19/1372 = 0.0138

Legend
- A. palliata range
- A. pigra range
- A. palliata coastline
- A. pigra coastline

Hurricane Category
- H1
- H2
- H3
- H4
- H5

0   125   250   500 Kilometers

Figure 4.1b: Total number of hurricanes that crossed through the ranges of *Alouatta pigra* and *Alouatta palliata* from 1851 to 2007.

Source: Data were obtained from the National Oceanic and Atmospheric Administration (NOAA) and plotted in ArcGIS 10.1 (ESRI, 2012). Political boundary data of all Central American countries consisted of datasets from the Digital Chart of the World (Defence Mapping Agency, 1992). These datasets have a standard 1:1,000,000 scale and were used as the base for coastline measures.

Discussion

The evolutionary history of *A. pigra* and *A. palliata* in Mesoamerica is not entirely understood. Colonisation of Mesoamerica by South American species took place after the formation of a land bridge between the two regions approximately 3.5 million years ago (Ford, 2006), and genetic data suggest that the two *Alouatta* species diverged from one another close to 3.0 Mya (Cores-Ortiz et al., 2007). Whether they colonised Mesoamerica before or after this speciation event, *A. pigra* is currently limited to a small geographic range in the Yucatan Peninsula compared to a relatively wide distribution of *A. palliata*. The aim of this paper was to investigate the potential role of environmental stochasticity in the geographic distribution and possibly speciation of *A. pigra* and *A. palliata* through: an examination of the forest characteristics that predict the presence of *A. pigra*; a comparison of the behavioural traits of *A. pigra* and *A. palliata* that might suggest an evolutionary history shaped by living in a stochastic environment; and finally a comparison of the prevalence of severe weather events, such as hurricanes, in the ranges of the two species.

In terms of patch characteristics, we found a higher relative abundance of
A. pigra in Belize in lowland broad-leaf moist and wet forest, in line with
previous reports that *A. pigra* inhabit primarily lowland riverine coastal forests
(Horwich and Johnson, 1986; Reid, 1997). The highest published densities
reported for *A. pigra* have been at the Community Baboon Sanctuary (CBS) near
the Belize River (257 ind/km², Ostro et al., 1999) and at Monkey River (102 ind/
km², Pavelka, 2003). Seasonally flooded forests create localised disturbance
which may improve soil quality and thus the quality of the howler food supply
(Peres, 1997). We also found a significant positive relationship between both
human settlements and patch size and the occurrence of *A. pigra* and between
the amount of edge and the relative abundance of *A. pigra*. This suggests that
A. pigra is tolerant of disturbance and may even prefer disturbed forest patches.
One reason for such a preference may be that disturbed areas are colonised
by fast growing pioneer species that invest little in chemical defence, thus
produce leaves high in protein and low in fibre and secondary compounds
(Coley, 1983). Folivores are able to take advantage of this and maintain a high
quality diet despite an overall reduction in stem density. We found this to be
the case following hurricane Iris in Monkey River where *A. pigra* shifted their
leaf consumption to include up to 75% *Cecropia peltata*, which had the highest
protein to fibre ratio and the third highest concentration of sugar of all ingested
species (Behie et al., 2014).

Living in disturbed forest patches may improve the food supply for folivores in
the long term, but immediately following a severe natural disaster there would
be substantial reductions in the available food supply. Many trees regenerate
new leaves immediately following a hurricane (Klinger, 2006; Zimmerman and
Covich, 2007; Waide, 1991), while other food sources such as fruit or flowers
take longer to return (Waide, 1991; Behie and Pavelka, 2005; Ratsimbazafy et
al., 2002). After Hurricane Iris hit Monkey River there was a 52 per cent loss of
major fruit trees and an 18 month absence in all fruit production forcing resident
A. pigra groups to rely on a completely folivorous diet (Behie and Pavelka, 2005;
Behie and Pavelka, in press). Following natural disturbances ring-tailed lemurs
(*Lemur catta*; LaFleur and Gould, 2009; Ratsambazafy et al., 2002), ruffed lemurs
(*Varecia v. editorium*; Ratzimbazafy, 2006) lion-tailed macaques (*Macaca silenus*;
Menon and Poirer, 1996) and black howlers (*Alouatta pigra*; Behie and Pavelka,
2005) altered their diets to include plant parts and species not previously
ingested, some of which were exotic to the region or were located outside of the
regular forest habitat.

Such changes in food supply would be expected to result in demographic and
behavioural changes to deal with an unpredictable and irregular food supply.
One such mechanism to cope with this would be to reduce group size to deal
with increased feeding competition. The average group size of black-and-

white lemurs decreased from seven individuals before Cyclone Gretelle to 2.5 individuals after (Ratsimbazafy et al., 2002) and after wildfires moved through Borneo, the number of gibbon groups of more than five members significantly decreased, while the number of pairs increased (O'Brien et al., 2004). Group size in *A. pigra* in Monkey River following Hurricane Iris fell from 6.32 individuals before the storm to less than five in the first year following the storm. This suggests that animals exposed to severe disturbance adapt in the short term by reducing group size.

If exposed to severe environmental change on a regular basis, it stands to reason that group size may be constrained, placing upper limits on how many animals could live in a group without suffering the ill effects of increased competition at times of food scarcity. This was suggested by Wright (1999) who noted that lemur populations affected by frequent cyclones lived in smaller groups than groups not impacted by severe weather. In our comparison with *A. palliata* we found that *A. pigra* have significantly smaller and less variable group sizes than do *A. palliata*. Mean group size in *A. palliata* is 15.37 (Glander, 1978; Estrada, 1982, 1984; Chapman, 1987; Larose, 1996; Stoner, 1996; Estrada et al., 1999; Serio-Silva et al., 1999; Solano et al., 1999; Rodriguez-Luna, 2003; Williams Guillen, 2003; Munoz et al., 2006; Asensio et al., 2007; Dunn et al., 2009; Dunn et al., 2010) and mean group size in *A. pigra* is 6.83 (Silver et al., 1998; Silver and Marsh, 2003; Pavelka and Knopff, 2004; Pozo-Montuy and Serio-Silva, 2006). Considering the hurricane activity that occurs in the range of *A. pigra*, low group size could very well represent an adaptation to living in a stochastic environment. James et al. (1997) studied groups of *A. pigra* living at Bermudian Landing, Belize that were subjected to population declines from hurricanes in 1931, 1954 and 1978 as well as from a yellow fever epidemic which occurred in 1971. Each of these severe weather events caused a drastic decline in population numbers, and although they have recovered, the average group size in this region is only 4.6 individuals, lower than most other species of howlers. In 1999 this number increased to between four to 10 individuals, which although larger, is still smaller than other howler species (Ostro et al., 1999). A similar decline in group size was recorded in Monkey River following Hurricane Iris where group size remains smaller than before the storm even after more than a decade has passed.

We also found differences in activity patterns between *A. pigra* and *A. palliata* during periods of fruit scarcity. The prolonged fruit shortage following Hurricane Iris resulted in differences in activity levels between periods when fruit was not available (2002–mid-2004; mean fruit intake 4.93%) and when it was available at close to pre-hurricane levels (mid-2004–2007; mean fruit intake 28.75%). When fruit consumption was absent or very low, animals spent more time inactive and less time feeding and locomoting, probably minimising energy expenditure in response to low energy intake. An increase in time spent inactive and a reduction in time spent feeding has also been seen in howler monkeys following translocation to an unfamiliar environment (Silver and Marsh, 2003) and in lemurs in response to unpredictable resource availability and dramatic and prolonged shortages in fruit production in Madagascar (Wright, 1999). In lemurs this is also associated with a lower basal metabolic rate, which varies in response to changing fruit production, allowing them to maximise energy conservation (Pereira, 1993; Jolly, 1966). Increasing inactivity in the howlers at Monkey River in response to prolonged fruit shortage likely serves a similar function allowing howlers to conserve energy at times when higher quality resources (e.g. fruit) are not ingested. This however is not the response seen in *A. palliata* who have been reported to increase travel time (Williams-Guillen, 2003) or ranging distance (Estrada, 1984) when fruit is less available. This may be reflective of increased scramble competition in the larger groups of *A. palliata*, however, it may also be a result of *A. palliata* living in regions that show more predictability in fruit production with less need to conserve energy for long periods of food scarcity.

This successful dispersal of *A. palliata* may have resulted in their outcompeting *A. pigra* in most areas, pushing them up into their currently restricted range of southern Mexico, Belize and northern Guatemala (Ford, 2006). If this is true, and *A. palliata* were able to outcompete *A. pigra* and push them out of many regions, then the current range of *A. pigra* must represent areas that are not tolerated by *A. palliata*. There is no denying that exposure to hurricanes would create a selective pressure for animals living in hurricane belts, an idea that Colin himself agrees may have influenced speciation and/or current biogeographical distributions. Data presented here show that significantly more hurricanes pass into the range of *A. pigra* than the range of *A. palliata*. This may have been one reason why *A. palliata* did not extend their range into the hurricane belt of the Yucatan Peninsula, leaving *A. pigra* as the only *Alouatta* species to colonise the area. This is supported by the fact that *A. pigra* are commonly found in and may actually prefer disturbed forests and live in small groups who exhibit energy conservation strategies with regards to their activity budgets. Such adaptations to environmental stochasticity are also seen in lemurs regularly exposed to cyclones in Madagascar (Wright, 1999) suggesting they are necessary adaptive mechanisms to cope with high degrees of environmental perturbations. *A. palliata* may be less able to tolerate hurricane activity and unable to colonise the hurricane belt of the Yucatan Peninsula. While there may be other reasons contributing to the range separation of *A. pigra* and *A. palliata*, we argue that the role of hurricane activity cannot be ruled out as a possible explanation for the current biogeographical separation and potentially of speciation in these two species and potentially other species exposed to severe weather conditions.

This is an idea rarely considered in theories of primate speciation or when considering current biogeographical ranges, but as severe weather events become more frequent and intense is one that warrants further investigation.

Acknowledgements

We wish to thank the Belize Government for granting us permission to conduct this research. The Monkey River Research Project could not be done without the assistance of guides and research assistants who aid in the collection of data and monitoring of the monkey population. Financial support for this research was received from The Natural Sciences and Engineering Council of Canada, The International Primatological Society, Sigma Xi, The Department of Anthropology, Faculty of Social Sciences and Graduate Studies at the University of Calgary. All data collection met the principles of the ethical treatment for animals and were approved by the Life and Environmental Sciences Animal Care Committee at the University of Calgary.

References

Asensio N, Cristobal-Azkarate J, Dias PAD, Vea JJ, Rodriguez-Luna E. 2007. Foraging habits of *Alouatta palliata mexicana* in three forest fragments. *Folia Primatol* 78:141–153.

Baumgarten A. 2006. The distribution and biogeography of Central American howling monkeys (*Alouatta pigra* and *A. palliata*). Masters Thesis, Louisiana State University.

Behie AM, Pavelka MSM. 2005. The short-term effect of Hurricane Iris on the diet and activity of black howlers (*Alouatta pigra*) in Monkey River, Belize. *Folia Primatol* 76:1–9.

Behie AM, Pavelka MSM. 2013. The interacting roles of nutrition, stress and parasitism in determining population density of howler monkeys living in a hurricane disturbed forest fragment. Chapter to be included in the book *Primates infragments: Complexity and resilence*. March LK, editor. Springer Press.

Behie AM, Kutz S, Pavelka SMS. 2014. Cascading effects of climate change: How do hurricane-damaged forests increase exposure to parasies. *Biotropica*. 46:25–34.

Behie AM, Pavelka MSM. In press. Fruit as a key factor in howler monkey population density: Conservation implications. In: Kowalewski M, Garber PA, Cortes-Ortiz L, Urbani B, Youlatos D, editors. *Howler monkeys: Examining the biology, adaptive radiation, and behavioral ecology of the most widely distributed genus of Neotropical primate.* New York: Springer Press.

Chapman CA. 1987. Flexibility in diets of three Costa Rican primates. *Folia Primatol* 49:90–105.

Coley PD. 1983. Herbivory and defensive characteristics of tree species in a lowland tropical forest. *Ecol Monographs* 53(2):209–234.

Cortes-Oritz L, Bermingham E, Rico C, Rodriguez-Luna E, Sampaio I, Ruiz-Garcia M. 2003. Molecular systematics and biogeography of the Neotropical monkey genus, *Alouatta. Mol Phyl Evol* 26:64–81.

Data East. 2012. XTools Pro for ArcGIS Desktop. 9.2. Data East, LLC.

Defense Mapping Agency. 1992. *Digital Chart of the World.* Fairfax, VA: Defense Mapping Agency.

Dunn JC, Asensio N, Arroyo-Rodriguez V, Schnitzer S, Cristobal-Azkarate J. 2009. The ranging costs of a fallback food: Liana consumption supplements diet but increases foraging effort in howler monkeys. *Biotropica* 43:612–618.

Dunn JC, Cristobal-Azkarate J, Vea JJ. 2010. Seasonal variations in the diet and feeding effort of two groups of howlers in different sized forest fragments. *Int J Primatol* 31:887–903.

Environmental Systems Resource Institute (ESRI). 2012. ArcGIS 10.1. Redlands, California: ESRI.

Estrada AR. 1982. Survey and census of howler monkeys (*Alouatta palliata*) in the rainforest 'Lox Tuxlas' Veracruz, Mexico. *Am J Primatol* 2:363–372.

Estrada A. 1984. Resource use by howler monkeys (*Alouatta palliata*) in the rainforest of 'Los Tuxlas', Veracruz, Mexico. *Int J Primatol* 5:105–131.

Estrada A, Anzures A, Coates-Estrada R. 1999. Tropical rainforest fragmentation, howler monkeys (*Alouatta palliata*), and dung beetles at Los Tuxlas, Mexico. *Am J Primatol* 48:253–262.

Ford SM. 2006. The biogeographic history of Mesoamerican primates. In: Estrada, A, Garber PA, Pavelka MSM, Luecke L, editors. *New perspectives in the study of Mesoamerican primates: Distribution, ecology, behavior, and conservation.* New York: Springer. pp. 81–120.

Glander KE. 1978. Howling monkey feeding behavior and plant secondary compounds: A study of strategies. In: Montgomery GG, editor. *The ecology of arboreal folivores*. Washington, DC: Smithsonian Press. pp. 561–573.

Groves CP. 2001. *Primate taxonomy*. Washington, DC: Smithsonian Books.

Horwich RH, Johnson ED. 1986. Geographic distribution of the black howler (*Alouatta pigra*) in Central America. *Primates* 2:53–62.

James R, Leberg PL, Quattro JM, Vrijenhoek RC. 1997. Genetic diversity in black howler monkeys (*Alouatta pigra*). *Am J Phys Anthropol* 102:329–336.

Jolly A. 1966. *Lemur behaviour: A Madagascar field study*. Chicago: University of Chicago Press.

Klinger R. 2006. The interaction of disturbances and small mammal community dynamics in a lowland forest in Belize. *J Anim Ecol* 75:1227–1238.

LaFleur M, Gould L. 2009. Feeding outside the forest: The importance of crop raiding and an invasive weed in the diet of gallery forest ring-tailed lemurs (*Lemur catta*) following a cyclone at the Beza Mahafaly Special Reserve, Madagascar. *Folia Primatol* 80:233–246.

Larose F. 1996. Foraging strategies, group size and food competition in the mantled howler monkey, *Alouatta pigra*. PhD thesis, University of Alberta.

Menon S, Poirer FE. 1996. Lion-tailed macaques (*Macaca silenus*) in a disturbed forest fragment: Activity patterns and time budget. *Int J Primatol* 17:969–985.

Milton K. 1980. *The foraging strategy of howler monkeys: A study in primate economics*. New York: Columbia University Press.

Milton K. 1981. Food choices and digestive strategies of two sympatric species. *Amer Nat* 117:496–505.

Munoz D, Estrada A, Naranjo E, Ochoa S. 2006. Foraging ecology of howler monkeys in a cacao (*Theobroma cacao*) plantation in Comalcalco, Mexico. *Am J Primatol* 68:127–142.

National Oceanic and Atmospheric Administration (NOAA). 2006. Historical North Atlantic and East-Central North Pacific tropical cyclone tracks, 1851–2005. www.csc.noaa.gov/hurricane_tracks (accessed September 2006).

O'Brien TG, Kinnaird MF, Nurcahyo A, Iqbal M, Rusmanto M. 2004. Abundance and distribution of sympatric gibbons in a threatened Sumatran rainforest. *Int J Primatol* 25:267–284.

Ostro LET, Silver SC, Koontz FW, Young TP, Horwich RH. 1999. Ranging behavior of translocated and established groups of black howler monkeys *Alouatta pigra* in Belize, Central America. *Biol Conserv* 87:181–190.

Pavelka MSM. 2003. Population, group, and range size and structure in black howler monkeys (*A. pigra*) at Monkey River in southern Belize. *Neotrop Primates* 11:187–189.

Pavelka MSM, Behie AM. 2005. The effect of hurricane Iris on the food supply of black howlers (*Alouatta pigra*) in southern Belize. *Biotropica* 37:102–108.

Pavelka MSM, Knopff KH. 2004. Diet and activity in black howler monkeys (*Alouatta pigra*) in southern Belize: Does degree of frugivory influence activity level? *Primates* 45:105–111.

Pavelka MSM, McGoogan KC, Steffens TS. 2007. Population size and characteristics of *Alouatta pigra* before and after a major hurricane. *Int J Primatol* 28:919–929.

Pereira ME. 1993. Seasonal adjustment of growth rate and adult body weight in ringtailed lemurs. In Kappeler PM, Ganzhorn JU, editors. *Lemur social systems and their ecological basis*. New York: Plenum Press. pp. 205–221.

Peres CA. 1997. Primate community structure at twenty western Amazonia flooded and un-flooded forests. *J Trop Ecol* 13:381–405.

Pozo-Montuy G, Serio-Silva JC. 2006. Comportamiento alimentario de monos aulladores negros (*Alouatta pigra Lawrence*, Cebidae) en habitat fragmentado en Balancan, Tabasco, Mexico. *Acta Zoologica Mexicana* 22:53–66.

Ratsimbazafy JH. 2006. Diet composition, foraging and feeding behavior in relation in habitat disturbance: Implications for the adaptability of ruffed lemurs (*Varecia V. editorium*) in Manombo forest, Madagascar. In: Gould L, Sauther ML, editors. *Lemurs: Ecology and adaptation*. New York: Springer Press. pp. 403–422.

Ratsimbazafy JH, Ramarosandratana HV, Zaonarivelo RJ. 2002. How do black-and-white ruffed lemurs survive in a highly disturbed habitat? *Lemur News* 7:7–10.

Reid FA. 1997. *A field guide to the mammals of Central America and Southeast Mexico*. New York: Oxford University Press.

Rodriguez-Luna E, Dominguez-Dominguez LE, Morales MJ, Martinez-Morales M. 2003. Foraging strategy changes in *Alouatta palliata mexicana* troop released on an island. In: Marsh LK, editor. *Primates in fragments: Ecology and conservation*. New York: Kluwer.

Rylands AB, Groves CP, Mittermeier RA, Cortés-Ortíz L, Hines JJH. 2006. Taxonomy and distribution of Mesoamerican primates. In: Estrada A, Garber PA, Pavelka MSM, Luecke L, editors. *New perspectives in the study of Mesoamerican primates: Distribution, ecology, behavior, and conservation.* New York: Springer. pp. 29–81.

Serio-Silva JC, Hernandez-Salazar LT, Rico-Gray V. 1999. Nutritional composition of the diet of *Alouatta palliata mexicana* females in different reproductive states. *Zoo Biol* 18:507–513.

Sheddon-Gonzalez A, Rodriguez-Luna E. 2010. Responses of a translocated howler monkey *Alouatta palliata* group to new environmental conditions. *Endang Sp Res* 12:25–30.

Silver SC, Marsh LK. 2003. Dietary flexibility, behavioural plasticity, and survival in fragments: Lessons from translocated howlers. In: Marsh LK, editor. *Primates in fragments: ecology and conservation.* New York: Kluwer.

Silver SC, Ostro LET, Yeager CP, Horwich R. 1998. Feeding ecology of the black howler monkey (*Alouatta pigra*) in northern Belize. *Am J Primatol* 45:263–279.

Smith JD. 1970. The systematic status of the black howler monkey, *Alouatta pigra* Lawrence. *J Mammal* 51:358–369.

Solano SJ, Martinez T, Estrada A, Coates-Estrada R. 1999. Uso de plantas como alimento por Alouatta palliata en un fragemento de selva en Los Tuxlas, Mexico. *Neotrop Primates* 7:8–11.

Stoner KE. 1996. Habitat selection and seasonal patterns of activity and foraging of mantled howling monkeys (*Alouatta palliata*) in north-eastern Costa Rica. *Int J Primatol* 17:1–30.

Waide RB. 1991. The effect of Hurricane Hugo on bird populations in the Luquillo Experimental Forest, Puerto Rico. *Biotropica* 23:475–480.

Williams-Guillen K. 2003. The behavioral ecology of mantled howling monkeys (*Alouatta palliata*) living in a Nicaraguan shade coffee plantation. Doctoral Dissertation, New York University.

Wright PC. 1999. Lemur traits and Madagascar ecology: Coping with an island environment. *Am J Phys Anthropol* 110:31–72.

Zimmerman JKH, Covich AP. 2007. Damage and recovery of riparian Sierra palms after hurricane Georges: Influence of topography and biotic characteristics. *Biotropica* 39:43–49.

5. Adolf Remane: Notes on his work on primates

Prof Ulrich Welsch

I met Colin Groves for the first time in the Anthropological Institute of the University of Zürich in 1964, where I did some of my PhD thesis research under the supervision of Professor Adolf Remane (1898–1976). The conversations I had with Colin were based on our common enthusiasm for morphology, comparative anatomy, phylogeny and theoretical ideas about the 'natural system', as I called it, following Professor Remane. Key elements of the natural system are the terms homology and analogy, which enabled scientists, even before the time of Charles Darwin, to establish the natural relationships among organisms. This concept was followed up much later with phylogenetic research. Based on my dealings with both scholars, I am convinced that there is a deep similarity between the minds of Adolf Remane and Colin Groves. Both were/are exceptionally gifted morphologists and both authored numerous high quality publications including entire books. In addition, primates were/are the main target of their interests. On a more personal note, to me both were/are very modest, yet at the same time show a sympathetic and friendly self-confidence, without a dogmatic attitude. Both scholars also worked/work with reliable and never failing consistancy, diligence and concentration, in a harmonious way that included uniting fieldwork with theory. They were/are also deeply interested in the historical dimensions of present-day concepts and had/have acquired a truly unusual knowledge in this field. Remane frequently read original texts (e.g. of Aristotle, Goethe and Cuvier), which is also the case of Colin whose own works are based in the classic texts of Buffon and many European authors of the nineteenth century. Both enjoyed/enjoy to share their knowledge freely with students and colleagues and in doing so they often showed/show a good sense of humour. Finally, both were/are outstanding university-academics, with a search for truth guiding their way of thinking.

Of course there are also differences of opinion between the two men. Remane condensed his experience and concepts in the book *Foundations of the Natural System, Comparative Anatomy and Phylogenetics* in 1952. I am sure that Colin Groves does not agree with every sentence in that book, but I am also sure that he has an understanding for Remane's arguments and logic. While Remane's interest in the scientific theory of taxonomy did not run very deep, I feel sure that he would have wholeheartedly joined in any discussion on Colin's *Primate Taxonomy* (2001) and that he would have followed in all probability Colin's 'Putting Primate Taxonomy into Practice' even if – deeply in his heart – he may have had reservations in one or the other case. The rest of this chapter will focus on the works of Robert Gustav Adolf Remane, which were published in German, thus underappreciated and largely unknown amongst many primatologists. As Colin himself is able to translate German works into English, it is fitting not only because the two men are so similar, but because Colin would appreciate the way that Remane's work was held back by the language in which it was written.

As alluded to above, Professor Robert Gustav Adolf Remane (1898–1976) was a multifaceted and stimulating zoologist. During his career he created a tremendous amount of scientific work including more than 300 publications ranging from unsurpassed monographs on single invertebrate groups (e.g. rotifers), to a broader field of marine biological and ecological topics, to more theoretical work on the foundations of the natural system, comparative anatomy and phylogeny and on practical and theoretical reflections on the phenomenon of homology. One important aspect in this coherent mosaic is Remane's primatological studies which dominated two periods of his career, the first at the beginning (1921–1928) and the second towards the end (1951–1965). His publications in this field concerned mainly the functional morphology of teeth and dentition of almost all extant and many extinct primate species and the methodological problems of hominoid phylogeny and theoretical problems of primate systematics.

Figure 5.1: Photograph of Professor Remane, at the age of 65.

Source: Given to author as a private gift from the Institute of Zoology in 1967. Photo was taken by the Institute in 1962.

It was the striking diversity of opinions on fossil teeth, which led him in 1921 to conduct his first thorough study of more than 900 dentition specimens of extant gibbons and apes. Apart from valuable, careful and very detailed descriptions, his work is always marked by a specific sober spirit, guided by an immense knowledge, and by enlightening comparative statements which always create a pleasant intellectual feeling while reading all his papers and articles, not unlike reading one of Colin's papers. He had a gift to not feel compelled to give all problems definitive answers, but was content to ask further questions based on preliminary conclusions. He found particular pleasure in finding the complexity of situations, for example of the possibility of reversibility of phylogenetic trends, of the simultaneous presence of very advanced specialisations and of primitive characters in one animal species (e.g. in *Tarsius* or in *Alouatta*).

Reading Remane's texts you always have the impression that you are not wasting your time with boring dental details but that you have gained new general biological insights, both on single primate species and on the interrelationships among primates. For example, in one instance he was able to compare the molar patterns of *Apidium*, *Oreopithecus*, *Pongo* and of human milk molars with interesting results in just a few sentences. The vividness of his thoughts was even more present when he spoke in the lecture hall. In his lectures, the wealth of his knowledge made it easy for him to reflect meaningfully on a variety of topics. He could easily discuss the specific morphological details of human canines or discuss topics considered by Georges Cuvier or Goethe, who on this or that subject had objected to Cuvier's viewpoint, then adding a sentence in ancient Greek, that this or that ambiguity had already been touched by Aristotle, all this quite naturally and without any pretentious attitude. One has to know that he was at home in the entire world of arts, philosophy and natural science in order to understand certain lines of argument in his work. In discussions he was unbeatable, logically thinking with an immense knowledge of all fields of zoology, anthropology, ecology, botany and philosophy, possibly only being rivalled by Colin himself. All his profound primatological – and other – work is free of any personal vanity and is marked by a rare solidity; it was created under often difficult circumstances which were caused by the absurd ups and downs of Central European history in the twentieth century and sometimes by strong personal discomfort due to migraines.

In order to understand Remane's complex theoretical work on the natural system, comparative anatomy and phylogenetics in depth, it is important to be familiar with the, in-part, pre-Darwinian developments of comparative anatomy, mainly in France, England and Germany. Georges Cuvier (1769–1832), Johann Wolfgang Goethe (1749–1832) and Carl Gegenbaur (1826–1903) were constants of orientation. For Cuvier he had a lifelong sympathy. In his lectures, which I had the privilege to hear from 1961–66, he additionally often referred to

Richard Owen (1804–1892), Etienne Geoffroy Saint-Hilaire (1772–1844), Ernst Haeckel (1834–1919) and William King Gregory (1876–1970), AS Romer (1894–1832) and to GG Simpson (1902–1984). Of course, he was fully aware of the unique significance of Charles Darwin (1809–1882) for the entire development of modern biology, but now and then slight 'mental reservations' became visible, especially against dogmatic followers of Darwin and those who claimed to be in the possession of 'the' Darwinian truth. Remane knew, that particularly eager protagonists could be good specialists, but could at the same time lack in depth of insight into the entire kingdom of organisms with its endless diversity, and contradictions. He was quite against absolute statements and also saw opaque spots in the 'synthetic evolutionary theory', which ideologically minded debaters did not want to see. The unobtrusive, ever present reference to the scientists of the past was not only an expression of personal modesty but was also of great educational significance for his students. Remane never saw natural science as an anti-thesis to other fields of the human mind such as art, philosophy or religion.

He had a unique gift to analyse and understand morphology, and he was open-minded to theoretical questions, but he could also become silent in view of the endless complexity of life. He was also an enthusiastic outdoor biologist with unbeatable knowledge of plants and animals, again similar to Colin's knowledge of most things flora and fauna.

His primatological work was based on the knowledge of several thousand teeth, which he studied personally in all important European museum collections and by careful and serious reading of all relevant literature. Gifted by an obviously inborn feeling for morphology, supported by a unique power of memory and guided by an unusual intelligence, he wrote more than 30 publications and handbook articles with primatological contents, including:

1. Individual topics, such as e.g. the unique morphology of the human canines (Remane, 1924a); or the critical discussion of R. Fourtaus' *Prohylobates tandyi* and *Dryopithecus mogharensis* (Remane, 1924b); or the interpretation of the dentition of *Oreopithecus* (Remane, 1952a) in which he showed that the attempt to derive the pattern of human upper molars from those of *Oreopithecus* and thus placing *Oreopithecus* into the hominids, is theoretically possible but only by help of a transitory form on paper and by using a specific variant of human molars; or the presumable phylogenetic position of *Gigantopithecus* (Remane, 1953), or the origin of bilophodont molars in Cercopithecidae (Remane, 1951); or aberrant morphologies of teeth and skulls.

2. Critical reviews, including those on primate systematics or the 'natural history' of primates (Remane 1956a, 1956b, 1960a, 1965), which due to their original and critical thoughts, are by no means outdated.

3. Handbook articles, in particular in the *Handbook of Primatology* (*Primatologia*) and comparable articles (Remane 1921a, 1921b, 1952b, 1955), which are not word-rich elaborations but concise intelligent compositions full of facts, accompanied by clear line drawings or photomicrographs. The facts are usually condensed into useful tables, e.g. on measurements and proportions of teeth. Already his first publication (1921) on teeth of gibbons and apes is based on the analysis of the skulls of 322 gorillas, 287 chimpanzees, 160 orangutans and 145 gibbons. Such a rich background renders particular weight on specific papers, for example on the dentition of *Oreopithecus* (Remane, 1952a) or *Australopithecus* (Remane, 1952b). This analysis also opened his eyes to the remarkable variability of morphological characters of primate teeth, which for him remained a life-long warning not to make definite statements on single fossil teeth or jaw fragments. He gave many important examples for the variability of skull and mandible shape, for example of *Pan*, of measurements of teeth and of crown patterns (Remane, 1952a), and he took this fact seriously. However, he analysed also single fossil teeth as far as one can go, but never crossed the borders of serious and objective science. The quantitative and qualitative analyses always comprises crown, roots and alveoli and usually details of maxilla and mandible. Of course most information can be extracted from the dental crown, for example patterns of cusps, crests and furrows, including patterns of attrition (Remane 1921ab). He was aware that different scientists use different methods, for example when measuring length and width of teeth, which can make it impossible to compare corresponding data. Therefore he suggested application of the same measuring techniques for primate teeth (Remane, 1927). In his contribution on methodic problems of hominid phylogeny II (Remane, 1954), he discusses not only the variability of all types of teeth in apes and man, but also of many cranial characters. Furthermore he carefully discusses pitfalls when not considering variability, which can lead to wrong simplifications and wrong phylogenetic reconstructions. Finally his discussion on species, subspecies and populations, also in regard of fossil findings, is of significance.

What makes Remane's texts and analyses always interesting and often helpful are his often surprising functional correlations, the comparative aspects and of course the always present phylogenetic background. Unfortunately it became a considerable drawback that Remane published almost exclusively in German. As he himself read with ease and pleasure in a multitude of languages including ancient Greek and Latin, French and English he possibly subconsciously expected a similar versatility in English and French-speaking scientists. This language barrier kept many of his publications unknown and underrepresented, but this is a real tragedy due to the fact that his work:

a) contains an invaluable treasure of facts (e.g. dental measurements and evaluations of crown patterns, illustrated by clear drawings)

b) opens eyes beyond the facts

c) includes, especially in his 1956, 1960 and 1962 handbook articles, all the older European and North-American literature (1900 until about 1960)

d) helps to avoid repetitive work

e) is written in a relatively simple, grammatically correct German.

What also makes such articles valuable even today is that the data can be used for studies not foreseen when they were compiled; primarily they can be transferred into computers for further mathematical evaluations.

When a character was particularly variable, for example the morphology of upper outer incisors or the molar crown patterns of chimpanzees or the morphology of human milk molars, Remane formed groups by which he tried to give some order to the variability. This, today, could be refined by computer analysis. But, it may be added, Remane always remarked, that the best discriminator for visible morphological characters are the human retina (with its many millions of neurons) and brain. In addition he used to say that although large numbers are important to give weight to statements, statements only make sense when the characters are evaluated and not only counted. In this he followed of course the great French comparative anatomists around 1800 and in the early years of the nineteenth century.

His statements were always sober and clear, e.g. 'fossil material too scanty for final statements', 'a well-founded evaluation can be made only when more fossil material is available', 'the XY-index as used in this study for the evaluation of fossil material is almost without any worth', in regard of the inner cusp of P_3: 'since there are regularly variations in regard of the presence of single structures in meristic organs (here the lower premolars) this question has no great significance', etc.

His analysis (1965) of teeth and the fragments of the mandibles of *Gigantopithecus blacki* are sober and careful and consider not only details of the Pleistocene strata but, in fairness, also the suggestions of other scientists in regard of the systematic position of this big ape (Remane et al., 1960; Remane, 1960a, 1965). He convincingly excludes *Gigantopithecus* from any lineage towards *Homo*, as had been suggested before. He carefully excludes *Gigantopithecus* from any close relationship with *Pongo* (and *Pan*); he describes similarities with *Gorilla* and overlap of teeth sizes with *Gorilla beringei*. The similarities with *Gorilla* are common primitive features. He clearly works out specific characters, for example the high hypsodont crowns and the specific morphology of the lower P3 and lower canines and concludes that *Gigantopithecus* was a herbivore and represents an own lineage of apes rooting probably in the 'Sivalik-pongids'.

Interesting and worth considering is his 1965 evaluation of the 'Sivalik-Pongiden' – the fossil remains of apes as found in the Sivalik hills (Remane, 1965). He discusses at length all findings and 'species' as described by Lydekker, Pilgrim, Gregory and Hellman, Lewis, von Koenigswald, Hooijer, and Dehm. He mentions among others that the size of the teeth varies from those of *Gorilla* (*Dryopithecus giganteus*) to those of a small *Pan*. The teeth are generally typical for apes. The 'species' are more similar among each other than among extant apes. Single teeth show rather specific agreements with those of *Pongo*, for example the P3 of *'Sivapithecus himalayensis'* (Pilgrim, 1927). Some of these P3-similarities are also to be found in *Indopithecus* (Hooijer, 1951). The M2 of *Dryopithecus giganteus* is marked by a pattern of furrows reminding those of *Pongo* molars. This and additional observations lead him to speculate that ancestors of *Pongo* may be present among the Sivalik pongids. These possible ancestors, however, have in general more primitive characters than modern *Pongo* (clearer defined cusps, bigger M3, more rounded symphysis). However, it would be premature to ascribe all Sivalik apes to *Pongo* and possible relatives. For example, *Dryopithecus pilgrimi* has rather narrow incisors as can be seen in *Gorilla* and *Dryopithecus fontani*. Individual teeth have a high length-width-index, as to be seen in *Gigantopithecus* and the closer relationship of *Homo*. Such individual evaluations take a large part of this contribution. Remane finally gives long arguments against a closer relationship between *Ramapithecus* and the *Homo*-lineage.

Today, Remane's most useful work may be his 209-page 'Zähne und Gebiß' (teeth and dentition) in *Primatologia* III/2, (1960b). It is probably the best introduction for a beginner in primate dental morphology and for the advanced beginner an always helpful reference. It deals with all genera and all important species. It includes much of his older publications, occasionally with more differentiated interpretations. Valuable is the chapter on deciduous teeth, which often give hints to the origin of structures on the permanent teeth. Unique are the cross-references and cross-comparisons among all primate groups, enlightening and convincing are the argumentations for homology or analogy. Fine examples of a sovereign mind are many specific paragraphs, for example on the detailed comparisons between *Pan* and *Homo*, and the parallel development of bilophodont molars in primates (e.g. in *Cebus*, Cercopithecoidea, *Symphalangus*, Archaeolemuridae and in part Indriidae). Almost every page has at least one clear illustration; the great trends in the development of primate dentitions are summarised in the last chapter, with summaries on the specific families and genera (e.g. *Hylobates*, *Gorilla*, *Pongo*, *Pan* and *Homo*).

References

Heydemann B. Zum Tode von Professor Dr.Dr.'.c. Adolf Remane (1977). *Faun-Ökol Mitt* 5:85–91.

Hooijer, DA. 1951. Questions relating to a new large anthropoid ape from the Mio-Pliocene of the Siwaliks. *Am J Phys Anthropol* 9: 79–95.

Pilgrim GE, 1927. A Sivapithecus palate and other primate fossils from India. *Memoirs of the Geological Survey of India (Palaeontologica Indica)*, NS(14):1–26.

Remane A. 1921. Beiträge zur Morphologie des Anthropoidengebisses. *Archiv f. Naturgeschichte, Abt. A* 87:1–179.

Remane A. 1921. Zur Beurteilung der fossilen Anthropoiden. *Centralblatt für Mineralogie, Geologie und Paläontologie* 11:335–339.

Remane A. 1924a. Einige Bemerkungen über *Prohylobates Tanyi* R. Fourtau und *Dryopithecus mogharensis* R. Fourtau. *Centralblatt f. Mineralogie, Geologie und Paläontologie* 7:220–223.

Remane A. 1924b. Einige Bemerkungen zur Eckzahnfrage. *Anthropolog Anz* 1:35–40.

Remane A. 1927. Methoden zur Untersuchung der Primaten. Zur Meßtechnik der Primatenzähne. *Handbuch biologischer Arbeitsmethoden*. Abt. 7:609–635 Berlin, Wien.

Remane A. 1951. Die Entstehung der Bilophodontie bei den Milchmolaren oft phylogenet. Vorstufe Cercopithecidae. *Anat Anz* 98:161–165.

Remane A. 1952a. Methodische Probleme der Hominiden-Phylogenie I. *Z Morph Anthrop* 44:188–200.

Remane A. 1952b. Der vordere Prämolar (P3) von *Australopithecus prometheus* und die morphologische Stellung des Australopithecinengebisses. *Z Morph Anthropol* 43:288–310.

Remane A. 1953. Die primitivsten Menschenformen (Australopithecinae) und das Problem der tertiären Menschen. *Schriften Naturwiss. Verein. Schleswig Holstein* 29:3–10.

Remane A. 1954. Methodische Probleme der Hominiden Phylogenie II. *Z Morph Anthrop* 46:225–268.

Remane A. 1955. Ist *Oreopithecus* ein Hominide? *Abh. Math.-Naturw. Kl. Akad. Wiss.* Mainz N.R. 12:469–497.

Remane A. 1956a. Paläontologie und Evolution der Primaten, besonders der Nicht-Hominoiden. *Primatologia* I(8). Basel, New York: S. Karger. pp. 267–378.

Remane A. 1956b. *Die Grundlagen des Natürlichen Systems, der Vergleichenden Anatomie und der Phylogenetik.* 2. Aufl. Leipzig: Akademische Verlagsgesellschaft.

Remane A. 1960a. Die Stellung von *Gigantopithecus. Anthropol Anz* 24:146–159.

Remane A. 1960b. Zähne und Gebiss. *Primatologia* III(2). Basel, New York: S. Karger. pp. 637–846.

Remane A. 1961. Probleme der Systematik der Primaten. *Z wiss Zool* 165:1–34.

Remane A. 1962. *Masse und Proportionen des Milchgebisses der Hominoidea.* Bibl. Primatl. 1. Basel, New York: S. Karger. pp. 229–238.

Remane A. 1965. Die Geschichte der Menschenaffen. Hrsg. Gerhard Heberer. *Menschliche Abstammungslehre.* Stuttgart: Gustav Fischer Verlag. pp. 249–309.

6. Retouch intensity on Quina scrapers at Combe Grenal: A test of the reduction model

Peter Hiscock and Chris Clarkson

Introduction

There have been extensive discussions about whether hominids other than *H. sapiens* had the cognitive capacity to plan and conceptualise elaborate tool forms in advance of use and to transmit those conceptual systems to others. Outside Africa such discussions have been most extensive for questions of what cognitive capacities were possessed by Neanderthals and how their approach to planning and tool use differed from the subsequent *H. sapiens*. A core concern of these questions has been how morphological variation is understood and how that variation can usefully be expressed in classificatory systems that are capable of revealing evolutionary change. Such considerations have been critical in biological debates about the nature of species as well as about specific species boundaries. Similar deliberations are pursued in studies of lithic artefacts, as archaeologists explore how morphological transformations within individual artefacts as well as evolutionary transitions in populations of artefacts, are represented in metrical indices and classifications.

One intensively debated issue in lithic analysis concerns whether conventional practices of analysing retouched flakes by classifying them into a number of tool types is valid or problematic, and whether those types represent tools of distinctly different designs or alternatively arbitrary divisions between objects that display continuous morphological variation. Inferences about these issues have formed the basis of different explanations for the Mousterian facies, and the opposing claims about whether Neanderthals conceived of a large number of tool designs or not (e.g. Bordes, 1972; Binford, 1973; Binford and Binford, 1966; Bourguignon, 1997; Dibble, 1984, 1988b; Dibble and Rolland, 1992; Rolland and Dibble, 1990; Hiscock and Attenbrow, 2005; Hiscock et al., 2009; Holdaway et al., 1996; Mellars, 1996; Turq, 1992, 2000). Debates about the nature of economy, technology and cognition in ancient hominids are both significant and exciting, but they rest on the accuracy and clarity of depictions of artefact patterning and the meaning of morphological and technological diversity.

Although much has been written on the characterisation of retouched flake variability in Middle Palaeolithic assemblages, key aspects of the archaeological patterns remain unresolved.

In recent decades, two different hypotheses have competed as the best explanation for the variability observed in Mousterian implements from the Dordogne. One model advocates the primacy of retouch intensity in models explaining morphological diversity in ancient tools, with some researchers arguing this is the sole significant factor creating typological variation. The second model argues that intensity of retouch is only one of many factors creating variation and that others are often more significant. In this chapter we test these competing models by presenting a detailed, quantitative analysis of the relationship between different types of Quina scrapers and the extent of retouching that they have undergone in one level of Combe Grenal.

The question of Quina scrapers and reduction

A well-known model for Middle Palaeolithic scraper reduction was proposed by Harold Dibble (1984, 1987a, 1987b, 1988a, 1988b, 1995). He hypothesised the transformation of scrapers from one typological class to another as they received additional reduction. Dibble argued that extent of reduction was the key factor causing differences between four implement classes with retouch onto their dorsal surface: (1) single-edged side scrapers with retouch on one lateral margin (Bordes types 9–11); (2) double scrapers with two separate retouched edges (Bordes types 12–17); (3) convergent scrapers which have two retouched edges that touch (Bordes types 8, 18–21); and (4) transverse scrapers which have retouch across the distal end of the flake (Bordes types 22–24). Examples of these classes are provided in Figure 6.1. Dibble interpreted these four kinds of implement as a result of different amounts of reduction, in which all specimens began as single scrapers, but with additional retouching were either transformed into transverse scrapers or alternatively into double and eventually convergent scrapers. This model, schematically shown in Figure 6.2, notionally positions each of these four classes along a continuum of greater or lesser amounts of retouch, and reveals the proposition that there were two branches along which individual scrapers could travel from the same starting point. Dibble (1988b: 49; 1995: 319) suggested that those individual single scrapers that were further retouched were either worked at the distal end to become transverse forms or on the second lateral margin to become double/convergent implements. He suggested that the sequence followed by any individual specimen may have depended on the shape of the flake, with short/broad flakes being worked into transverse scrapers while longer, narrow flakes were retouched laterally to become double and convergent scrapers. However, Dibble (1995: 319) argued that much of the variation between implement classes, and specifically the morphology diagnostic of different types, was a product of differences in the level of retouching and the length of time they had been used and maintained: single scrapers had undergone little retouching while both transverse and convergent scrapers were more intensively retouched. In a series of papers he argued that the smaller average size of convergent and transverse scrapers, both in absolute terms and relative to their platform size, was evidence that this model was correct for Quina assemblages from southwest France and elsewhere.

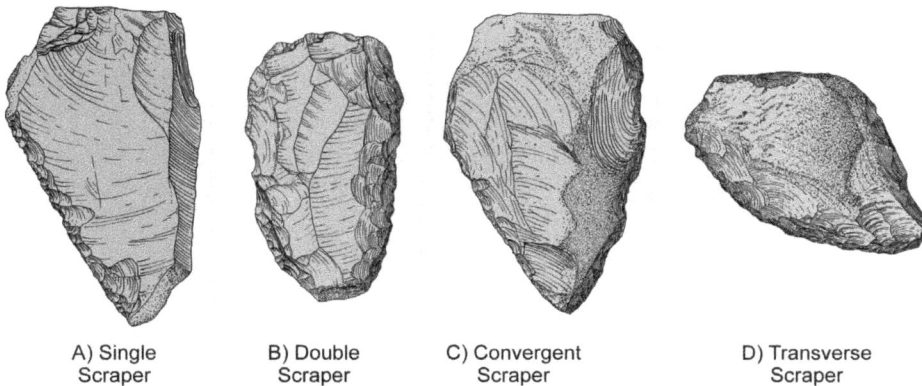

| A) Single Scraper | B) Double Scraper | C) Convergent Scraper | D) Transverse Scraper |

Figure 6.1: Examples of specimens classified into each of the four scraper classes: A) Single scraper, B) Double scraper, C) Convergent scraper, and D) Transverse scraper. All specimens are from Combe Grenal, Layer 21.

Source: Authors' original work depicting artefacts from Combe General.

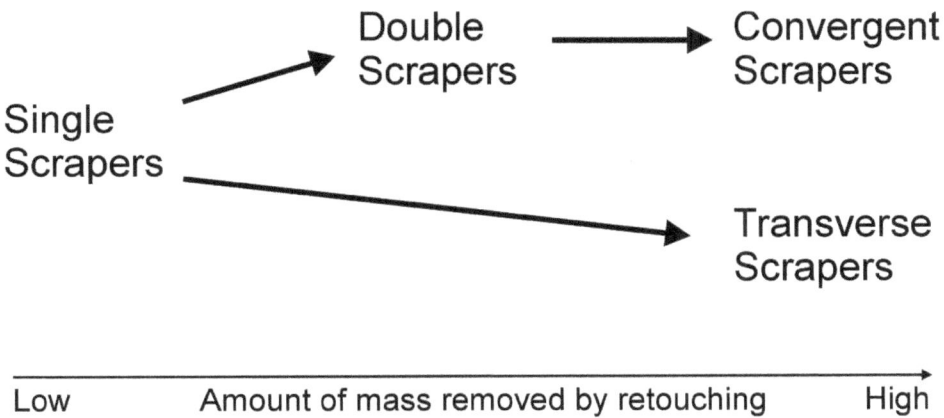

Low   Amount of mass removed by retouching   High

Figure 6.2: Diagrammatic representation of the staged reduction model proposed by Dibble.

Source: Based on Dibble (1987: 115).

This conclusion implies and necessitates particular interpretations of Mousterian variability. For example, although Dibble believed that the typology of Bordes still had value, he argued that the different types were a continuum created by differing extents of edge resharpening and that the implement classes therefore represented coherent stages in the continuum of retouch (Dibble and Rolland 1992: 11). Consequently Dibble (1988b: 52) stated that traditional implement types were best employed as a proxy for the extent of tool maintenance/ resharpening in archaeological assemblages.

One interpretation that has followed from Dibble's model of flake retouching is that differences between implement types and industries reflect differences in the intensity of tool use (and by implication the nature of land use); they do not reflect mental constructions of Neanderthals, nor do they reveal specific differences in design or function. For instance, Dibble and Rolland (1992: 17) argued that the production of industries dominated by convergent or transverse scrapers were a consequence of economic practices that encouraged more intensive tool use, such as the intensive maintenance of tools during cold palaeo-climatic phases in which there were long winter residence and patterns of settlement based on the interception of migratory herds, situations in which provisioning of stone for tools could have proved difficult. They contrast this with the contexts of industries dominated by side scrapers (and denticulates), which they hypothesised resulted from less intensive tool and site use that occurred under milder climatic phases in which Neanderthal economy was focused on the pursuit of dispersed, mobile game. The value of these kinds of interpretations depends on the veracity of the characterisation of traditional implement types as comparable units primarily reflecting differences in the extent of tool resharpening.

A number of commentaries and further studies have followed the publication of Dibble's model, many supporting his argument of the value of traditional implement types for studies of the extent of implement reduction (e.g. Gordon, 1993; Holdaway et al., 1996). However, significant reconsiderations of the factors involved in implement creation have been offered. The most potent is the proposition that the extent of retouching is not a function of the intensity of edge maintenance alone, but was often a reflection of the size and morphology of the flake to which retouch had been applied. For instance, Dibble (1991: 266), Gordon (1993: 211), and Holdaway and others (1996) all argued that larger flakes typically had greater potential for edge resharpening, and consequently in extensively reduced assemblages those larger specimens received more retouching than smaller ones. One result of the continued reduction of larger specimens, but not smaller ones, is that extensively retouched flakes were sometimes still larger when discarded than less extensively retouched ones made on smaller flakes (Dibble, 1991). While this proposition has been applied to notched types (e.g. Holdaway et al., 1996; Hiscock and Clarkson, 2007), its implications for the interpretation of other implement types and for the value of typology as a measure of the extent of retouching has received less attention. One obvious implication is that the amount of retouching applied to a specimen cannot be judged by its size (Dibble, 1991), a realisation that encouraged the development and growth of several methods for measuring retouch intensity on Middle Palaeolithic tools (see Dibble, 1995; Hiscock and Clarkson, 2005). However, the existence of a strong relationship between flake form and retouch has been argued to create problems for the interpretation of implement types as reduction stages.

For example, if retouching is a response to flake morphology and there is variation in the size and shape of flakes being retouched, an almost inevitable reality in most prehistoric contexts, then the amount of mass removed during retouching may vary substantially between specimens assigned to any implement type. This appears to be the case in the data presented by Dibble (1987b: 113) for the La Quina scrapers, which display extraordinarily high levels of intra-type variability in reduction measures, such as flake area/ platform area ratios which show coefficients of variation of 125% for single scrapers, 49% for double scrapers, 91% for convergent scrapers and 182% for transverse scrapers. Although Dibble still found statistically significant differences between the means of these four implement classes, the measured variability probably reflects very great differences in the amount of retouching between specimens in a single implement class. In such circumstances the value of conventional types as units measuring the extent of reduction may be questioned, and Hiscock (1994) argued that analysts would be better able to discuss differences in amounts of retouching if they focused on measuring the manufacture of individual specimens rather than merely the contrast between types (also Hiscock and Clarkson, 2005; Clarkson and Hiscock, 2008).

Furthermore, many researchers have argued that intensity of retouch is not the most important factor affecting the form of retouched flakes. For instance, Kuhn (1992) has argued that if flake form played a significant role in determining the position and amount of retouch on each object then typological composition is not principally affected by the intensity of tool use and so industrial variation may not directly correspond to different patterns of settlement and mobility. Instead, Kuhn argues the typological composition of an assemblage would reflect the size and shape of available flakes, which in turn would reflect the form and availability of raw material and the tactics of core reduction. While raw material procurement and core reduction may also be linked to economic and settlement patterns, the connection with the abundance of each implement type would be remote and indistinct. While Kuhn did not deny the proposition that intensity of retouch may be an indicator of settlement/mobility systems, he argued that types are not reliable indicators of intensity of retouch, and that archaeologists will require dedicated measurements of retouch intensity prior to developing inferences about the land-use from lithic artefacts.

Long-term archaeological research in southwest France has yielded much evidence for the complex articulation of core reduction systems and the patterns of retouched tools made on the flakes produced in those systems (e.g. Bisson, 2001; Bourguignon, 1997; Bourguignon et al., 2004; Thiébaut, 2003; Turq, 2000; Verjux, 1988; Verjux and Rousseau, 1986), reinforcing the possibility that flake form may have an important role in the construction of morphological diversity amongst Mousterian implements. Many discussions of flake-retouch relationships have posited a simple relationship between flake elongation and the position of retouch, suggesting that long flakes were often retouched on their lateral margins, whereas short, wide flakes were often worked at the distal end (e.g. Bordes, 1961: 806, 1968: 101; Turq, 1989; Mellars, 1992). A number of researchers have argued that the flakes on which single and transverse scrapers were made are very different, and that regular production, and/or selection, of flakes with particular characteristics was a significant factor in the formation of the typological composition of any assemblage (e.g. Turq, 1989, 1992). As a consequence, Turq (1989) argued that there were clear morphological discontinuities in the form of single and transverse scrapers in the Dordogne, evidence that would not be conformable with Dibble's reduction hypothesis. The hypothesised connection of flake form and systems of core reduction has also been argued to be evidence for deliberate and planned acts of selection/ production (e.g. Boëda, 1988; Turq, 1989, 1992).

Some models of the way the morphology of flakes strongly influenced the nature, and typological category of implements have hypothesised complex interactions between multiple characteristics that affected the nature of retouching. An example is Alain Turq's proposal that scrapers in Quina industries reflected a regular pattern of flake selection and retouching. He suggested that transverse scrapers, unlike single side scrapers, were made on flakes that were thick relative to their length and ventral surface area; a proposition that would account for differences between types in the relationship of platform and ventral areas, which Dibble (1984, 1987a, 1987b, 1995) had employed as evidence for different degrees of reduction. Furthermore, Turq argues that scrapers were typically made on flakes with asymmetrical cross-sections and retouch was located on the flake margin furthest from the maximum thickness (Turq, 1992: 75). In a diagram, presented here as Figure 6.3, Turq (1992: 77) implied that the potential for resharpening was related to the asymmetry of each flake selected for retouching, with symmetrical ones having little mass removed before steep retouching came close to reaching the thickest part of the flake while asymmetrical flakes could have considerably more mass removed through retouching before reaching the same state. This proposition linked variation in scraper morphology with flake morphology as well as extent of reduction, implying that flake shape and selection were the proximate factors creating variation in both the location/orientation of retouch and the amount of mass removed by retouching on different specimens, and consequently the typological category into which each specimen was placed. Turq's model not only contrasts with Dibble's in the emphasis given to flake form rather than extent of reduction, but also implies that there may be a great deal of difference in the extent of reduction of specimens with similar cross-sections and placed in the same typological category.

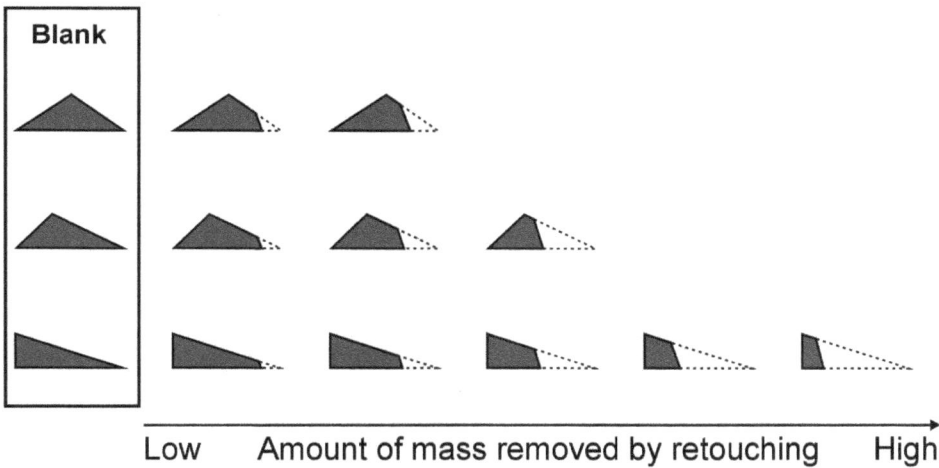

Figure 6.3: Notional illustration of the relationship of blank cross-section and extent of reduction for dorsally retouched Quina scrapers.

Source: After Turq (1992: Figure 6.2).

Considered in this way, the distinctions between two different models of Quina scraper variability are clear. On the one hand Dibble's 'scraper reduction model', posits a single branching scheme and asserts that traditional typological categories represent different points/stages along a continuum of greater or lesser amounts of retouch. The frequency of specimens in each type may therefore be used as a proxy for the intensity of reduction that an assemblage has undergone. From this perspective intensity of reduction is the primary cause of typological variation, and although differences in flake morphology exist, their effect on typological variation is minimal. Consequently typological diversity through time and space can be directly interpreted as a result of access to raw material and settlement/economic activities. On the other hand, what Hiscock and Clarkson (2008) have called the 'blank-retouch interaction hypothesis' proposes that traditional typological categories represent complex patterns of morphological variation created by several factors, particularly differences in the distribution and intensity of retouch in response to flake form. Distinctions between conventional implement types may therefore have little coherent covariation with intensity of retouch, and should not necessarily be treated as representing different points along a reduction continuum. This hypothesis implies that the frequency of specimens in each type may not be a reliable indicator of the intensity of retouching that an assemblage has undergone, and that typological diversity through time and space is difficult to directly interpret in terms of settlement/economic activities. Instead, this hypothesis asserts that Borde's typology reflects morphological patterns created by a constellation of factors including flake morphology, material cost, tool design, as well as amount of uselife/resharpening, and that the resulting typological patterns are not necessarily sensitive to variation in the intensity of retouch.

Our goal here is to examine the applicability of these two opposing models to one Quina assemblage, recovered from Layer 21 in Combe Grenal. Although these models predict different behavioural processes, they both invoke extent of retouching as a mechanism constructing morphological variation; the two models differ in the way retouching is articulated to other technological and economic factors. We emphasise that there is no reason to expect that one will inevitably be the most appropriate in all situations. It is possible for the 'reduction hypothesis' to be correct for some assemblages and the 'blank-retouch interaction hypothesis' to be correct for others. In this way these opposing models are not competitors in a search for a universal truth but are actually expressions of the variable operation of multiple factors that may have created morphological variation in Mousterian implements. Consequently our examination of these two models for Layer 21 at Combe Grenal is not a test of the general veracity of either model, but actually an assessment of what kinds of processes were operating in the Neanderthal technological system in the Perigord Noir at the time that layer formed.

Our approach to measuring the extent of retouching

Our sample of artefacts comes from Combe Grenal, excavated by François Bordes (1972) and now held at the Musèe National de Prèhistoire des Eyzies. The following analysis uses measurements of complete dorsally retouched flakes from Layer 21, a Quina level. Technological cores and unretouched flakes, broken specimens, and a small number of burins, end scrapers, Mousterian points and a truncated-faceted piece were excluded from the analysis. For this paper our sample consists of 306 objects, representing all specimens in each of the major typological categories: single scrapers (N = 172), double scrapers (N = 28), convergent scrapers (N = 40), transverse scrapers (N = 66). The number of specimens in each of the Bordes type classes used in our analysis is listed in Table 6.1.

Table 6.1: Sample of complete retouched flakes from Layer 21 used in this analysis, presented by implement type.

Implement types	N
Single	
9 Single straight scraper	50
10 Single convex scraper	112
11 Single concave scraper	10
Double	
12 Double straight scraper	14
13 Double straight-convex scraper	7
14 Double straight-concave scraper	2
15 Double convex scraper	4
16 Double concave scraper	1
Convergent	
8 Limace	2
18 Straight convergent scraper	3
19 Convex convergent scraper	5
21 Dejete scraper	30
Transverse	
22 Straight transverse scraper	14
23 Convex transverse scraper	49
24 Concave transverse scraper	3

Source: Author's data.

Our analysis of these implements employs two measures of the position of retouching on each specimen and the amount of mass removed through retouching (Figure 6.4). The first is a version of the Geometric index of unifacial reduction (GIUR), a measure we have experimentally verified (Hiscock and Clarkson, 2005, 2009; Clarkson and Hiscock, 2008). Our experiments showed that scar height ratios, taken at multiple points around a retouched flake, yield an average GIUR value which has a non-linear relationship with the mass removed by retouching (Hiscock and Clarkson 2005: 1019). Experimental retouching of flakes demonstrated that there was a strong log-linear relationship between the calculated Kuhn GIUR and the percentage of original flake weight that has been lost (Figure 6.5). This relationship appears to hold irrespective of whether retouching is applied to the lateral or distal margin (Hiscock and Clarkson,

2005), or to one or more than one edge (Clarkson and Hiscock, 2008; Hiscock and Clarkson, 2009). For instance, when we experimentally retouched flakes on one lateral margin, producing items similar to single side scrapers, there was a strong positive relationship between the index value and the mass removed by retouching ($r = 0.933$, $r^2 = 0.871$). When we experimentally retouched flakes on two lateral margins the Kuhn GIUR was still strongly and significantly correlated with the proportion of mass lost from each flake ($r = 0.88$, $r^2 = 0.778$). We have argued elsewhere that while variations in the GIUR/mass-lost relationship occurred as a consequence of differences in the shape and size of flake blanks, a strong relationship exists for most flakes that are dorsally retouched, and consequently we take the Kuhn GIUR to be a reliable measure of the extent of dorsal, unifacial retouch in most instances, including the specimens discussed in this analysis, irrespective of the nature of the flake (Hiscock and Clarkson, 2005: 1022). Furthermore, the high coefficient of determination (r^2) allows us to use the regression line and 95% confidence intervals shown in Figure 6.5 to estimate the approximate amount of mass removed during retouching.

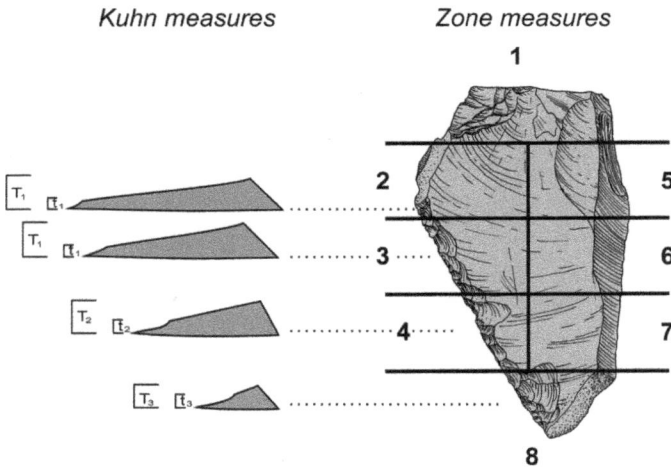

Figure 6.4: Illustration of the measurements of reduction used: multiple values of Kuhn's (1990) unifacial reduction index and a count of the number of zones which have been retouched.

Source: Hiscock and Clarkson (2005).

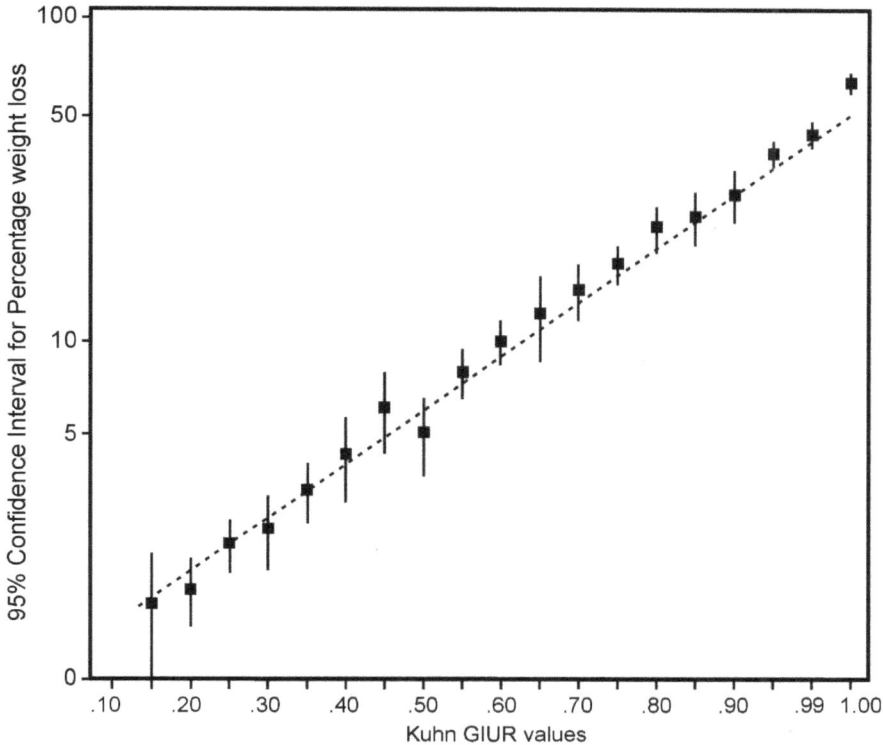

Figure 6.5: Relationship between Kuhn GIUR (as 0.05 intervals) and the percentage of original flake mass lost through retouching (shown with 95% confidence intervals) in the experimental dataset (Hiscock and Clarkson 2005). Broken line is the regression line (r = 0.933, r² = 0.871) published by Hiscock and Clarkson (2009).

Source: Hiscock and Clarkson (2009).

A second measure of retouching was the distribution of retouch on the margins of each flake. This measure provides an indication of the lateral expansion of retouching around the specimen; complementing the GIUR, which measures how far retouch has penetrated into the centre of a flake (Hiscock and Attenbrow, 2005: 59). This 'Retouched zone index' was obtained by observing which of eight zones, illustrated in Figure 6.4, were retouched. The zones were defined in terms of five equal divisions of the percussion length, but with the left and right margins being separated to create eight locations (proximal, distal, three zones on the right margin and three on the left). The face on which scars occurred was not relevant for this measure, giving retouched flakes values between 1 and 8 zones. This recording system was not only used to measure the amount of retouch around the flake margin, but also served as a way to compare the location of retouch on different specimens.

Other measures of flake retouching, such as Clarkson's (2002) invasiveness index or Holdaway, McPherron and Roth's (1996) surface area/platform thickness ratio were considered to be of lesser value on the steeply, unifacially retouched flakes in our sample and are not presented here. Although we have previously expressed doubt about the sensitivity of Dibble's (1987) surface area/platform area ratio as a measure of reduction we have calculated this below as a comparison to published data that has been used to discuss models of Quina retouch.

The extent of retouching and implications for reduction

With these measurements of the amount of retouching, we are able to evaluate whether the different implement categories (single, double, convergent, transverse scrapers) actually represent clusters of specimens that have been reduced to different extents, as hypothesised by Dibble. Descriptive statistics for the reduction indices in our sample, presented in Table 6.2, show a pattern somewhat similar to that reported by Dibble (1987: 113) for the La Quina site, and which he used in support of his reduction model. For instance, the mean surface area/platform area values are higher for single scrapers than double and convergent ones, and transverse scrapers display the smallest mean; with the means being very similar to those Dibble found at La Quina. This offers support for the proposition that, on average, single scrapers were less reduced that the other three scraper categories. Average values for the Kuhn GIUR and retouched zone index could also be used to suggest that single scrapers were on average less reduced than double or convergent scrapers; giving support to the idea of a single-double-convergent sequence of scraper transformations in Layer 21. ANOVA treatment of our data reveals statistically significant differences between the implement classes in the Kuhn GIUR ($F = 5.485$, d.f. $= 4$, $p = 0.001$, with the index broken into five groups: 0.01–0.19, 0.20–0.39, 0.40–0.59, 0.60–0.79, and 0.8–1.0) and in the retouched zone index ($F = 10.112$, d.f. $= 7$, $p < 0.001$); but not in the surface area/platform area ratio ($F = 0.295$, d.f. $= 4$, $p = 0.881$), with the index broken into five groups: 0.01–4.99, 5–9.99, 10–49.99, 50–99.99, and >100. These statistics all indicate that there is patterned variation in the central tendencies for retouching intensity between the four classes of implement.

However, the relationship of transverse and single scrapers is not consistent with the predictions of Dibble's reduction model. Differences in mean Kuhn GIUR alone ($t = 3.239$, d.f. $= 236$, $p = 0.001$) conform with the predictions of Dibble's (1987) model, although the question of how to interpret the large variation in each class is discussed below. Average surface area / platform area values were not significantly different for transverse and single scrapers ($t = -0.892$, d.f. $= 208$, p = 0.373), and the retouched zone index indicates that transverse scrapers have significantly less extensively retouched margins than single scrapers (t = -4.185, d.f. = 97.8, p < 0.001), a finding that is not compatible with Dibble's model in which the addition of distal retouch converted single scrapers into transverse ones. These statistics imply a difference between single and transverse scrapers in intensity and location of reduction, but not necessarily as sequential stages as Dibble argued in model of his single-transverse sequences.

Table 6.2: Descriptive statistics for the Kuhn GIUR, Retouched zone index, and surface area/platform area ratio of four implement classes in Layer 21 of Combe Grenal.

	Kuhn GIUR	Retouched zone index	Surface area/platform area
Single (N = 150)	0.49 ± 0.20 0.15–1.00	4.05 ± 1.42 1–8	14.59 ± 31.10 1.0–297.4
Double (N = 23)	0.60 ± 0.16 0.31–0.91	6.61 ± 1.12 2–8	11.86 ± 7.20 2.9–31.2
Convergent (N = 31)	0.61 ± 0.16 0.26–0.98	6.26 ± 1.59 2–8	11.81 ± 20.32 1.2–105.8
Transverse (N = 60)	0.61 ± 0.21 0.17–1.00	2.97 ± 1.78 1–7	10.79 ± 17.78 0.5–102.7

Note: Top line is mean and standard deviation, lower line is the minimum and maximum value.

Source: Authors' calculation.

These data document differences between these implement classes in the average degree of reduction, but such differences may not constitute evidence of the transformation of specimens from one implement class to another. An examination of the variation found within each implement class reveals that the assemblage from Layer 21 does not conform to Dibble's reduction model. Each of implement class displays high levels of variation in the reduction measures. In particular single and transverse scrapers show large ranges of reduction indices. For example, on single scrapers the coefficient of variation for Kuhn GIUR is 41% and for the retouched zone index it is 35%, while transverse scrapers have a coefficient of variation for Kuhn GIUR of 34% and for the retouched zone index 60%. This indicates that each of those typological groupings contain specimens with very different levels of retouch. Using the Kuhn GIUR to estimate the proportion of original flake mass removed through retouching shows that single scrapers lost 2–66% of their weight, double scrapers 3–30%, convergent scrapers 5–35%, and transverse scrapers 4–66% of blank weight. When intensity of reduction is expressed in this way it is clear that Dibble's reduction models do not account for all of the specimens in Layer 21. For instance, some single scrapers are extensively reduced; some more than twice as reduced as any double or convergent scrapers. The existence of single scrapers with very high amounts of mass removed through retouching, and that were not converted into double or convergent forms, demonstrates that specimens typologically classified as single scrapers were not all 'early stage', with only little retouch. Conversely the existence of double, convergent and transverse scrapers with less than 5–10% of mass removed through retouching, representing the initial creation of the edge and perhaps one resharpening episode, demonstrates that such forms were not always more heavily retouched than single scrapers. Similarly, many transverse scrapers were not noticeably more reduced than many single scrapers, as might be expected if they were created at a later stage. However, other transverse scrapers have been extensively retouched, probably losing more than 50% of their original mass. This illustrates that the intensity of reduction within each implement class is highly variable. Further evidence for this within class variation in retouching intensity, and its implications, is provided in the following sections.

Single Scrapers

The striking characteristic of single scrapers in Layer 21, besides the strong pattern of retouch positioned on one lateral margin, is the great difference in the extent of reduction that different specimens had undergone. Some of the variation in the extent of retouching is displayed by the retouched zone index. The majority of single scrapers were retouched along much of one lateral margin, resulting in retouch scars in four or five zones (Figure 6.6). However, a few specimens had retouch restricted to a small portion of the lateral margin, only one or two zones; and some specimens also had small occurrences of retouch elsewhere on the flake, in more than five zones. The distribution of retouch around the flake margin was clearly related to blank characteristics, such as edge angle, cross-section and distribution of cortex.

Another dimension of retouch intensity, measured by the Kuhn GIUR, also displays extreme variation. Figure 6.6 shows a histogram of the abundance of specimens with different levels of the Kuhn GIUR. Almost 20% of single scrapers had a GIUR less than 0.3, equating to less than about 5% of the original flake mass removed by retouching. Most single scrapers had GIUR values of 0.3–0.8, representing about 5–20% of mass loss. Some single scrapers, about 10% of those in Layer 21, had GIUR values greater than 0.8, representing retouch that removed approximately 30% to more than 60% of the original mass. While conversion of GIUR values to mass lost through retouching in this way is only an estimate, it expresses the large differences in retouch intensity that are evident on different single scrapers.

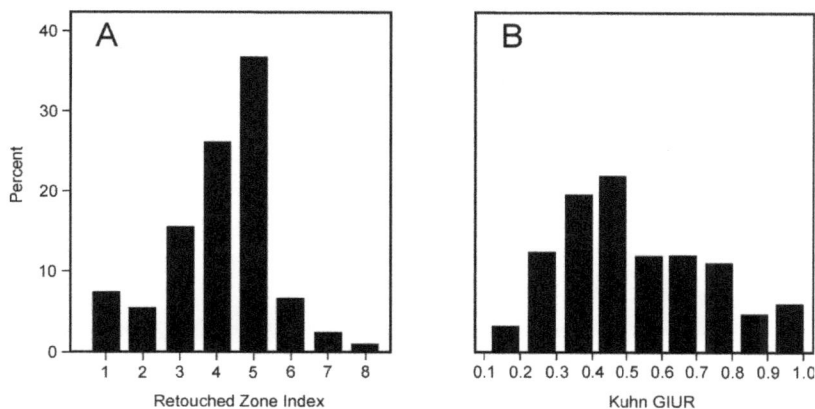

Figure 6.6: Histogram of the Kuhn GIUR values for single scrapers from Layer 21.

Source: Authors' calculations.

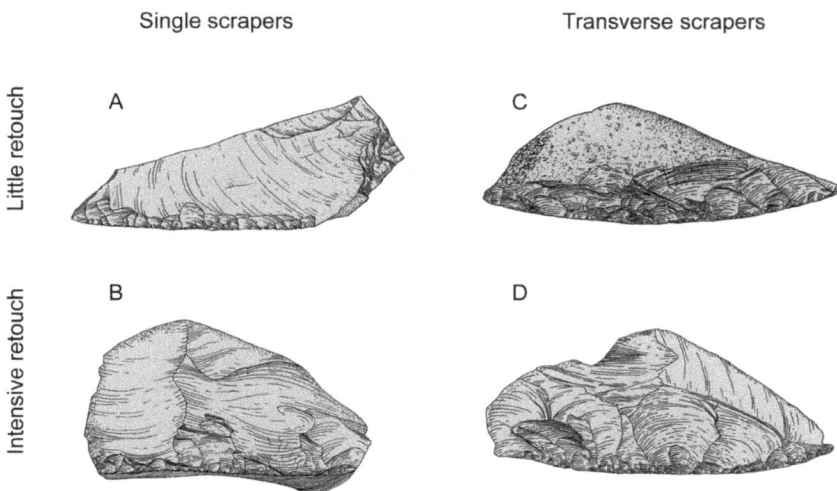

Figure 6.7: Examples of different levels of reduction on scrapers: A-B = Single scrapers, C-D = Transverse scrapers; A and C = little mass removed, B and D = extensive mass removed. A = Single scraper with a Kuhn index of 0.44; B = Single scraper with a GIUR index of 1.00. C = Transverse scraper with a Kuhn index of 0.56; D = Transverse scraper with a Kuhn index of 0.91.

Source: Authors' calculations.

Differences in the extent of retouching on single scrapers are illustrated in Figure 6.7, which presents two single scrapers: one with a small amount of material removed by retouching and the other with a large amount. The first specimen is a long flake with a series of small retouch scars, mostly about 3 mm

117

long, on three zones of the left lateral margin (Figure 6.7A). The Kuhn GIUR of 0.44 recorded for this specimen, in association with low unretouched edge angles of 20–25° in the retouched zones, is consistent with less than 5–10% of the original flake mass being removed by retouching. The other specimen (Figure 6.7B) was the remnant of a wide, thick flake which has been extensively reduced through the removal of large flakes from along the entire right lateral margin (retouch in five zones). This specimen has a GIUR of 1.00, with the retouch scars having removed the thickest part of the flake; a pattern consistent with the removal of approximately 45–65% or more of the original flake mass by retouching. Together these two illustrations exemplify the different levels of reduction present amongst single scrapers in Layer 21.

In conjunction with the statistics, these specimens demonstrate that some single scrapers in Layer 21 were minimally retouched while others were heavily retouched. The heavily retouched specimens, as indicated by the GIUR, typically have retouch scars only on one lateral margin and had always been a single scraper throughout the retouching process. The evidence from such specimens shows that some single scrapers became very intensively retouched but that the level of reduction did not alter their typological status.

Double and convergent scrapers

Double and convergent scrapers are almost certainly made from single scrapers that had appropriate sizes and shapes, since one retouched margin must have been created before the other. The higher mean and minimum GIUR values for both classes, in comparison to single scrapers, are consistent with that interpretation, but do not prove it. However, the evidence for Layer 21 does not conform to Dibble's proposed single-double-convergent sequence of type stages. We have already discussed the observation that single scrapers were sometimes very intensively retouched and so specimens in that typological class do not always represent a stage of minimal reduction. This demonstrates that double/convergent scrapers are not always highly retouched and single scrapers were not always minimally retouched.

Furthermore, in the collection from Layer 21 there is no difference in the intensity of reduction of the double and convergent scrapers. There is no significant difference between these two classes for any measure of retouching intensity: Kuhn GIUR (t = -0.566, d.f. = 66, p = 0.573), retouched zone index (t = 0.995, d.f. = 66, p = 0.323), and platform surface/platform area index (t = -0.010, d.f. = 52, p = 0.992). The ranges and distribution of values are also comparable for these measures; evidence that indicates specimens in both groups show varied but comparable levels of retouch intensity. Since convergent scrapers in this assemblage are not more reduced than double scrapers the notion that double scrapers were regularly converted into convergent scrapers is unlikely to be correct.

Instead, it seems likely that double and convergent scrapers are made on different kinds of flakes. A number of features of the flake are preserved on these retouched specimens and show statistically significant differences between the two classes. For example, the mean platform thickness of double scrapers is significantly lower than for convergent scrapers (t = -3.318, d.f. = 49, p = 0.002). This evidence indicates that the relative positioning of retouched edges in double and convergent scrapers, leading them to be assigned to different types, may reflect the knapper's response to dissimilarities in flake form rather than extent of retouch. Hence, it is possible to conclude that more specimens classified as double and convergent scrapers were reworked single scrapers, but that many of the convergent scrapers are not more intensively retouched than double scrapers, the typological distinction largely reflects the influence of the different flake blanks from which they were made.

Transverse scrapers

Transverse scrapers also display large differences in the extent of retouching. Nearly 60% of transverse scrapers had a GIUR less than 0.6, probably indicating less than 10% mass lost through retouching; but 20% of specimens had values of 1.0, indicating they had more than 40–50% of their initial mass removed. These differences can also be illustrated using specific implements as exemplars (Figure 6.7). For instance, Figure 6.7C shows a transverse scraper made on a primary decortication flake, which has had a series of small retouch scars at the distal end. The Kuhn GIUR of 0.56 recorded for this specimen, in association with a low unretouched edge angle of 34° at the distal end, is consistent with less than 10% of the original flake mass being removed by retouching. In contrast, another transverse scraper shown in Figure 6.7D had a series of large flake scars at the distal end, with retouch removing the thickest part of the flake along one half of the edge to give a GIUR of 0.91. This pattern is consistent with the removal of at least 30–35% of the original flake mass by retouching. These two illustrations exemplify the different levels of reduction present amongst transverse scrapers in this layer.

Large differences in retouch intensity between specimens classified as transverse scrapers also reflect flake characteristics: specimens with GIUR of less than 0.6 have, on average, significantly smaller platform thickness (t = 2.094, d.f. = 59, p = 0.041), smaller flake thickness (t = 3.616, d.f. = 64, p = 0.001) and lower unretouched edge angles (t = 3.240, d.f. = 64, p = 0.002). Reduction intensity was therefore connected to the size and morphology of flakes, with larger flakes being more extensively retouched. However, despite the great variation in retouch between specimens in Layer 21, retouch intensity did not alter the typological status of transverse scrapers. That inference is inconsistent with the notion that transverse scrapers were once single side scrapers that had subsequently had additional retouch added to the distal end. Instead this evidence indicates that many or all transverse scrapers had always been transversely retouched, throughout their entire production and re-sharpening history.

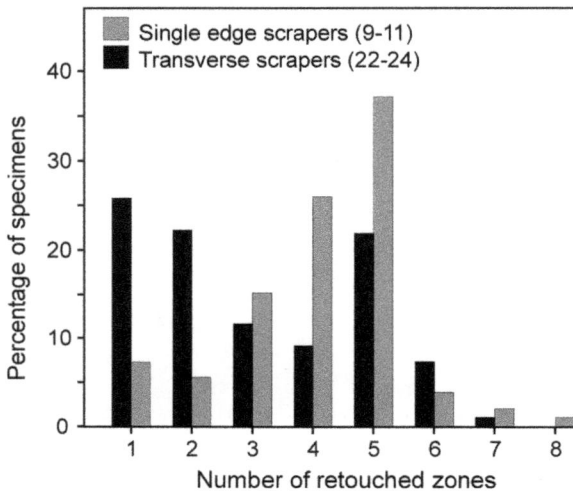

Figure 6.8: Histogram showing differences in the distribution of retouch on specimens classified as single scrapers and transverse scrapers.

Source: Author's data.

This conclusion is reinforced by information about the distribution of retouch around the perimeter of flakes (Figure 6.8). Distribution of retouch around flake perimeters is not consistent with all transverse specimens originally being single scrapers. More than 50% of transverse scrapers have retouch only toward the distal end (<3 retouched zones). These specimens were never single scrapers, and we conclude that at least half the transverse scrapers began as transverse scrapers. Those with 4–6 retouched zones may once have been single side scrapers that had retouch added to the distal end, or they may have begun as transverse scrapers that subsequently had retouch added to a lateral margin. While it is possible that in Layer 21 Dibble's hypothesised transformation of single into transverse scrapers sometimes occurred, this must have been infrequent compared to the common process creating transverse scrapers, in which knappers began to retouch at the distal end and continued to retouch in that location.

The initiation and maintenance of restricted patterns of retouch, at either the distal end or on a margin, probably reflects the influence of blank form. Transverse and single scrapers were regularly made on different flake blanks. For example, flakes which received retouch at their distal end (transverse scrapers) were thicker (t = 1.929, d.f. = 236, p = 0.055) and had higher unretouched edge angles (t = 2.249, d.f. = 99, p = 0.027) than those worked only the lateral margin. Differences in flake shape and thickness are therefore hypothesised to have been factors affecting the decision of knappers to begin working flakes laterally or distally.

A retouching scheme for Layer 21

The evidence presented here is consistent with a retouching scheme that is more elaborate and less stage-based than the one proposed by Dibble (1984, 1987b, 1995). Our interpretation of the retouching processes that created typological scraper groups in Layer 21 of Combe Grenal is represented in Figure 6.9. Most frequently, single side scrapers were retouched only on one margin for their entire history of production and maintenance. Some of those specimens were discarded after only a small amount of retouching but others were very intensively retouched on the same margin but remained, in typological terms, single scrapers. Some single side scrapers were retouched on additional margins to produce specimens classified as either double or convergent scrapers. Single scrapers were typically converted into either a double scraper or a convergent scraper, but there is little evidence for double scrapers being reworked to form convergent ones. Double and convergent scrapers have comparable levels of retouch, and are not sequential stages of retouch; they represent alternative strategies applied to single scrapers with subtly different sizes and shapes. The choice of whether to continue retouching one margin or to begin working a second, and in the latter case to retouch parallel or converging edges, appears to be related to differences in flake size and morphology. The precise interaction of blank form and retouch intensity will be pursued in future publications.

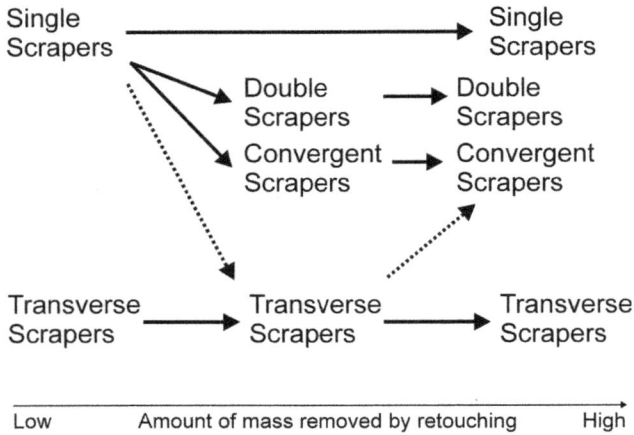

Figure 6.9: Illustration of the typological status and reduction history of flakes retouched to different degrees, using the same graphical conventions as Figure 6.1.

Source: Author's original figure.

Retouching of flakes to produce transverse scrapers appears to have been largely separate to patterns of lateral retouching leading to single, double and convergent scrapers (see Figure 6.9). Some transformations of single side scrapers to transverse scrapers, or the reworking of transverse scrapers into double/convergent scrapers, may have occurred but in Layer 21 this was infrequent. Our interpretation of the evidence is that the majority of transverse scrapers were distally retouched throughout their 'life-span' and they had never been single side scrapers. Transverse scrapers therefore principally represent the result of a parallel pattern of production that is separate from, and constitutes an alternative to, the retouching strategy that created single side scrapers.

Implications for the interpretation and analysis of Mousterian variability

We note that while our analysis demonstrates the non-sequential structure of Quina scrapers at Combe Grenal Layer 21, and the incorrectness of Dibble's reduction model for that assemblage, the application of our analytical approach to Denticulate retouched flakes in Layers 11–12 revealed that notched tool types did follow a sequence as predicted by the reduction model (Hiscock and Clarkson, 2007). This contrast shows that models of tool production may be correct for one site, or one level within a site, but not another, and arguments that one model is universally correct are not valuable. Additionally this finding reinvigorates questions of whether there is a distinct difference in tool production between Quina and Denticulate industries.

Furthermore, our interpretation of Quina scraper retouching in Layer 21 of Combe Grenal has a number of implications. Evidence presented indicates implement production was not simply a series of stages, it was multi-linear, and flake morphology was an important influence on the pattern of retouch distribution and intensity. Dibble's model of the implement classes, as stages of reduction, is not a viable depiction of the retouching technology represented in Layer 21. Our reconstruction of retouching processes conforms to many of the propositions contained in the 'blank-retouch interaction hypothesis' advocated by Turq and others.

Each of the Bordes' types examined here had multiple histories of retouching. Within each implement type some specimens received little retouch while others were intensively retouched. Perhaps retouch sometimes changed a specimen to such an extent that the type into which it would be classified was altered, but many specimens remained typologically stable even though they received additional retouch. Because different specimens belonging to each type had different histories, in particular very different amounts of retouch, the Bordes typology is not a reliable system for measuring retouch intensity. In assemblages such as this, Bordesian types tend to record the pattern and character of retouch preserved on flakes at the time they were discarded, but intensity of retouch cannot be accurately inferred from the type classification alone. Consequently, studies of spatial and temporal changes in retouch intensity will be more reliable when made on the basis of dedicated and experimentally verified systems of measurement, such as the Kuhn GIUR. In this context it is likely that robust models of land use and provisioning will not be able to be built on typological counts alone, and that long standing questions about the relationship between industrial variation and broader-scale organisational patterns can only be based on understandings of the technological processes that underpin differences in retouched flakes.

This is the first time that detailed quantitative measurements of reduction intensity have been used to assess the applicability of Dibble's 'reduction model' to Quina assemblages, and to do so by not only employing the same measures as Dibble had used but also other independent, and we argue more accurate, measurements of the retouch intensity. Our analysis reveals that the traditional typological groups are a complex product of multiple processes and not principally a signal of differing levels of retouch, even though retouch intensity is undoubtedly one of a number of factors creating morphological variation between specimens. Our conclusion that at Combe Grenal, and perhaps for many Mousterian assemblages, there is a strong interaction between blank and the nature of retouch which knappers applied to the blank also invites consideration of broad questions about the interpretation of implement patterns. The retouching scheme we have inferred for Layer 21 implies that sequential transformation of retouched flakes from one implement type to another was rare, and instead there were multiple, albeit branching, pathways of reduction. Demonstrating that retouching often maintained relatively stable tool forms need not indicate that Mousterian knappers had a specific design. In the instance of Combe Grenal the strong connection of retouch location, form and extent with blank form may provide a mechanism for creating regular and stable implement shapes over the reduction process, even if no well-defined, formal design was in place. Habitual application of production rules to blanks of different shapes may maintain stability in the appearance and location of a retouched edge during extended reduction. However, as Turq, Boëda and others have proposed, such production rules connecting blank form and retouching process to produce regularity in implement form might be considered a kind of design system for Middle Palaeolithic hominids. Debates of how we should think of goal-oriented behaviours in the Mousterian, and indeed the nature of technological and cogntive processes that were involved and their articulation with economic and ecological contexts, still require exploration in the quest to understand the construction of morphological diversity in Middle Palaeolithic implements. The evaluation of what constitute meaningful and valuable units of measurement, and how they may or may not be connected to traditional implement types, is not resolved; on the contrary, this discussion is merely beginning.

Acknowledgements

We offer this paper in reference to Colin Groves' significant work on classification, units of measurement, and of course hominin evolution. Both authors have enjoyed our interactions with Colin at all points through our careers, and we thank him for his support and engagement.

We acknowledge and appreciate the permission to examine the collection granted by Dr J-J Cleyet-Merle, the Director of the Musèe National de Prèhistoire des Eyzies. We thank Dr Cleyet-Merle and the Musèe National de Prèhistoire des Eyzies for providing their facilities for the prolonged duration of our project. For assistance and discussions at the Musèe National de Prèhistoire des Eyzies we thank Alain Turq, Andre Morala, and Jean-Philippe Faivre. Additionally we thank both Harold Dibble and Shannon McPherron for their advice and their generous help. A preliminary version of this paper analysing a sample of artefacts from Layer 21 was published as Hiscock and Clarkson (2008), but this paper presents an analysis of all relevant specimens from the layer and supersedes that earlier depiction. This research was funded by an Australian Research Council Discovery Grant (DP0451472 – A reappraisal of Western European Mousterian tools from Australian perspectives).

References

Binford LR. 1973. Interassemblage variability – the Mousterian and the 'functional' argument. In: Renfrew C, editor. *The explanation of culture change*. London: Duckworth. pp. 227–254.

Binford LR. 1989. Isolating the transition to cultural adaptations: An organizational approach. In: Trinkaus, E, editor. *The emergence of modern humans: Biocultural adaptations in the later Pleistocene*. Cambridge: Cambridge University Press. pp. 18–41.

Binford LR, Binford SR. 1966. A preliminary analysis of functional variability in the Mousterian of Levallois Facies. *Am Antiq* 68:238–295.

Bisson MS. 2001. Interview with a Neanderthal: An experimental approach for reconstructing scraper production rules, and their implications for imposed form in Middle Palaeolithic tools. *Cambridge Archaeol J* 11:165–84.

Boëda E. 1988. Le concept laminare: Rupture et filiation avec le concept Levallois. In: Kozlowski J, editor. *L'Homme Neanderthal*, vol. 8, *La Mutation*. Liège: Etudes et Recherches Archéologique de l'Université de Liege (ERAUL). pp. 41–60.

Bordes F. 1968. *The Old Stone Age*. New York: McGraw-Hill.

Bordes F. 1972. *A tale of two caves*. New York: Harper and Row.

Bordes F. 1961. *Typologie du Paléolithique Ancien et Moyen*. Mémoires de l'Institut Préhistoriques de l'Université de Bordeaux 1, Delmas: Bordeaux.

Bordes F, De Sonneville-Bordes D. 1970. The significance of variability in Palaeolithic assemblages. *World Archaeol* 2:61–73.

Bourguignon L. 1997. Le Moustérien de type Quina: Définition d'une nou-velle entité technique. Thèse de Doctorat de l'Université de Paris X, Nanterre.

Bourguignon L, Faivre J-P, Turq A. 2004. Ramification des chaînes opératoires: Une spécificité du Moustérien? *Paléo* 16:37–48.

Clarkson C. 2002. An Index of Invasiveness for the measurement of unifacial and bifacial retouch: A theoretical, experimental and archaeological verification. *J Archaeol Sci* 29:65–75.

Clarkson C, Hiscock P. 2008. Tapping into the past: Exploring the extent of Palaeolithic retouching through experimentation. *J Lithic Tech* 33:1–15.

Close A. 1991. On the validity of Middle Paleolithic tool types: A test case from the Eastern Sahara. *J Field Archaeol* 18:256–264.

Debenath A, Dibble HL. 1994. *Handbook of Paleolithic typology*, Volume One. *Lower and Middle Paleolithic of Europe*. Philadelphia: University Museum, University of Pennsylvania.

Dibble HL. 1984. Interpreting typological variation of Middle Paleolithic scrapers: Function, style, or sequence of reduction? *J Field Archaeol* 11:431–436.

Dibble HL. 1987a. Reduction sequences in the manufacture of Mousterian implements of France. In: Soffer, O, editor. *The Pleistocene Old World regional perspectives*. New York: Plenum Press. pp. 33–45.

Dibble HL. 1987b. The interpretation of Middle Paleolithic scraper morphology. *Am Antiq* 52:109–117.

Dibble HL. 1988a. Typological aspects of reduction and intensity of utilization of lithic resources in the French Mousterian. In: Dibble H, Montet-White A, editors. *Upper Pleistocene prehistory of Western Eurasia*. Philadelphia: University Museum, University of Pennsylvania. pp. 181–194.

Dibble HL. 1988b. The interpretation of middle Paleolithic scraper reduction patterns. In: *L'Homme de Néandertal*, vol. 4, *La Technique*. Actes du Colloque International de Liége, L'Homme de Neandertal. pp. 49–58.

Dibble HL. 1991. Rebuttal to Close. *J Field Archaeol* 18:264–269.

Dibble HL. 1995. Middle Paleolithic scraper reduction: Background, clarification, and review of evidence to date. *J Archaeol Meth Th* 2:299–368.

Dibble HL, Rolland N. 1992. On assemblage variability in the Middle Paleolithic of Western Europe: History, perspectives, and a new synthesis. In: Dibble HL, Mellars P, editors. *The Middle Paleolithic: Adaptation, behavior, and variability*. Philadelphia: University Museum, University of Pennsylvania. pp. 1–28.

Gordon D. 1993. Mousterian tool selection, reduction and discard at Ghar, Israel. *J Field Archaeol* 20:205–218.

Hiscock P. 1994. The end of points. In: Sullivan M, Brockwell S, Webb A, editors. *Archaeology in the north*. Darwin: North Australia Research Unit, The Australian National University. pp. 72–83.

Hiscock P. 2004. Slippery and Billy: Intention, selection and equifinality in lithic artefacts. *Cambridge Archaeol J* 14:71–77.

Hiscock P, Attenbrow V. 2005. *Australia's Eastern Regional Sequence revisited: Technology and change at Capertee 3*. British Archaeological Reports. International Monograph Series 1397. Oxford: Archaeopress.

Hiscock P, Clarkson C. 2005. Experimental evaluation of Kuhn's Geometric Index of Reduction and the flat-flake problem. *J Archaeol Sci* 32:1015–1022.

Hiscock P, Clarkson C. 2007. Retouched notches at Combe Grenal (France) and the Reduction Hypothesis. *Am Antiq* 72:176–190.

Hiscock P, Clarkson C. 2008. The construction of morphological diversity: A study of Mousterian implement retouching at Combe Grenal. In: Andrefsky W, editor. *Artifact life-cycle and the organization of lithic technologies*. Cambridge: Cambridge University Press. pp. 106–135.

Hiscock, P, Clarkson C. 2009. The reality of reduction experiments and the GIUR: reply to Eren and Sampson. *J Archaeol Sci* 36:1576–1581.

Hiscock P, Turq A, Faivre J-P, Bourguignon L. 2009. Quina procurement and tool production. In: Adams B, Blades B, editors. *Lithic materials and Paleolithic societies*. Blackwell. pp. 232–246.

Holdaway S, Mcpherron S, Roth B. 1996. Notched tool reuse and raw material availability in French Middle Paleolithic sites. *Am Antiq* 61:377–387.

Kuhn S. 1990. A geometric index of reduction for unifacial stone tools. *J Archaeol Sci* 17:585–593.

Kuhn S. 1992. Blank morphology and reduction as determinants of Mousterian scraper morphology. *Am Antiq* 57:115–128.

Mellars P. 1965. Sequence and development of Mousterian traditions in south-west France. *Nature* 205:626–627.

Mellars P. 1986. A new chronology for the French Mousterian period. *Nature* 322:410–411.

Mellars P. 1988. The chronology of the south-west French Mousterian: A review of the current debate. In: Otte M, editor. *L'Homme de Néanderthal*, vol. 4: *La technique*. Liege: Etudes et recherches Archaéologiques de l'Université de Liége. p. 97–120.

Mellars P. 1992. Technological change in the Mousterian of southwest France. In: Dibble HL, Mellars P, editors. *The Middle Paleolithic: Adaptation, behavior, and variability*. Philadelphia: University Museum, University of Pennsylvania. pp. 29–43.

Mellars P. 1996. *The Neanderthal legacy*. New York: Princeton University Press.

Rolland N. 1981. The interpretation of Middle Paleolithic variability. *Man* 16:15–42.

Rolland N. 1988. Observations on some Middle Paleolithic time series in southern France. In: Dibble H, Montet-White A, editors. *Upper Pleistocene prehistory of Western Eurasia*. University Museum Monograph 54. Philadelphia: University Museum, University of Pennsylvania. pp. 161–180.

Rolland N, Dibble HL. 1990. A new synthesis of Middle Paleolithic assemblage variability. *Am Antiq* 55:480–499.

Thiebaut C. 2003. L'industrie lithique de la couche III du Roc de Marsal: Le probléme de l'attribution d'une série lithique au Moustérien à Denticulés. *Paléo* 15:141–168.

Turq A. 1989. Approche technologique et économique du facies Moustérien de type Quina: Etude préliminaire. *Bull Soc Préhist Fr* 86:244–56.

Turq A. 1992. Raw material and technological studies of the Quina Mousterian in Perigord. In: Dibble HL, Mellars P, editors. *The Middle Paleolithic: Adaptation, behavior, and variability*. Philadelphia: University Museum, University of Pennsylvania. pp. 75–85.

Turq A. 2000. Paléolithique inférieur et moyen entre Dordogne et Lot. *Paléo*, Supplément 2.

Verjux C. 1988. Les Denticules Mousteriens. In: *L'Homme de Néandertal*, vol. 4, *La Technique*. Actes du Colloque International de Liége, L'Homme de Neandertal. pp. 197–204.

Verjux C, Rousseau D-D. 1986. La retouche Quina: Une mise au point. *Bull Soc Préhist Fr* 11–12:404–415.

7. What are species and why does it matter? Anopheline taxonomy and the transmission of malaria

Robert Attenborough

Introduction

By the mid-twentieth century, taxonomy had, like many things Victorian, become unfashionable. When Washburn (1951: 298) argued for a 'new physical anthropology', he described the discipline's dominant approach of that period as 'static, with emphasis on classification based on types', though he did qualify that characterisation as 'oversimplified'. The change he wanted to encourage was one of emphasis, bringing genetics in without totally rejecting systematics. Even so, it must have seemed to many readers that a viewpoint centred on taxonomy would be allied, not to the new, but to the 'old physical anthropology' (see also Fuentes, 2010; Little and Collins, 2012).

From this period onwards, an antipathy to taxonomy was almost palpable in some quarters, even if manifested more in neglect than in critique. The typological mentality underlying the enterprise was seen as akin to that of stamp-collecting; it was predicated on a static, almost pre-Darwinian view of biological variation; too much was left to the subjective judgment of the taxonomist; some industrious practitioners had taken 'splitting' to absurd and chaotic extremes; the rules of nomenclature were arcane and obfuscatory. Above all, the work necessary and sufficient to label the entities out there in nature had essentially been done by then. Further fiddling with categories and names was mere finicky detail; and when it led to changes in an approved taxonomic name, that was more a nuisance than a scientific advance.

Since the 1980s or so, however, the tide has been turning, even if taxonomists of some groups (e.g. plants) still fear that their trade is itself an endangered species. Colin Groves must long have been amongst those who sensed that earlier critiques had thrown out both baby and bathwater. In much of his work from the 1960s to the present, he has addressed taxonomic issues both directly (e.g. Groves, 2001a; Groves and Grubb, 2011) and indirectly, through the taxonomic underpinning he has brought to other work (e.g. Groves, 1989; Groves, 2008). I draw on some of his arguments here.

It says much for the adaptability of Linnaean taxonomy that a pre-evolutionary system for classifying biological diversity should have weathered the intellectual shocks of Darwinism, Mendelian genetics, 1930s population genetics, and in our own time molecular genetics. A further shock came from within taxonomy itself, in the form of a challenge to make taxonomy conform more rigorously to phylogeny. The cladistic school of taxonomy emerged in 1950 with the publication in German of Hennig's major theoretical work; though by 1966, when it appeared in English, other pioneers were already thinking along comparable lines (Cain and Harrison, 1960; Groves, 2001a). Hennig insisted that taxonomic groups should be monophyletic, which many traditional taxonomic groups were not; and he even wanted to have each taxonomic group's rank linked systematically to its antiquity, at least within higher-order taxa. Cladistic taxonomy was indeed a 'bombshell' and initiated a 'scientific revolution' (Groves 2001a: 8, 18).

Since the mid-century 'new synthesis' of evolution with genetics (Huxley, 1942), one particular taxonomic rank – the species – has been a focus of special interest and intensive analysis amongst cladistic and more traditional taxonomists. Why are species special? Part of the answer, as Groves puts it, is that species are 'kinds of animals (and other organisms)' as the lay public generally understands them (2001a: 26). A happy example was provided by Mayr from his fieldwork in the Arfak Mountains of the Vogelkop Peninsula, now in West Papua, Indonesia. He found that local people had 136 vernacular names for the 137 bird species that museum taxonomists recognised as occurring in the area. Their classification conflated just two of the museum taxonomists' species. This, said Mayr, was 'an indication that both groups of observers deal with the same, non-arbitrary discontinuities of nature' (1963: 17). Indeed: the observation vividly illustrates why species are special, and not purely artificial. But Mayr did not claim – nor would Godfrey and Marks (1991) for example, nor Groves – that it provides all the answers to the role of species in evolutionary theory, or solves all the taxonomists' day-to-day practical problems.

So what, then, are species, and why does it matter? The double-barrelled question comes *verbatim* from Colin Groves himself in recent conversation. But the first part is also the question with which he began a much earlier major theoretical work (Groves, 1989). He has reviewed this long-standing concern several more times, both in detail (Groves, 2001a: Part I) and more succinctly (Groves and Grubb, 2011: 1–10); see also Groves (2001b, 2004, 2012). In this chapter, I start from the solid ground that he established in these reviews (confounding his lament to Mittermeier and Richardson [2008: ii] that 'nobody ever reads Part I' of *Primate Taxonomy*). Then I shall set off in a quite different direction.

What are species?

In introducing the taxonomy of the primates, Groves (2001a) reviewed 11 main concepts of what a species is (amongst sexually reproducing organisms). Importantly, he divided them into two groups: theoretical concepts, dealing with 'what a species is in essence'; and operational concepts dealing with 'how you can recognize one when you meet one' (2001a: 26). Here I only consider one concept from each group.

Mayr's Biological Species Concept (BSC) originated with the neo-Darwinian 'new synthesis', and became part of the prevailing orthodoxy along with it, cited innumerable times now. Under this theoretical concept, species are 'groups of actually or potentially interbreeding natural populations which are reproductively isolated from other such groups' (Mayr, 1963: 19). Reacting against the preceding Typological Species Concept, Mayr and like-minded thinkers were concerned to emphasise species as units of evolution – populations or sets of populations sharing a common gene pool. Reproductive continuity was central under the BSC. Morphology and phenotype played no part in defining species, although Mayr did allow that 'where the taxonomist applies morphological criteria, he uses them as secondary indications of reproductive isolation' (1963: 16–17).

The BSC still fits well with modern evolutionary thinking, and remains important at that level. Groves cited it approvingly in earlier work (e.g. Groves, 1989: 1–3). But his position has become more sceptical (e.g. Groves, 2001a; Groves and Grubb, 2011); and this has much to do with the weaknesses of the BSC as a guide to a working taxonomist. For one thing, if the test is which sets of populations are 'actually or potentially interbreeding' and which are 'reproductively isolated' in nature, that raises visions of the taxonomist as naturalist field-worker, binoculars at the ready, watching interbreeding not happening. Not only is such negative evidence hard to gather and inherently unlikely to convince; often the reality is that the taxonomist works mainly in the museum or the laboratory, on preserved specimens rather than observations in life, providing evidence that bears only inferentially at best on reproductive isolation.

Furthermore, even field observational evidence cannot determine which non-interbreeding populations might potentially interbreed (Groves, 2001a: 26–27; Groves and Grubb, 2011); nor, as Mayr himself rightly insisted (1963: 92), can evidence from captivity settle the question either. The classic, common instance of this problem arises where populations or sets of populations are allopatric. Populations might be separated by geographical barriers or unsuitable habitat, so that there is little or no actual interbreeding. But could the separated populations potentially, naturally, interbreed? The BSC provides no clear, evidence-based way of answering that question.

Groves also draws attention to a simpler, more factual problem with the BSC. With extensive genetic evidence from wild populations now available, we can see that good species, recognised by everyone as such, do actually interbreed, at least on occasion. There are both primate and ungulate examples of this (Groves, 2001a; Groves and Grubb, 2011: 2).

Groves concludes that, for all the great merits of Mayr's work, the BSC is irreparably flawed as a basis for practical taxonomy. Even taxonomists professing to implement it have often only paid it lip service, and 'it can be claimed that the BSC had made very little difference in how practicing taxonomists actually practiced' (Groves, 2001a: 27; Groves and Grubb, 2011). One effect that it did have, however, was an ill-effect: a bias arose amongst some workers under the BSC against the recognition of allopatric species, which tended to be 'lumped', even where sympatric species may have been correctly diagnosed (Groves, 2001a: 27).

Amongst the 10 concepts competing to succeed the BSC, Groves' reviews identify a clear winner: the operational Phylogenetic Species Concept (PSC), due principally to Cracraft (e.g. 1983), though further developed by others (Groves, 2001a: 30–32; 2004). Species are seen as evolutionary lineages – units of evolution and biodiversity. The concept is 'based on the results of evolution (on pattern), not on the processes by which these results may or may not have come about' (Groves, 2001a: 31). Cracraft's formal definition of the PSC includes reference to 'patterns of ancestry and descent', a reference which for practical purposes Groves strips out. Like reproductive isolation, it would lead us off towards the untestable. For operational purposes, he defines a species simply as: 'a diagnosable entity' (Groves, 2001a: 32). 'Diagnosable' here means 'identifiable 100% of the time, having fixed genetic differences from all others' (Groves, 2001a: 313). This is still a demanding definition. Species diagnoses on this basis may admittedly be based on plausible assumptions as to the heritable basis of diagnostic characters (rather than clear demonstrations), and on the indications of small samples as to their prevalence. But at least there is a reasonably objective basis on which to reach a conclusion that can be defended as the best supported one in the current state of knowledge. And it is a concept that is compatible with Hennigian cladistics (Nixon and Wheeler, 1990). Groves does not claim that there are no uncertainties or drawbacks to the PSC; only that they are fewer and less serious than for the competing concepts.

Put in a nutshell, 'the advantage of the PSC is that it depends entirely on the evidence to hand; there is no extrapolation' (Groves and Grubb, 2011: 1). And 'this is as close as we can come to putting a finger on the units of biodiversity.

The next level … is where the excitement begins for many workers … But first we have to determine what the units actually are' (Groves and Grubb, 2011: 2). I agree; and I propose to proceed on that basis.

Why does it matter?

Does all this make a difference? Indeed it does. It has been precisely the PSC's power to make a real difference to accepted classifications that has made it unwelcome where a high value is placed on taxonomic stability. The difference it can make, simply in terms of the number of species recognised, is illustrated by Cracraft's own work as an avian taxonomist. Cracraft (1992) applied the PSC to the birds of paradise (Paradisaeidae) with dramatic results. A family previously thought to comprise 40–43 species could now boast some 90 of them. The contrast with the taxonomic stability found in the New Guinea example cited earlier – Mayr's from the Arfak Mountains – is instructive. If one's first thought is that perhaps Mayr's finding was simply an outrageous fluke, a second thought, and a third one, show that it is more complex.

While Mayr supplied few details, Bulmer's (1970) ethnotaxonomic research elsewhere in New Guinea provided a case study with more nuance. His work was amongst the Kalam (now the standard spelling, though Bulmer spelt it 'Karam': Pawley, 2011) of the Schrader Mountains. Bulmer estimated that only about 60% of Kalam terminal taxa for vertebrates correspond well with species as recognised zoologically. This does not, however, conflict with his main argument, that Kalam are like zoologists in being 'concerned with, and to a large extent aware of, the discontinuities which define biological species, even where their folk-taxa do not correspond one-to-one to these' (1970: 1082). The mismatches are explained partly as cases where Kalam are less familiar with those forms, and partly as either Kalam 'lumping' of zoological species (e.g. five microhylid frog species which, unlike all other frogs, they regard as inedible) or Kalam 'splitting' (e.g. where mature male birds of paradise, with their different plumage and behaviour, are placed in a different terminal taxon from females and immature males, even though Kalam know that this is what they are).

The essence of both Mayr's and Bulmer's examples, however, was that they were local. Presumably the species in question were mostly or all sympatric. Cracraft, on the other hand, in surveying a whole radiating family dispersed across the broad New Guinea region, was assessing the taxonomic status of population sets that included many instances of allopatry. What Cracraft had done was not primarily to collect new specimens, and certainly not to find new evidence about reproductive isolation. It was primarily to restore species status to numerous 'operational taxonomic units' (Groves, 2001a: 7), which under the BSC had been regarded as subspecies: allopatric, by definition, therefore. 'Restore' because, ironically, many of these taxonomic units had had species status until it became the trend under the BSC to relegate allopatric species to subspecies status where they were similar enough descriptively for that to be plausible. Cracraft's treatment of the evidence was quite traditional, analysing morphological variation in relation to geographical distribution, though his analysis was guided by the criteria of the PSC. The shock was in the outcome. Although Cracraft actually recognised fewer 'terminal taxonomic units' than the 100–115 that had been recognised under the BSC, he more than doubled the number recognised at species, as opposed to subspecies, level. Such a large disturbance of the previously accepted order can make readers uncomfortable. We ourselves are, after all, a classifying species, and sometimes classifications and their anomalies are strongly marked culturally (Douglas, 1966).

The PSC has brought similar, sometimes equally dramatic, changes to mammalian taxonomy. Madagascan primates supply several examples. Louis and others (2006) reviewed the genus *Lepilemur* (sportive lemurs), a group of 'superficially indistinguishable' (p. 2) primates, and argued on the basis of both molecular and phenotypic data that the true species diversity of the genus was 22, double that previously recognised. Andriantomphohavana and others (2007) similarly reviewed the taxonomy of the genus *Avahi* (woolly lemurs), supporting the five species already recognised, and proposing the elevation of two subspecies to species status and the recognition of a further entirely new species. In both genera, the taxa in question are separated by rivers as well as sheer distance: see also this volume (Chapter 15) for a catarrhine case.

A recent and dramatic illustration of the difference that a different species concept can make – and the controversy it can cause – comes from Groves' own work. Groves and Grubb (2011) and Groves and Leslie (2011) have presented a revised scheme for the classification of the ungulates in which, amongst other things, they recognised 279 extant bovid species where only 143 had been recognised previously. This was too much for Zachos and others (2013), who launched a strongly worded critique of authors promoting 'species inflation and taxonomic artefacts'; to which Groves (2013) duly responded.

Their exchange is instructive. The parties are agreed on certain points. Sometimes, traditional classifications do underestimate diversity at the species level. This diversity is worth uncovering and recognising. In this task, genetic data can usefully supplement morphological data. A categorical taxonomic system is not fully adequate to reflecting a continuous evolutionary process, so whether a speciation process is sufficiently advanced to justify formal recognition for an incipient species may sometimes be moot. There can be real-world consequences to recognising more or fewer species.

But for Zachos et al., the splitting of, for example, one klipspringer species (*Oreotragus oreotragus*) into 11 species is simply unacceptable: 'spectacular … taxonomic inflation' (Groves and Grubb, 2011: 275–279; Zachos et al., 2013: 3). Their principal criticisms, across this and other examples, appear to be two: that the PSC has been inappropriately applied; and that there are insufficient data. Despite passing mention of the Genetic Species Concept ('a group of genetically compatible interbreeding natural populations that is genetically isolated from other such groups'), Zachos and others make no clear case as to what species concept or what method of implementing one they would have preferred to see. The very fact of major divergence from tradition in the resulting number of species recognised seems to be a fault perceived in the PSC. As for the data, Groves (2013) concedes the small samples but points out that no more data were available; and he argues that one should draw the conclusion that follows from the available data, while remaining open to testing it against more data once available. Zachos and others make similar criticisms in relation to other examples, e.g. the recognition of six mainland serow (*Capricornis*) species in place of one (Groves and Grubb, 2011: 255–261; Zachos et al., 2013: 3), even though the data are more abundant in that case. Groves, lacking their trust in traditional species diagnoses, is correspondingly more willing to advance claims based on the PSC.

All these examples – birds of paradise, lemurs, klipspringers and serows – concern instances where authors applying the PSC have recognised more species than previously. This does not in itself show that the higher number is either correct or incorrect. Many but not all are taxa previously deemed to be subspecies. All, I believe, are allopatric – often not even parapatric but well separated geographically – with respect to other populations with which they had been or might have been considered conspecific. Allopatry is not invariably central, however, to cases where species diagnosis is affected by subtleties of detail and concept. Among insects and some other invertebrates, sympatric sibling or cryptic species have long been reported, separated reproductively, genetically, ecologically or chronologically (Mayr, 1963). Behaviour and ecology are often important as distinguishing characters, where the morphological differences are slight.

Overall, the species concept deployed not only makes a difference – potentially a large difference – to the classification one comes up with; it is also a difference that matters greatly to taxonomists. To Groves, for one, it matters to make the most scientific, most evidence-based estimate one can of the species-level diversity in nature; and, even if the answer that comes up is unexpected or inconvenient, to take it seriously. But, taxonomists apart, does it matter more widely?

There is at least one further, and very practical, respect in which it matters. Conservation work risks being misdirected unless it is operating on the best available understanding of the evolutionary lineages it is dealing with. Without that understanding, the rich biodiversity of what survives may be under-estimated; but so may the extent and nature of the need for action. The lineages and habitats in direst danger of extinction – or conversely, on a conservation triage approach, those viable enough to benefit from intervention – may not be correctly identified to conservation agencies. Thus, excessive 'lumping' carries an undisputed risk of concealing conservation needs; while Zachos and others add a countervailing concern that excessive 'splitting' may provide additional targets for taxonomically minded trophy hunters and collectors.

In the remainder of this chapter, I shall attempt to explore a quite different practical reason why it is helpful to do as Groves and like-minded scholars have done with the taxa of interest to them: to consider and refine species concepts very carefully and to work from them towards the best attainable empirical determinations and characterisations of species.

Anopheline taxonomy

Flies of the family Culicidae (Diptera) – mosquitoes – are divided into two subfamilies, Culicinae and Anophelinae; and Anophelinae into three genera, including *Anopheles* (Krzywinski and Besansky, 2003). By virtue of the blood meals that they imbibe from their hosts – mammal, bird and reptile – the females of many culicid species transmit infections between them, including, in the human case, arboviruses, filariasis and malaria. Because it is only anophelines, specifically *Anopheles*, of certain species only, that transmit human malaria, there is a disproportionate focus on anophelines in the literature; and they will be my focus too. The number of anopheline species recognised, including cryptic species, has increased substantially in recent years. Below I explore some implications of this development. My debt especially to three recent reviews (Beebe et al., 2013; Sinka et al., 2012; White et al., 2011), which draw on much wider literatures than I shall do directly, will be very apparent.

Anopheles is a very speciose genus, with a near worldwide distribution, containing six subgenera and hundreds of species. Four of the subgenera are endemic to South America, where the genus is likely to have originated, at least 50 million years ago (Reidenbach et al., 2009). Their current distribution includes many regions where malaria does not normally occur endemically nowadays, such as southern Australia and northern Europe – though islands in the Remote Pacific, beyond Buxton's Line which runs east and south of Vanuatu, remain *Anopheles*-free. In the 1980s, around 400 species of *Anopheles* were recognised (Bruce-Chwatt, 1985). Now, after much further work in genetics to supplement the morphology, 465 species are formally recognised and there are also over 50 unnamed members of species complexes (Sinka et al., 2012). Only a minority of all anopheline species have the capacity to transmit the five malaria parasites that normally infect people (*Plasmodium falciparum, P. vivax, P. malariae, P. knowlesi and P. ovale*), and fewer still do so frequently.

The global distribution of *Anopheles* species involves complex regionally contrasted patterns of allopatry and sympatry. The southwest Pacific region – lying between Weber's Line in the west (running through the Moluccas) and Buxton's Line in the east, and centred on the islands and archipelagos of New Guinea, the Bismarcks, the Solomons, Vanuatu and Australia – may serve to illustrate this complexity. At best current reckoning, at least 56 *Anopheles* species occur in this region (Beebe et al., 2013; Foley et al., 2007). Most of these belong to one of five species groups endemic to the Australian Faunal Region (the *An. punctulatus, An. longirostris, An. lungae, An. bancroftii* and *An. annulipes* groups). There are also four ungrouped endemic species, and eight species from the Oriental Faunal Region which have dispersed eastwards into the Moluccas or further into the Australian Faunal Region. There are elements of allopatry in their distribution: for example, the *An. lungae* group occurs only in the Solomons including Bougainville, whereas the *An. longirostris* and *An. bancroftii* groups occur only in New Guinea, and the *An. annulipes* group mainly in Australia though with two representatives in New Guinea. On the other hand, *An. farauti* (in the *An. punctulatus* group) occurs throughout the tropical parts of the region, though rarely far from the coast, and is thus sympatric with many other species (Beebe et al., 2013).

The best studied species group in the region is the *An. punctulatus* group (subgenus *Cellia*). In the 1980s this group had five recognised member species, of which just three – *An. punctulatus, An. koliensis* and *An. farauti* (or *An. farauti* 1) – were known from Papua New Guinea. Research undertaken around this time was reported in these terms (e.g. Attenborough et al., 1997; Charlwood et al., 1986). Since the 1990s it has become apparent that the situation is more complex. The biodiversity of this species group, and the extent to which it is made up of morphologically similar cryptic species, is only now becoming fully appreciated (Beebe et al., 2013). This is because genetic data have allowed further species-level distinctions to be made where morphological data alone are uninformative or (as it turns out) unreliable. The *An. punctulatus* group now consists of 13 species. Most of the more recently diagnosed ones were previously included in *An. farauti*, and some still await a formal name. They have been distinguished via observations on chromosomal inversions, allozyme variation, variation under species-specific genomic DNA probes, and PCR-RFLP variation in ribosomal DNA (rDNA), especially its internal transcribed spacer 2 (ITS2) region. Species status for these taxa was further confirmed in some cases by cross-mating experiments demonstrating sterility, and by lack of hybridity at the rDNA locus even in large samples of field-collected specimens. The 13 species appear to be organised as two main clades: most of the *An. farauti*-like species in one; the *An. punctulatus*-like species plus *An. farauti* 4 in the other; with the position of *An. koliensis* indeterminate (Beebe et al., 2013).

At a broad level, there is much co-occurrence amongst the *An. punctulatus* group species. Three of the 13 species occur in the Moluccas; four in New Britain; formerly five in the Solomons (including Bougainville), though fewer now; 11 in one or another part of the New Guinea island. In more fine-grained biogeographical terms, there is less true sympatry of these species, especially in New Guinea where different species occupy different regions; and there is also some ecological separation e.g. in breeding habitats. Nonetheless, there is some true sympatry: for example, *An. koliensis, An. hinesorum* (formerly *An. farauti* 2) and *An. farauti* 4 have not only overlapping distributions, especially in New Guinea's inland northern lowlands, but also similar larval habitats, including transient ones created by human and pig activity. They are also not reliably distinguishable on morphological criteria only. The latter two are thus instances of cryptic species-level biodiversity uncovered through genetic research. That they are – unlike the newly recognised vertebrate species discussed earlier – sympatric is very interesting. Beebe and others (2013) do not explicitly discuss the species concept that has been applied in diagnosing these and other species in the group; but their methods and findings appear to conform to the PSC.

Similar statements could be made about the other four *Anopheles* species groups endemic to the region, in which some 25 new species overall have emerged through recent research. These groups are, however, all less well known than the *An. punctulatus* group.

For a second example, the most thoroughly researched of all, I turn to sub-Saharan Africa and the *An. gambiae* species group. For this group we can see research well advanced in some directions that currently still remain in the future for the southwest Pacific anophelines. The major African malaria vector *An. gambiae* was originally taken to be a single species. Then cross-mating experiments showed that sterile male progeny resulted from crossing, first western and eastern coastal saltwater breeding populations with each other or with freshwater breeding populations, and then certain freshwater populations with each other (White et al., 2011). We now have an *An. gambiae* species complex consisting of some seven well recognised species: the western coastal species *An. melas*; the eastern coastal species *An. merus*; a third salt-tolerant species, *An. bwambae*, found near hot springs in Uganda; at least two widespread freshwater species, *An. gambiae sensu stricto* and *An. arabiensis*, which are extensively but not wholly sympatric; and two allopatric species of more restricted distribution, both originally included in *An. quadriannulatus* but now separated, again on the strength of crossing experiments, as

An. quadriannulatus A and B (White et al., 2011) (see also maps of Sinka et al., 2012). Fixed genetic differences between these species were discovered, first karyologically, as banding patterns and paracentric chromosomal inversions, and then at the DNA level, again primarily in the intergenic spacer of rDNA. The species in the *An. gambiae* complex remain morphologically indistinguishable (White et al., 2011): that is, no morphological differences that are fixed and therefore diagnostic have been identified. Here, then we have a species group smaller than the *An. punctulatus* one but similarly complex, with at least seven, sometimes sympatric, well studied sibling members clearly recognised as such despite their phenotypic similarity. Species hybrids have been found in the wild and may be fertile if female; but are extremely rare and do not alter the diagnosis of these good species, whose phylogeny and history of ecological interaction with humans are discussed by White and others (2011).

The complexity goes to another level too in this case. A great deal of complexly patterned chromosomal and molecular variation has been discovered within *An. gambiae s.s.* (hereafter, *An. gambiae*), and these two modes of variation are not simply related to each other. It transpires that the molecular level is the more fundamental one reproductively. The molecular forms labelled M and S, identified by fixed single-nucleotide differences in rDNA, have been widely considered to be examples of 'incipient speciation', but recently some workers have gone further and accorded them species status, as *An. coluzzii* and *An. gambiae* respectively (Coetzee et al., 2013). Though sympatric, they are ecologically differentiated at a micro level, especially in larval habitat. S characteristically breeds in ephemeral rain puddles and its larvae grow fast. M tends to breed in longer-lasting artificial habitats associated with irrigated agriculture; its larvae grow more slowly and are outcompeted by S in the absence of predators, but are better at predator avoidance – they become more inactive in their presence (White et al., 2011). A recent study in Burkina Faso confirmed strong assortative mating of the forms, whereby: first, most mating swarms were of one form only, temporally or spatially separated from the other; second, even in mixed-form swarms, a large majority of pairs collected were of the same form; and third, even in the tiny number of mixed-form pairs, all the females had sperm of their own form in their spermothecae, not of the males they were caught with, presumably having mated recently with males of their own form (Dabiré et al., 2013). This supports the evidence of Pennetier and others (2010) for close-range mate-type recognition, based specifically on auditory flight-tone matching via difference tones. It also supports their recognition as species. Despite the existence of pre-mating barriers and probably post-mating barriers too, hybridisation of M and S does occur, at low but non-negligible and regionally variable levels, and genetic evidence indicates continuing gene flow between forms (White et al., 2011). Most authors prior to Coetzee and others (2013) recognised a strong evolutionary differentiation but did not go beyond calling the forms 'incipient species'. As both Groves (2013) and Zachos and others (2013) would presumably agree, in the continuous process whereby species come into existence, there will be instances in which it is moot where we draw any categorical line; and this appears to be one of them. We seem to have an excellent example of speciation under way; and interestingly it is happening sympatrically. The possible future discovery of further complexity and differentiation is not ruled out.

In the two limited examples of anopheline taxonomy chosen for discussion here, then, we can see that, although neither the PSC nor any of its competitor concepts is explicitly invoked, molecular technology and more in-depth research tend to uncover more complexity and greater biodiversity, both unambiguously at the species level and emergent amongst sets of populations, both allopatric and sympatric. Positive and statistically convincing identification of reproductive isolation mechanisms, often by experimentation, has more often been practical with insects than with birds or mammals, but is compatible with the PSC; and the primary criterion of 'fixed genetic differences' (White et al., 2011: 114) captures its essence.

The transmission of malaria

There are, then, many more anopheline species, and differentiable subpopulations within the formally named species (not always meeting the criteria for subspecies), than were recognised a few years or decades ago. Probably very few concerns have arisen that any anopheline taxa might need conservation. But there is a different practical reason why we should be interested in getting anopheline taxonomy right, and specifically in not underestimating their species biodiversity. That lies in the potential that a better understanding of malaria's vectors should have in combatting the transmission of the *Plasmodium* parasite. One simple illustration lies in a mistargeted malaria control campaign in Vietnam against a non-vector species misidentified as a vector (Krzywinski and Besansky, 2003; van Bortel et al., 2001). The more fine-grained our knowledge of each species or form, even those that are cryptic, the better guided our interventions should be. As Beebe and others (2013) put it: 'effectiveness of malaria control interventions depends on the biology of the [*Anopheles*] species present.'

Given the multiplicity of the factors required for human malaria transmission, there are many points at which the cycle might in principle fail or be interrupted. Depending on her proclivities and opportunities, a mosquito's first blood meal might or might not be from a human host; that blood meal might or might not be infected with the gametocytes of one of the plasmodia that cause human malaria; the mosquito might or might not be a competent vector (i.e. susceptible to infection); she might or might not live long enough for the malaria sporozoites to develop and reach her salivary glands; her next blood meal may or, depending on similar factors, may not be from another human host, and may or may not start up a new infection in that person; the mosquito may or may not proceed to high levels of reproductive success.

Mosquito survivorship, density and anthropophily are the basis of vectorial capacity, which is thus a matter of numbers and probabilities. Since the proportion of anophelines that have the potential to transmit a malarial infection is typically only a few per cent, quantitative reductions in abundance and biting rates, however these arise, may bring appreciable gains. Current efforts to achieve this amount principally to indoor residual insecticide spraying and (often long-lasting) insecticide-treated bed-nets. Used effectively, these can lead to large reductions in biting rates. A genetic study in Equatorial Guinea has shown large reductions in anopheline populations subject to these measures (Athrey et al., 2012). In addition, larval control measures have, but only lately, been shown to be effective where coverage of larval habitats is high enough (Tusting et al., 2013). And anti-malarial medications, where effective, must reduce the opportunities for mosquitoes to be infected.

The key point here is that the different variables affecting malaria transmission frequently vary in a species-specific way, and sometimes in a population-specific way. *Anopheles* species are not all equal in their importance for malaria epidemiology; nor are their roles as malaria vectors simply a function of their global distribution or local abundance. And, as White and others (2011: 112) say, 'the rare species that possess all four of these traits [strong preference for human blood, physiological competence to parasite infection, long life, high population density] are not clustered phylogenetically but rather are interdigitated with nonvector species in four of six subgenera and even in sibling species complexes'. This statement applies globally (Sinka et al., 2012), in Africa (White et al., 2011), and in the Pacific (Beebe et al., 2013). It has long been known that anopheline species vary in their ecology, demography and behaviour as well as their distribution, in ways that affect malaria transmission patterns and the overall importance of each as malaria vectors (Bruce-Chwatt, 1985). But this is true of newly distinguished *Anopheles* species and populations, too. Important questions now for renewed malaria control or even local elimination efforts include: what are the details of these patterns of variation? And how can malaria transmission interventions best be designed on the basis of that knowledge? I only discuss, selectively, the first question here.

Some 70 *Anopheles* species, worldwide, out of ~500, can transmit the *Plasmodium* parasites; and Sinka and others (2012) designate 41 of those as dominant vector species or species complexes (DVS). In Africa, they designate three species (*An. gambiae* and *An. arabiensis*, both in the *An. gambiae* complex; plus *An. funestus*) as the most dominant of the continent's DVSs; and a further three species (including two more members of the *An. gambiae* complex, *An. melas* and *An. merus*) and one species complex as secondary DVSs. The recognition by some authors of *An. coluzzii* (see above) brings the list of most dominant DVSs to four.

Of the 56 species in the Pacific region, Beebe and others (2013) divide the 38 which occur in New Guinea, the Bismarcks, the Solomons and/or Vanuatu into four vector status categories: 3 primary vectors (all in the *An. punctulatus* group); 18 secondary vectors (including 4 in the *An. punctulatus* group); 8 possible vectors (including 1 in the *An. punctulatus* group), pending more extensive research; and 9 non-vectors (including 5 in the *An. punctulatus* group).

The three primary Pacific vectors of malaria go by the names longest-known in the *An. punctulatus* group: *An. farauti s.s.*, *An. koliensis* and *An. punctulatus* itself. All are widespread in the region and can be locally abundant. Some key features of their species-specific distribution, ecology and behaviour, relevant to their vectorial capacity, are summarised in broad terms in Table 7.1.

Table 7.1: Key features of the primary Pacific malaria vectors.

Species	Distribution	Environment	Breeding sites	Longevity	Anthropophily
An. farauti	Moluccas New Guinea Bismarcks Solomons Vanuatu N. Australia	Seldom far from coasts	Small ground pools to large coastal swamps & lagoons. Larvae tolerate brackish water	Variable	Adaptable & variable: readily feeds on humans but in many places also pigs, dogs & probably native birds & mammals; in the Solomons strongly anthropophilic
An. koliensis	New Guinea New Britain Buka Formerly most of Solomons	Inland lowlands & river flood plains to 300 m a.s.l.	Wheel tracks, drains, natural pools, swamps	Medium	Prefers humans; will feed on pigs & dogs
An. punctulatus	Moluccas New Guinea Bismarcks Buka Formerly Solomons	Lowlands, foothills, mountain valleys. Clay soils, perennial rainfall	Rock pools in & near rivers & streams, wheel ruts, foot & hoof prints, pig wallows, transient water. Eggs can survive desiccation	Most long lived of its species group	Most anthropophilic of its species group

Source: Summarised from Beebe and others (2013).

Relative to the primary vectors, other species play smaller roles, or no role, in malaria transmission, for reasons generally related to the criteria set out by White and others (2011). At the extreme, several New Guinea species including *An. farauti* 5 are rare, *An. clowi* so rare that it has only been found twice since 1946; *An. rennellensis* only occurs on Rennell Island (Solomons) where there is little or no malaria transmission; the three members of the *An. lungae* complex (endemic to Solomons) bite humans but have never been found infected with human malaria parasites, and at least one of them is short-lived, so their capacity to transmit malaria is unconfirmed; *An. irenicus* (formerly *An. farauti* 7) is not only restricted to Guadalcanal (Solomons) but also has never been recorded as biting humans, despite local abundance as indicated by larval collections. Thus these species and some others play nil, negligible or unproven roles in malaria transmission (Beebe et al., 2013).

This leaves, however, a number of secondary vectors, including newly recognised species, which may play significant malaria transmission roles on top of those of the primary vectors, at least locally where they occur or are abundant. Within the *An. punctulatus* group, for instance, *An. farauti* 6 is quite common in the cool moist highlands valleys of New Guinea over 1000 m a.s.l., to which it appears adapted; and it probably plays a major role in the now worsening problem of highlands malaria transmission. *An. farauti* 4, *An. hinesorum* (formerly *An. farauti* 2) and *An. koliensis* all transmit malaria, but are hard to distinguish morphologically, and are all sympatric in lowland New Guinea; so more field research is still required to characterise sharply the abundance, ecology and vectorial properties of the first two especially. *An. hinesorum* in New Guinea readily bites humans but is highly zoophilic in Buka, Bougainville and the Solomons. Similarly, mosquitoes of the *An. longirostris* complex transmit malaria but have been found to be zoophilic in some areas and anthropophilic in others; whether reflecting species-specific behavioural differences between the cryptic species in this complex remains to be seen following further research (Beebe et al., 2013).

As with taxonomy, so also with vectorial capacity, more research has been undertaken in Africa than in the Pacific. Within the morphologically homogeneous *An. gambiae* complex, *An. gambiae* is usually considered the most anthropophilic, though there are grounds to see the situation as more complex. *An. gambiae* thrives in many different environments, but appears specialised in its association with humans in all those environments and at all stages of its life cycle (White et al., 2011). Ayala and Coluzzi (2005) argue that *An. gambiae* is a very young species (or, now, species pair), descended from an *An. quadriannulatus*-like ancestor; and thence ultimately, like other complex members, from an *An. arabiensis*-like ancestor. The proposed selection pressures were principally those produced by human population density and environmental impacts, beginning in the African late Neolithic, less than 4,000 years ago; resulting, for example, in heliophilic larvae. Along with its anthropophily, *An. gambiae* is also endophagic and endophilic, with implications for intervention (White et al., 2011). The highly anthropophilic incipient species M and S, or *An. coluzzii* and *An. gambiae*, show a partial but marked ecological differentiation as described above, with M predominant in more urbanised, more polluted environments with longer-lasting, more predator-infested breeding habitats; whereas S predominates in more rural settings (Kamdem et al., 2012).

Where sympatric, *An. gambiae* and *An. arabiensis* compete, with *An. gambiae* prevailing in rainforests and other relatively well watered habitats, and *An. arabiensis* in drier ones. *An. arabiensis* possibly dispersed from the Middle East over 6,000 years ago, probably as a zoophilic and exophilic species, acquiring anthropophily secondarily, and being now second in that respect only to *An. gambiae* in East Africa – though it remains zoophilic and exophilic in Madagascar, perhaps on account of historically lower population density (Ayala and Coluzzi, 2005).

At the opposite extreme of anthropophily, still within the same superficially homogeneous species complex, *An. quadriannulatus* is generally reported to be highly zoophilic and therefore a non-vector of human malaria. It has never been found naturally infected with *Plasmodium falciparum* malaria, though it has been shown in the laboratory to be a competent vector of it (White et al., 2011).

Another pressure driving the recent and rapid radiation of the *An. gambiae* complex has apparently been the adaptation permitting larval physiological tolerance of brackish water – twice independently, with one lineage, more closely related to *An. quadriannulatus*, leading to *An. bwambae* and the western coastal species *An. melas*, and the other, more closely related to *An. gambiae*, leading to the eastern coastal species *An. merus*. This adaptation would appear to be the dominant differentiating factor for these species, given their still saltwater-focused distribution (Ayala and Coluzzi, 2005).

Anopheline mosquitoes' role as malaria vectors brings them no known evolutionary advantage. That there are variations amongst species, populations and individual mosquitoes in genes conferring immune resistance to malarial infection might suggest that there is some evolutionary cost. Laboratory genetic lines of *An. gambiae* more refractory to infection are able to kill many immature *Plasmodium* parasites – though *P. falciparum* least, and African *P. falciparum* least of all. These findings may not be transferable to field-collected mosquitoes, whose median parasite density is typically very low. But if they are transferable, an evolutionary 'arms race' might be hypothesised, whereby *An. gambiae* has evolved a degree of resistance to infection with the *Plasmodium* species and strains that naturally infect it, but in turn those plasmodia most exposed to this selection pressure have evolved the ability to evade the mosquitoes' immune defences. White and others (2011) review the now substantial evidence now available on immune gene variation affecting their vector competence. The net result is great variation, even within malaria-endemic zones, between and within species, in the likelihood that an individual mosquito's bite will be infective.

The processes that direct that individual mosquito's bite to a human host, rather than some other vertebrate (for which it may not be infective), are also crucial epidemiologically; and they vary according partly to anopheline species or population. As a variable, however, anthropophily can be problematic. Mosquito preferences for particular host species do not necessarily translate directly into biting rates on those species, as hosts may differ in their numbers, accessibility, presence in particular micro-environments, and defensive behaviour. Thus true host preferences unbiased by these factors are generally unmeasurable in uncontrolled field conditions. Consequently, descriptions such as anthropophilic or zoophilic are frequently based on less than satisfactory evidence. Nonetheless, it is clear that anophelines do have preferences amongst hosts, that these preferences vary between species, and that they play a part in explaining the different biting rate patterns of different species: indeed population differences within *An. gambiae* have been demonstrated (Lefèvre et al., 2009). Of the multiple cues thought to guide mosquitoes' activation, anemotaxis, close-range approach to hosts, and landing behaviour, some are non-specific (warmth, humidity, carbon dioxide); but others are species-specific – human or cattle odours, for example, eliciting different responses according to the anthropophily/zoophily of the anopheline species. Amongst anthropophilic species there is also differential attraction to different individual humans, and one factor which exacerbates this in *An. gambiae* – one further risk factor for malaria, in other words – is beer consumption (Lefèvre et al., 2010).

Field estimates of population density are generally derived from mosquito landing rates at human, animal or artificial baits, and are therefore subject to the extent of their attraction to those baits. The abundance of more zoophilic mosquitoes is likely to be underestimated in situations where most of the data come from human landing rates. Nonetheless, despite biases affecting the estimates, species differences in population density are clearly real and sometimes very large.

An illustrative picture of some more recently diagnosed anopheline species, reviewed in relation to White and others' (2011) four criteria for vectorial efficiency, is presented in Table 7.2. As this table shows, there are potentially major malariological implications to making taxonomic distinctions amongst anophelines, including between species previously not distinguished.

Table 7.2: Selected contrasts in variables relevant to malaria transmission between selected cryptic species in the *An. punctulatus* and *An. gambiae* species complexes.

Species complex	Species	Anthropophily/ zoophily*	Estimated population density*	Longevity	Vector competence
An. punctulatus (Pacific)	*punctulatus*	Most anthropophilic in complex but ranges from total to modest dependence on human blood	High	Long-lived	Yes
	farauti	Highly variable	High	Variable	Yes
	hinesorum	Variable geographically	Variable geographically		Yes
	irenicus	Highly zoophilic	Common locally		No?
An. gambiae (Africa)	*gambiae*	Highly anthropophilic	High		Yes
	arabiensis	Second most anthropophilic	High		Yes
	quadriannulatus	Highly zoophilic			Shown in lab only

Note: *See text for discussion of difficulties in making unbiased estimates of these variables.

Source: Summarised from Beebe and others (2013); White and others (2011).

Research on anopheline demography, ecology and behaviour clearly has important potential implications – not pursued here – for interrupting malaria transmission. Suffice it to say that, to the (substantial) extent that these characteristics vary along taxonomic lines, especially species-specific lines, a fine-grained taxonomy of anophelines too has an important role in fresh approaches in the field to the still huge problem of malaria. The recent recognition of *An. coluzzii* as a species distinct from *An. gambiae* only strengthens this point.

To counterbalance optimism with a necessary caution, it also needs to be noted that we also have evidence for: phenotypic adaptability according to circumstances; variation within currently recognised species and even incipient species due either to such adaptability or genetic differentiation of sub-populations; and rapid changes in these traits due to either evolutionary or behavioural change.

Conclusion

The circle that I have attempted to square in this piece has been to seek a perspective, from my own angle as a biological anthropologist interested above all in one particular species (ourselves), on a central interest of Colin Groves: that is, on species concepts, species diagnosis, species biodiversity, and indeed the origin of species. My approach has been to review, first, species concepts, and thence, selected recent developments in the taxonomy of *Anopheles*, the mosquito genus that, by transmitting malaria, still wreaks enormous havoc upon human life and health. This review supports the proposition that a fine-grained taxonomy, based on the PSC criterion of fixed inherited differences, and including recognition of cryptic and incipient species that are barely distinguishable or indistinguishable morphologically, is an important prerequisite of further fundamental biological research on these mosquito populations. Optimum practical intervention also depends upon it: in this case, not in a conservation context but to improve human health in the tropical Western Pacific, sub-Saharan Africa and other places still greatly afflicted by this scourge.

Acknowledgements

This piece is written out of my appreciation of an almost career-long collaboration with the acute, polymathic, formidably industrious, frequently humorous, ever equable Colin Groves, my closest colleague in the School of Archaeology and Anthropology at The Australian National University. To Tom Burkot of James Cook University, I owe not only my long-ago introduction to malarial entomology but also my extreme gratitude for valuable guidance before this piece was written and a critical reading of it afterwards. Nigel Beebe of the University of Queensland and Nora Besansky of the University of Notre Dame were also most helpful entomologically. Andrew Pawley of ANU drew my attention to a most useful reference. Any remaining errors are of course my own responsibility. I thank the volume editors, Alison Behie and Marc Oxenham of ANU, for their advice and forbearance.

References

Andriantomphohavana R, Lei R, Zaonarivelo JR, Engberg SE, Nalanirina G, McGuire SM, Shore GD, Andrianasolo J, Herrington K, Brenneman RA et al. 2007. *Molecular phylogeny and taxonomic revision of the woolly lemurs, genus* Avahi *(Primates: Lemuriformes)*. Lubbock, TX: Museum of Texas Tech University.

Athrey G, Hodges TK, Reddy MR, Overgard HJ, Matias A, Ridl FC, Kleinschmidt I, Caccone A, Slotman MA. 2012. The effective population size of malaria mosquitoes: Large impact of vector control. *PLoS Genetics* 8(12):e1003097.

Attenborough RD, Burkot TR, Gardner DS. 1997. Altitude and the risk of bites from mosquitoes infected with malaria and filariasis among the Mianmin people of Papua New Guinea. *Trans R Soc Trop Med Hyg* 91:8–10.

Ayala FJ, Coluzzi M. 2005. Chromosome speciation: Humans, *Drosophila*, and mosquitoes. *Proc Natl Acad Sci U S A* 102(Suppl 1):6535–6542.

Beebe NW, Russell TL, Burkot TR, Lobo NF, Cooper RD. 2013. The systematics and bionomics of malaria vectors in the southwest Pacific. In: Manguin S, editor. Anopheles *mosquitoes: New insights into malaria vectors*. New York: Intech Publishing.

Bruce-Chwatt LJ. 1985. *Essential Malariology*. London: William Heinemann Medical Books.

Bulmer R. 1970. Which came first, the chicken or the egg-head? In: Pouillon J, Maranda P, editors. *Échanges et Communications: Mélanges offerts à Claude Lévi-Strauss à L'Occasion de son 60ème Anniversaire*. The Hague: Mouton.

Cain AJ, Harrison GA. 1960. Phyletic weighting. *Proc Zool Soc Lond* 135(1):1–31.

Charlwood JD, Graves PM, Alpers MP. 1986. The ecology of the *Anopheles punctulatus* group of mosquitoes from Papua New Guinea: A review of recent work. *P N G Med J* 29(1):19–26.

Coetzee M, Hunt RH, Wilkerson RC, della Torre A, Coulibaly MB, Besansky NJ. 2013. *Anopheles coluzzii* and *Anopheles amharicus*, new members of the *Anopheles gambiae* complex. *Zootaxa* 3619(3):246–274.

Cracraft J. 1983. Species concepts and speciation analysis. In: Johnston RF, editor. *Current Ornithology*. New York: Plenum Press. pp. 159–187.

Cracraft J. 1992. The species of the birds-of-paradise (Paradisaeidae): Applying the phylogenetic species concept to a complex pattern of diversification. *Cladistics* 8:1–43.

Dabiré KR, Sawadodgo S, Diabate A, Toe KH, Kengne P, Ouari A, Costantini C, Gouagna C, Simard F, Baldet T et al. 2013. Assortative mating in mixed swarms of the mosquito *Anopheles gambiae s.s.* M and S forms, in Burkina Faso, West Africa. *Med Vet Entomol* 27:298–312. 10.1111/mve.2013.27. issue-3.

Douglas M. 1966. *Purity and danger: An analysis of concepts of pollution and taboo*. London: Routledge & Kegan Paul.

Foley DH, Wilkerson RC, Cooper RD, Volovsek ME, Bryan JH. 2007. A molecular phylogeny of *Anopheles annulipes* (Diptera: Culicidae) *sensu lato*: The most species-rich anopheline complex. *Mol Phylogenet Evol* 43(1):283–297.

Fuentes A. 2010. The new biological anthropology: Bringing Washburn's New Physical Anthropology into 2010 and beyond – the 2008 AAPA luncheon lecture. *Yearb Phys Anthropol* 53:2–12.

Godfrey L, Marks J. 1991. The nature and origins of primate species. *Yearb Phys Anthropol* 34:39–68.

Groves CP. 1989. *A theory of human and primate evolution*. Oxford: Clarendon Press.

Groves CP. 2001a. *Primate taxonomy*. Washington, DC: Smithsonian Institution Press.

Groves CP. 2001b. Why taxonomic stability is a bad idea, or why are there so few species of primates (or are there?). *Evol Anthropol* 10:192–198.

Groves CP. 2004. The what, why, and how of primate taxonomy. *Int J Primatol* 25:1105–1126.

Groves CP. 2008. *Extended family: Long-lost cousins. A personal look at the history of Primatology*. Arlington, VA: Conservation International.

Groves CP. 2012. Species concept in primates. *Am J Primatol* 74:687–691.

Groves CP. 2013. The nature of species: A rejoinder to Zachos et al. *Mamm Biol* 78(1):7–9.

Groves CP, Grubb P. 2011. *Ungulate taxonomy*. Baltimore: Johns Hopkins University Press.

Groves CP, Leslie DM. 2011. Family Bovidae (Hollow-horned ruminants). In: Wilson DE, Mittermeier RA, editors. *Handbook of the mammals of the world*. Barcelona: Lynx Edicions.

Huxley JS. 1942. *Evolution: The modern synthesis*. London: Allen & Unwin.

Kamdem C, Tene Fossog B, Simard F, Etouna J, Ndo C, Kengne P, Boussès P, Etoa F-X, Awono-Nkondjio C, Besansky NJ et al. 2012. Anthropogenic habitat disturbance and ecological divergence between incipient species of the malaria mosquito *Anopheles gambiae*. *PLoS ONE* 7(6):e39453.

Krzywinski J, Besansky NJ. 2003. Molecular systematics of *Anopheles*: From subgenera to subpopulations. *Annu Rev Entomol* 48:111–139.

Lefèvre T, Gouagna L-C, Dabiré KR, Elguero E, Fontenille D, Costantini C, Thomas F. 2009. Evolutionary lability of odour-mediated host preference by the malaria vector *Anopheles gambiae*. *Trop Med Int Health* 14(2):228–236.

Lefèvre T, Gouagna L-C, Dabiré KR, Elguero E, Fontenille D, Renaud F, Costantini C, Thomas F. 2010. Beer consumption increases human attractiveness to malaria mosquitoes. *PLoS ONE* 5(3):e9546.

Little MA, Collins KJ. 2012. Joseph S. Weiner and the foundation of post-WWII human biology in the United Kingdom. *Yearb Phys Anthropol* 55:114–131.

Louis EE, Engberg SE, Lei R, Geng H, Sommer JA, Andriantompohavana R, Randria G, Prosper, Ramaromilanto B, Rakotoarisoa G et al. 2006. *Molecular and morphological analyses of the Sportive Lemurs (family Megaladapidae: genus Lepilemur) reveals 11 previously unrecognized species.* Special Publications. Lubbock, TX: Museum of Texas Tech University.

Mayr E. 1963. *Animal species and evolution.* Cambridge, Mass: Belknap Press of Harvard University Press.

Mittermeier RA, Richardson M. 2008. Foreword. In: Groves CP, *Extended family: Long-lost cousins. A personal look at the history of Primatology.* Arlington, VA: Conservation International.

Nixon KC, Wheeler QD. 1990. An amplification of the phylogenetic species concept. *Cladistics* 6:211–223.

Pawley A. 2011. *A dictionary of Kalam with ethnographic notes.* Canberra: Pacific Linguistics.

Pennetier C, Warren B, Dabiré KR, Russell IJ, Gibson G. 2010. 'Singing on the wing' as a mechanism for species recognition in the malarial mosquito *Anopheles gambiae*. *Curr Biol* 20:131–136.

Reidenbach KR, Cook S, Bertone MA, Harbach RE, Wiegmann BM, Besansky NJ. 2009. Phylogenetic analysis and temporal diversification of mosquitoes (Diptera: Culicidae) based on nuclear genes and morphology. *BMC Evol Biol* 9:298.

Sinka ME, Bangs MJ, Manguin S, Rubio-Palis Y, Chareonviriyaphap T, Coetzee M, Mbogo CM, Hemingway J, Patil AP, Temperley WH et al. 2012. A global map of dominant malaria vectors. *Parasit Vectors* 5(69):1–11.

Tusting LS, Thwing J, Sinclair D, Fillinger U, Gimnig J, Bonner KE, Bottomley C, Lindsay SW. 2013. Mosquito larval source management for controlling malaria. *Cochrane Database of Syst Rev* 2013(8).

van Bortel, W, Harbach, RE, Trung HD, Roelants, P, Backeljau, T, Coosemans M. 2001. Confirmation of *Anopheles varuna* in Vietnam, previously misidentified and mistargeted as the malaria vector *Anopheles minimus*. *Am J Trop Med Hyg* 65(6):729–732.

Washburn SL. 1951. The new physical anthropology. *Trans N Y Acad Sci* 13(7):298–304.

White BJ, Collins FH, Besansky NJ. 2011. Evolution of *Anopheles gambiae* in relation to humans and malaria. *Annu Rev Ecol Evol Syst* 42:111–132.

Zachos FE, Apollonio M, Barmann EV, Festa-Bianchet M, Golich U, Habel JC, Haring E, Kruckenhauser L, Lovari S, McDevitt AD, Pertoldi C, Rossner GE, Sanchez-Villagra MR, Scandura M, Suchentrunk F. 2013. Species inflation and taxonomic artefacts – a critical comment on recent trends in mammalian classification. *Mammalian Biology* 78:1–6.

PART III

8. Lamarck on species and evolution

Marc F Oxenham

Introduction

For the last decade I have lived three doors down from Colin's office, in the bowels of the AD Hope building at The Australian National University. However, my first contact with Colin was indirect, in as much as he was an examiner of my Honours thesis ('Progress and Evolution: A re-evaluation of some ideas, devices and scholars in the study of human evolution to 1950') back in 1995. Those familiar with the Australian and New Zealand university systems will realise that Honours is where undergraduates start to play with the big kids, and examiner's reports can be a rude entree to the world of academia. Colin's positive comments inspired me to think about publishing from my Honours research, and while my PhD quickly got in the way of that, this chapter is in fact a reworked early version of a paper drafted just after my Honours year. While my research interests have developed in very different directions since then, I have maintained a strong 'armchair' interest in the history of evolutionary thought and hope this contribution will excite others to revisit the works of the early evolutionary theorists. Much of my discussion concerns the nature of and the role of species in Lamarck's theory of evolution. Many readers are no doubt aware of Colin's own particular interest in species (see the discussion by Robert Attenborough, this volume, particularly with respect to the phylogenetic species concept) and while Colin may not necessarily agree with my interpretations of Lamarck on species, I am sure he would see it a most appropriate topic for this volume.

In popular and scientific mythology Darwin is reified as the founder of modern evolutionary theory and Lamarck lampooned as that 'odd chap' who believed in the inheritance of acquired characteristics. While not of concern here, the inheritance of acquired characteristics (what is now termed transgenerational epigenetic inheritance) would now seem to be a reality (see Morris, 2012 for an overview of recent developments in this area). The purpose of this chapter is not to bring Darwin 'down a peg', and in fact Darwin hardly gets a mention, but to entreat the reader to see Lamarck in an alternative light.

The role and influence of the French naturalist Jean-Baptiste Pierre Antoine de Monet, Chevalier de Lamarck (1744–1829) with regard to the development of evolutionary theory has been extensively researched over the past 50 years or so (Cannon, 1959; Gillispie, 1959; Lovejoy, 1959; Burkhardt, 1977, 1984; Hull, 1984; Lovtrup, 1987; Corsi, 1988) and is not pursued here. In this chapter I wish to review Lamarck's ideas on the nature of species, which directly relates to his evolutionary model, and then re-evaluate what I will argue are misinterpretations of two central aspects of this model: first, the view that Lamarck's theory was strictly vertical in nature and lacked a crucial horizontal component; secondly, that his evolutionary model is best viewed as a collection of multiple, independent lineages and is inconsistent with a theory of descent. Both of these themes are generally considered evidence of pivotal differences between Lamarck's and Darwinian, or modern, evolutionary theory. While I am not proposing that Darwin's theory of evolution be seen as resting on the foundations of Lamarck's theorising, I am asking that Lamarck at least be given a fair go in light of a close reading of his actual works.

Zoological philosophy

Coming from a background in botany Lamarck assumed the position of professor of invertebrates at the Museum National d'Histoire Naturelle in Paris in 1793 at 50 years of age. At the turn of the nineteenth century Lamarck made a seemingly abrupt ideological change with his conversion to evolutionism (Lamarck, 1800, 1802). Less than a decade later he became the first scholar (Lamarck, 1809) to publish a detailed theory of bio-evolution or transmutation. While neither of these terms were used by Lamarck, contemporary synonymous concepts such as 'changed', 'converted', 'mutation' and 'transformed' were used in their stead. Lamarck was also to a large extent responsible for integrating the threads of an emerging nineteenth century bio-evolutionary theory with the notion of progress.

Lamarck's most famous work, published in 1809 (the year of Darwin's birth), *Zoological Philosophy*, was not only a treatise on evolution but a system of biology treating three broad areas of study: (1) zoological classification and evolution; (2) the nature and causes of life; and (3) the nature and causes of intelligence, emotions and so forth. While the first part of this work dealt specifically with his evolutionary ideas, some sections on the origin of life in the second part of the work are important in understanding Lamarck's evolutionary model.

In his own lifetime Lamarck's views were essentially either ignored or actively disparaged (Cannon, 1959; Bowler, 1984, 2003; Burkhardt, 1984; Hull, 1984), although he received support in some quarters, the French naturalist Henri de Blainville for instance (see Appel, 1987). Perhaps one reason for Lamarck's dearth of support was related to his lack of compunction in seeing humanity as a creature of the evolutionary process. Cannon (1959) has also suggested that

Georges Cuvier, a younger highly influential contemporary of Lamarck's, was the central cause of Lamarck's problems whilst living and dead. During his time at the Museum of Natural History Lamarck came into conflict with Cuvier over both geological gradualism and transmutation, both concepts to which Cuvier was totally antagonistic. The power and influence Cuvier wielded during these years (Coleman, 1964) did nothing for Lamarck's cause. Cannon (1959) has even suggested that nineteenth and twentieth century interpretations of Lamarck and his ideas stem from Cuvier's reinterpretations or misrepresentations. Indeed, in Cuvier's *Biographical Memoir* (Cuvier, 1831: 434) of Lamarck, he contrasted him to men of true genius:

> [Those], with minds not less ardent, nor less adapted to seize new relations, have been less severe in scrutinizing the evidence; with real discoveries with which they have enriched science, they have mingled many fanciful conceptions; and, believing themselves able to outstrip both experience and calculation, they have laboriously constructed vast edifices on imaginary foundations, resembling the enchanted palaces of our old romances, which vanished into air on the destruction of the talisman to which they owed their birth.

Notwithstanding, Cuvier was one man and insufficient to the task of countering all materialist thought at the time (see Corsi, 1988). Burkhardt (1984) has outlined three additional reasons why Lamarck was so unsuccessful in his own time. In brief these are the materialist overtones of his work; his reputation for wild speculation; and the fact that he was 'unable to cultivate a circle of capable naturalists willing to champion his views' (Burkhardt, 1984: xxxiv). Moreover, less than three decades after his death, Lamarck was to present 'a serious public relations problem for Darwin and the Darwinians' (Hull, 1984: xlvi). Lamarck was perceived as Darwin's scientific precursor in a sense. Lyell even went as far as to describe Darwin's theory of evolution as a modification of Lamarck's views, much to the annoyance of Darwin (Hull, 1984). A perception of Darwinian evolution as Lamarckian evolution revisited (Lovtrup, 1987) would not have been an idea that would have sat well in the Darwin camp. Unfortunately, only Lamarck's mechanism for change survives as his legacy to the history of evolutionary thought. Furthermore, it is unlikely that Lamarck would recognise what was understood and presented as Lamarckism after his death (see Bowler, 1992).

Lamarck on species

Lamarck's particular understanding of species impacts on the two principal themes of this chapter and it is necessary to outline his ideas on this subject.

Lamarck argued for the ability of species to change, and while not in itself ground breaking at the time, he presented his case by positing that species form a continuum:

> I do not mean that existing animals form a very simple series, regularly graded throughout; but I do mean that they form a branching series, irregularly graded and free from discontinuity, or at least once free from it. For it is alleged that there is now occasional discontinuity, owing to some species being lost. (Lamarck, 1809: 37)

In his view species changed or transformed very gradually, that is by way of extremely small changes or micro-mutations over time. Because of this Lovtrup (1987) has argued that Lamarck was in fact the first micro-mutationist. Lamarck needed evolution to be gradual to fit with his understanding of deep geological time and views of environmental change.

> As compared to the periods which we look upon as great in our ordinary calculations, an enormous time and wide variation in successive conditions must doubtless have been required to enable nature to bring the organization of animals to that degree of complexity and development in which we see it at its perfection. (Lamarck, 1809: 50)

Lamarck, like those after him who subscribed to the view of gradual species change, had a problem with species recognition. How does one separate a temporally and physically continuous entity into discrete units or species? Lamarck's solution was unique: species for him did not actually become extinct, although he noted the exception of recent cases of non-natural human induced extinctions. The fossil evidence in his time indicated many forms with no known contemporary representatives. Lamarck forwarded two explanatory arguments. The first, it was likely the living counterparts existed in the vast unexplored regions of the earth; the second and more important, species did not become extinct but they change over time by way of the accumulation of tiny mutations into new or different forms.

> May it not be possible ... that the fossils [apparently representing extinct species] ... belonged to species still existing, but which have changed since that time and become converted into the similar species that we now actually find. (Lamarck, 1809: 45).

Species for Lamarck could encompass both broad levels of variation and temporal depth. In modern evolutionary terms Lamarck was describing the model of phyletic (anagenetic) evolution. This particular conceptualisation of what defined species allowed him to see them both change through time and retain their essential identity.

There is, perhaps, an additional reason for Lamarck's subscription to this particular notion of species. It may have been a reaction to an interpretation of extinction events as the result of the non-viability of transmutated organisms. Such a view, if sustained, would constitute direct evidence against evolution. There is some evidence for this view when one considers opposition to transmutationism from Cuvier. Coleman (1964) and Bowler (1984) have noted that Cuvier, through his studies in comparative anatomy, had come to the conclusion that biological organisms were too complex to hold to the view transmutation could result in different viable organisms. For Cuvier species were fixed (implying species had a creationist and/or saltational origin). Lamarck saw transumtation occurring very gradually and there was certainly no place for useful monsters (the expected common result of large mutations as opposed to tiny unnoticed ones) in his scheme.

Vertical and lateral evolution

A number of scholars have either denied (Mayr, 1982, 1991) the idea of lateral, as opposed to vertical, evolutionary change in Lamarck's evolutionary model, or else seen this idea as having limited significance (Ruse, 1981, 1982; Bowler, 1984). It is Darwin who is generally credited with this ostensibly novel dichotomisation of evolutionary theory. For example, Mayr (1991: 17) argues that:

> For Lamarck, evolution was a strictly vertical phenomenon, proceeding in a single dimension, that of time. Evolution for him was a movement from less perfect to more perfect, from the most primitive infusorians up to the mammals and man.

and

> The problem of how these new species and incipient species came into being was clarified for Darwin by the Galapagos mockingbirds. These specimens showed that new species can originate by what we now call geographical (or allopatric) speciation...By this thought Darwin founded a branch of evolutionism which, for short, we might designate as horizontal evolutionism, in contrast with the strictly vertical evolutionism of Lamarck. (Mayr, 1991: 20)

Moreover, both Bowler (1984, 2003) and Ruse (1981, 1982) have tended to stress the idea that Lamarck was redeveloping the Medieval classificatory construct of a scale of being or *scala natura*, albeit a dynamic scale rather than a static one. Traditionally this scale concept was a hierarchical device that encompassed all life from the worms in the ground through to the angels themselves and even God. Each life form was ordered and ranked in such a manner as to create a continuous, unbroken but graduated chain of life (Hodgen, 1964).

Indeed, the concept of a scale of life is prominent in Lamarck's work. He outlined 14 classes of animals which were ranked or arranged from the single celled *infusorian* through to the most 'perfect' class *mammals*. Notwithstanding, Lamarck's understanding and use of the *scala natura* concept needs some explanation. For instance, what did Lamarck mean by the concept arrangement? Arrangement related to the order of his scale, in distinction to classification which referred specifically to the divisions within it. For example, Lamarck (1809: 56) defined arrangement as:

> [A]n order in that list [referring to his classes] which represents as nearly as possible the actual order followed by nature in the production of animals; an order conspicuously indicated by the affinities which she has set between them.

However, this was not a simple continuous unilinear scale with evolution being represented by change over time (refer to the Lamarck, 1809: 37 quotation above) and Lamarck's views can be seen to be quite modern. Most evolutionary biologists today would accept some sort of loose macro-historical trend of less to more complex, mediated by the observation that the evolution of *B* from *A* is constrained by the nature of *A* itself (see Mayr, 1982; Eldridge, 1985). Further, there are controversies and difficulties with current species concepts (e.g. Paterson, 1981; Eldridge, 1985; Tattersall, 1986), a topic to which Colin has also made important contributions (e.g. Groves, 2001). The point is that Lamarck was not simply reviving the *scala natura* in newer and more dynamic garb.

If not a simple unilineal scale what then was Lamarck's evolutionary scheme? In a summary of his own evolutionary views Lamarck notes:

> Nature has produced all the species of animals in succession, beginning with the most imperfect or simplest, and ending her work with the most perfect, so as to create a gradually increasing complexity in their organization; these animals have spread at large throughout all the habitable regions of the globe, and every species has derived from its environment the habits that we find in it and the structural modifications which observation shows us. (Lamarck, 1809: 126)

Two themes are clearly presented here by Lamarck: the first is an implicit law of nature which causes the *vertical* progression from simple to complex organisation, based on the *scala natura* already spoken of; the second theme is the *lateral* secondary transformation of species due to the effects of environment, habit and so forth. Although this bi-directional scale of vertical continuity and lateral change was a major intellectual conceptualisation of the Renaissance theistic model of Adamitic origins and subsequent diffusion (see Hodgen, 1964), it is Lamarck who was responsible for appropriating it into an evolutionary model.

Lamarck's advocacy of lateral, or non-progressive, evolutionary change is further illustrated in his evolutionary diagram (see Figure 8.1). For instance there is an initial lateral branching from the worms into an *annelids–cirrhipedes–molluscs* evolutionary sequence on the one hand and an *insects–arachnids–crustaceans* sequence on the other. There is a similar major branching event at the reptiles. Further, there are a series of lateral bifurcations from an ancestral amphibian stock. That this is the manner in which Lamarck viewed evolutionary history, and not simply an interpretation seen through the lens of a modern understanding of evolution, is evident in his own comments concerning this diagram:

> It is there shown that in my opinion the animal scale begins by at least two separate branches, and that as it proceeds it appears to terminate in several twigs in certain places. (Lamarck, 1809: 178)

The importance of this concept in his evolutionary model is further reinforced by other such specific references in his work:

> As we continue to examine the probable origin of the various animals, we cannot doubt that the reptiles, by means of two distinct branches, *caused by the environment*, have given rise, on the one hand, to the formation of the birds and, on the other hand, to the amphibian mammals, which have in their turn given rise to all the other mammals. (Lamarck, 1809: 176; italics added)

The environment was the causal agent for Lamarck's other secondary, or lateral, component of his evolutionary model. It is evident that environmental influence played an important role in influencing evolutionary direction. Lamarck devoted a chapter (VII: 106–127) to his secondary evolutionary causal agent.

> It is obvious then that as regards the character and situation of the substances which occupy the various parts of the earth's surface, there exits a variety of environmental factors which induces a corresponding variety in the shapes and structures of animals, *independent of* that special variety which necessarily results from the progress of the complexity of organisation in each animal. (Lamarck, 1809: 112)

Lamarck (1809: 127, italics added) went on in concluding his chapter on the role of environmental influence to state that:

> [I]t is not the shape either of the body or its parts which give rise to the habits of animals and their mode of life; but that it is, on the contrary,

the habits, mode of life *and all the other influences of the environment* which have in course of time built up the shape of the body and of the parts of animals. With new shapes, new faculties have been acquired, and little by little nature has succeeded in fashioning animals such as we actually see them.

It is clear that not only should Lamarck be given credit for the first comprehensive development of this vertical–lateral dichotomy within an evolutionary model, but that it was a fundamental component of his theory. Lamarck's answer to the riddle of specific diversity was a function of his lateral thinking.

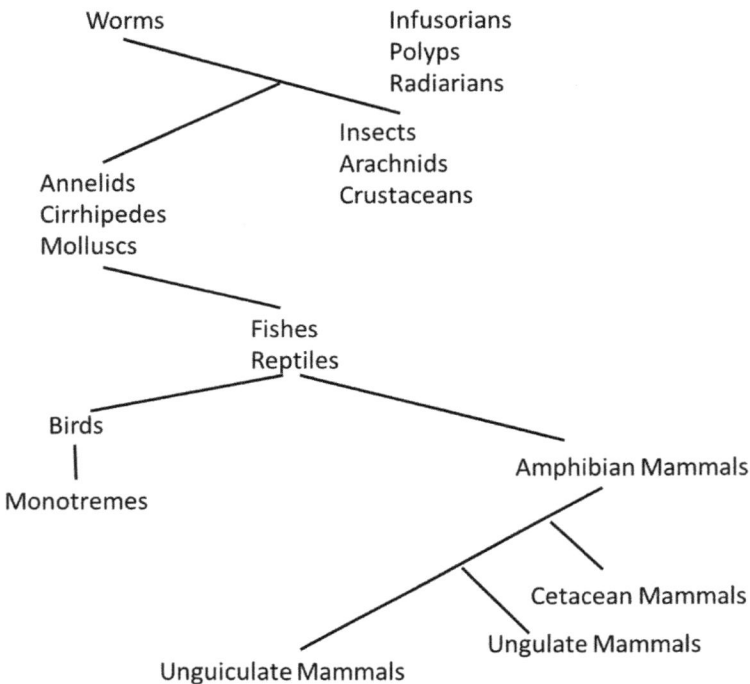

Figure 8.1: Diagram 'Showing the Origin of the Various Animals'.

Source: After Lamarck 1809: 179.

The multiple independent lineage view

Giving precedence to Lamarck for first developing a vertical–lateral evolutionary model would not be accepted by Ruse (1981, 1982) and Bowler (1984), both of whom have argued that Lamarck's evolutionary model cannot be viewed as encompassing a theory of common descent. In fact Bowler claimed that the *crucial* difference between modern, or Darwinian, evolutionary theory and Lamarck's evolutionary views was that Lamarck '…did not suppose all forms alive to have evolved from a common ancestry' (Bowler, 1984: 80). For Lamarck complexity equated with temporal depth of lineage. Humans, for example, are the longest lived lineage which arose at the earliest period in time, relative to other lineages, from a separate spontaneous generation event (see Bowler, 1984 Fig. 10; 2003 Fig. 9).[1] Lamarck is interpreted to be advocating multiple parallel lineages through time, each lineage having its roots in a separate and progressively later point of spontaneous generation (Bowler, 1984; Ruse, 1981, 1982). The most primitive species belong to the youngest lineages and in fact spontaneous generation is still occurring (Bowler, 1984; Ruse, 1981, 1982). However, I would argue that only the last point is partially correct.

Lamarck (1809: 247) devotes an entire chapter to the topic of spontaneous generation, which in summary is that:

> [I]t appears to me certain that nature does herself carry out spontaneous or direct generations, that she has this power, and that she utilises it at the anterior extremity of each organic kingdom, where the most imperfect living bodies are found; and that it is exclusively through their medium that she has given existence to all the rest.

Note that in Lamarck's scheme there are only two organic kingdoms: the plants and animals. This is Lamarck's principal view regarding the origins and subsequent development or evolution of plant and animal life. However, Lamarck does pose the question that is apparently the cause of the (mis)interpretation of his main thesis representing multiple and independent evolutionary lineages through time:

> [I]s it certain that she [nature] does not give rise to similar generations at any other point of these scales? (Lamarck 1809: 247)

When posing this question Lamarck (1809: 247) notes that he had hitherto held the view that:

> [I]n order to give existence to all living bodies, it was enough for nature to have formed directly the simplest and most imperfect of animals and plants.

It is in this context that he (1809: 247) expands on his question:

---

1 Bowler (1984: fig. 10; 2003: fig. 9) does not provide an argument or evidence for this particular interpretation. Presumably his view is based on arguments by Ruse (e.g. 1982) who produces a very similar schematic (1982: fig. 1.4) to Bowler. As will be come apparent later in the chapter, Ruse (1982), has misinterpreted Lamarck on this issue.

Why indeed should nature not give rise to direct generations at various points in the first half of the animal and plant scales, and even at the origin of certain separate branches of these scales? Why should she not establish, in favourable circumstances, in these diverse rudimentary living bodies, certain physical systems of organisation, different from those observed at the points where the animal and vegetable scales appear to begin?

Having posed these questions they are then put aside as aspects worthy of further investigation, but not in any way central to his main thesis. He (1809: 248, italics added) goes on to conclude this section thus:

Whether the kind of direct generations, here referred to, do or do not actually take place, *as to which at present I have no settled opinion*, it seems to me certain at all events that nature actually carries out such generations at the beginning of each kingdom of living bodies, and that she could never, except through this medium, have brought into existence the animals and plants which live on our earth.

Lamarck spent the majority of the first part of his book arguing that the dual agents of natural progress (vertical change) and environmental influence (lateral branching) acting on spontaneous generation events at the base of the plant and animal kingdoms are sufficient causes in and of themselves for producing the present variety and complexity of life. For Lamarck spontaneous generation is clearly important for the establishment of the animal and plant kingdoms, while the environment played a crucial role in subsequent branching events.

Not only has the direct formation of the simplest living bodies actually occurred [spontaneous generation], as I am about to show, but the following principles proves that such formations must still be constantly carried out and repeated where the conditions are favourable, in order that the existing state of things may continue. (Lamarck, 1809: 245)

The maintenance of the existing order Lamarck refers to is a reference to animals such as his infusorians (see Figure 8.1), which would become extinct, and thus disrupt the existing state of things, without continuous acts of spontaneous generation. In referring to the ephemeral and seasonal nature of these simplest of animals he goes on to state:

[H]ow fragile their existence, from what or in what way do they regenerate in the season when we again see them? Must we not think that these simple organisms, these rudiments of animality, so delicate and fragile, have been newly and directly fashioned by nature rather than have regenerated themselves? (Lamarck, 1809: 245)

With the exception of the generation of the original plant and animal progenitors, these acts of continuous spontaneous generation were not seen by Lamarck as the starting points for new independent lineages. Except for fleetingly toying with the possibility of such a scenario, the idea of multiple separate and independent lineages was certainly not part of Lamarck's evolutionary model as substantively outlined in *Zoological Philosophy*.

Ruse (1981: 10) has argued that (and it is worth quoting him at length):

> [O]ne must note that Lamarck's theory was in no way a theory of common descent, supposing that all organisms descended from one or a few common origins. We know that he thought simple forms of life are constantly being spontaneously generated through the action of heat, light, electricity and moisture on the inorganic world (*Philosophie*: 236–248). Then organic development continues on essentially the same path it started on. Lamarck believed that lions and so on, if destroyed, would be replaced in the course of time (*Philosophie*: 187). There is therefore no reason to believe, for example, that today's mammals and today's fish have common ancestors- they are merely at different stages on the scale of being [see Figure 8.2].

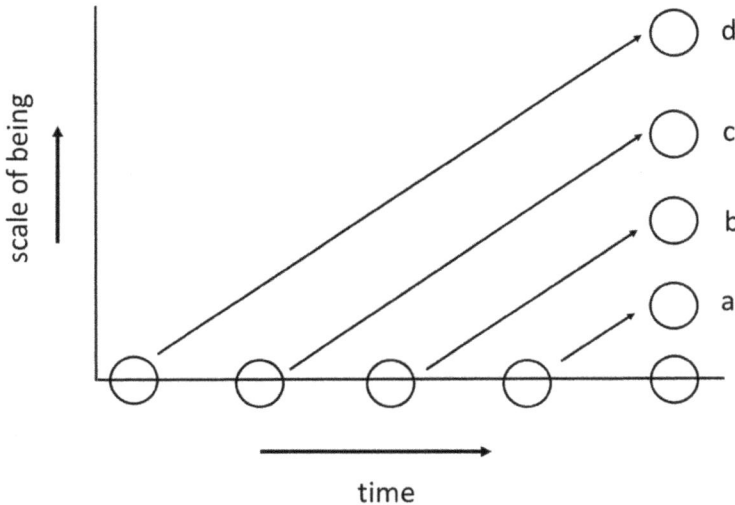

Figure 8.2: This is a modification of Ruse (1982: 8) following Bowler's (1984: Figure 10) interpretation of Lamarck's scheme. The circles along the X axis represent the spontaneous generation events of the 'simplest' organisms, while the circles on the right hand side of the Y axis represent progressively more complex (d being the most complex) organisms that have unilinearly evolved from separate original spontaneous generation events.

Source: After Ruse (1982).

Whilst I have already addressed the majority of points raised in this statement, it is necessary to examine the reference to the lions and other animals, as this is clearly an important component of Ruse's argument for multiple, independent lineages. Lamarck has been taken out of context, and this is not a piece of evidence supporting the multiple independent lineage interpretation. In fact, this quote is taken from the introductory section of part two of *Zoological Philosophy*, dealing specifically with the physical causes, effects and manifestations of life, and not his evolutionary model per se.

> After recognising the necessity for these acts of direct creation [spontaneous generation], we must enquire which are the living bodies that nature may produce spontaneously, and distinguish them from those which only derive their existence indirectly from her. Assuredly the lion, eagle, butterfly, oak, rose, do not derive their existence immediately from nature; they derive it as we know from individuals like themselves who transmit it to them by means of reproduction; and we may be sure that if the entire species of the lion or oak chanced to be destroyed in those parts of the earth where they are now distributed, it would be long before the combined powers of nature could restore them. (Lamarck, 1809: 186–187)

Lamarck is making two important points here, neither of which can be attributed to subscription to a model of multiple, independent lineages: (1) complex animals (e.g. lions and oaks) derive [descend] from other complex forms and are not the product of spontaneous generation; (2) enormous periods of time are involved in the complex process of evolution.

A final point before concluding concerns Bowler's (1984, 2003) view that the multiple, independent lineage model is the only one able to explain why, in Lamarck's world view, the *scala natura* is still visible. Lamarck provided two mechanisms to explain the continued existence of his scale. The first relates to his ideas on extinction: generally it does not happen. His understanding of the concept of species, examined previously, did not encompass the idea of extinction and this is why we still see life at all levels of organisational complexity. Secondly, and related to the first point, is the role of environmental influence again. In outlining his aims for the first part of *Zoological Philosophy* Lamarck (1809: 15, italics added) noted:

> I shall [also] show the influence of environment and habit on the organs of animals, as being the factors which *favour* or *arrest* their development.

An example of this *arrest* in development can be seen in the way he dealt with an anti-evolutionary argument that used mummified Egyptian cats amongst other things. It was argued that as these cats, which were several thousand years old, were essentially identical to modern forms this proved that species did not change. Lamarck invoked the environmental argument to claim that the climate and environment in Egypt had not changed to any degree over the past several thousand years and thus one would not expect to see any change in Egyptian cats over this time. Progress, his vertical evolutionary component, would tend to cause all life to advance toward perfection. However, environmental influence, his lateral component, would serve to redirect and also arrest progressive advance. Lamarck's understanding of species and the role of the environment in his evolutionary model were sufficient in themselves to explain the apparent preservation of the arrangement of life as diagrammatically represented in Figure 8.1.

Conclusions

Ruse (1981) remarked that Lamarck was a very confusing writer and suggested this may have been because Lamarck was confused himself. True, *Zoological Philosophy* is written in a generally unclear and confusing style, but it is also apparent that he was struggling with a number of novel ideas and concepts. Even without an equivalent set of 'Darwin note books' it is clear that *Zoological Philosophy* has been through a number of drafts and alterations in theoretical orientation. Lamarck even noted in the preface to this work that this was a new, corrected and enlarged version of *Recherches sur les corps vivants* (Lamarck, 1802). Perhaps it is the clutter of these vestiges of changes in point of view that facilitated some of the misinterpretations of his work dealt with in this chapter.

Notwithstanding such concerns, Lamarck clearly and successfully grappled with the concepts of vertical and lateral evolutionary change. The view that he supported and promulgated a model of multiple, independent lineages all catalysed with their own independent spontaneous origin events is clearly not supported by a close reading of Lamarck's own words. Moreover, the development of the vertical and lateral dichotomy in Lamarck's model prefigured, at least, its appearance in Darwin's (1859) published model half a century later. What Lamarck gave us, whether anyone was listening or not, was a phylogentic model with two fundamental bifurcations, plant and animal, these in turn provided a multitude of environmentally induced branching events: a model of vertical and lateral evolutionary change.

Some 200 years after Lamarck published *Zoological Philosophy*, Colin Groves (Groves and Grubb, 2011; Gippoliti and Groves, 2012) was criticised for his own views on species (e.g. Zachos and Lovari, 2013). While I am sure Colin will be vindicated with time and, indeed, has made a stellar progress on this front (e.g. Gippoliti et al., 2013), I am not so sure about Lamarck.

References

Appel TA. 1987. *The Cuvier-Geoffroy debate: French biology in the decades before Darwin*. New York: Oxford University Press.

Bowler PJ. 1984. *Evolution: The history of an idea*. Berkeley: University of California Press.

Bowler PJ. 1992. Lamarckism. In: Keller EV and Lloyd EA, editors. *Keywords in evolutionary biology*. Cambridge, MA: Harvard University Press. pp. 188–193.

Bowler PJ. 2003. *Evolution: The history of an idea*. Third edition. Berkeley: University of California Press.

Burkhardt RW Jr. 1977. *The spirit of system: Lamarck and evolutionary biology*. Cambridge: Cambridge University Press.

Burkhardt RW Jr. 1984. The zoological philosophy of J.B. Lamarck. In: Lamarck JB. *Zoological philosophy: An exposition with regard to the natural history of animals*. Translation of the 1809 edition by H Elliot. Chicago: The University of Chicago Press. p. xv–xxxix.

Cannon HG. 1959. *Lamarck and modern genetics*. Manchester: Manchester University Press.

Coleman W. 1964. *Georges Cuvier, zoologist: A study in the history of evolution theory*. Cambridge, MA: Harvard University Press.

Corsi P. 1988. *The age of Lamarck: Evolutionary theories in France 1790–1830*. Berkeley: University of California Press.

Cuvier G. 1831 [1984]. Biographical Memoir of M. De Lamarck. In: Lamarck JB. *Zoological philosophy: An exposition with regard to the natural history of animals*. Translation of the 1809 edition by H Elliot. Chicago: The University of Chicago Press. pp. 434–453.

Darwin C. 1859 [1985]. *The origin of species by means of natural selection*. Edited with an introduction by J.W. Burrow. Penguin Books.

Eldredge N. 1985. *Unfinished synthesis: Biological hierarchies and modern evolutionary thought*. New York: Oxford University Press.

Gillispie C. 1959. Lamarck and Darwin in the history of science. In: Glass B, Temkin O, and Straus W Jr., editors. *Forerunners of Darwin: 1745–1859*. Baltimore: Johns Hopkins University Press. pp. 265–291.

Gippoliti S, Cotterill FPD, Groves CP. 2013. Mammal taxonomy without taxonomists: A reply to Zachos and Lovari. *Hystrix* 24(2):145–147.

Gippoliti S, Groves CP., 2012. "Taxonomic inflation" in the historical context of mammalogy and conservation. *Hystrix* 23(2):8–11.

Groves CP. 2001. *Primate taxonomy*. Washington, DC: Smithsonian Institution Press.

Groves CP, Grubb P. 2011. *Ungulate taxonomy*. Baltimore: Johns Hopkins University Press.

Hodgen MT. 1964. *Early anthropology in the sixteenth and seventeenth centuries*. Philadelphia: University of Pennsylvania Press.

Hull DL. 1984. Lamarck among the Anglos. In: Lamarck JB. *Zoological philosophy: An exposition with regard to the natural history of animals*. Translation of the 1809 edition by H Elliot. Chicago: The University of Chicago Press. p. xl–lxvi.

Lamarck JB. 1800. Introductory Lecture for 1800. In: Lamarck JB. *Zoological philosophy: An exposition with regard to the natural history of animals*. Translation of the 1809 edition by H Elliot. Chicago: The University of Chicago Press. pp. 407–433.

Lamarck JB. 1802. *Recherches sur L'organisation Des Corps Vivants*. Paris: Maillard.

Lamarck JB. 1809 [1984]. *Zoological philosophy: An exposition with regard to the natural history of animals*. Translated by H Elliot. Chicago: The University of Chicago Press.

Lovejoy AO. 1959. The argument for organic evolution before the origin of species, 1830–1858. In: Glass B, Temkin O, and Straus W Jr, editors. *Forerunners of Darwin: 1745–1859*. Baltimore: Johns Hopkins University Press. pp. 356–414.

Lovtrup S. 1987. *Darwinism: The refutation of a myth*. New York: Croom Helm in association with Methuen.

Mayr E. 1982. *The growth of biological thought: Diversity, evolution, and inheritance*. Cambridge, MA: Belknap Press.

Mayr E. 1991. *One long argument: Charles Darwin and the genesis of modern evolutionary thought*. Allen Lane: The Penguin Press.

Morris KV. 2012. Lamarck and the Missing Lnc. *The Scientist* 26(10):29–33.

Paterson HEH. 1981. The continuing search for the unknown and unknowable: A critique of contemporary ideas on speciation. *S Afr J Sci* 77:113–119.

Ruse M. 1981. *The Darwinian revolution: Science red in tooth and claw.* Chicago: University of Chicago Press.

Ruse M. 1982. *Darwinism defended: A guide to the evolution controversies.* London: Addison-Wesley Publishing.

Tattersall I. 1986. Species recognition in human paleontology. *J Hum Evol* 15:165–175.

Zachos FE, Lovari S. 2013. Taxonomic inflation and the poverty of the Phylogenetic Species Concept – a reply to Gippoliti and Groves. *Hystrix* 24(2):142–144.

9. Naming the scale of nature

Juliet Clutton-Brock

Introduction

As an archaeozoologist and mammalogist, I spent 30 years in the Osteology Room of the British Museum (Natural History), later named the Natural History Museum, and for a number of those years the visits of Colin and Phyll Groves enlivened the Mammal Section, not only with taxonomic discussions but also with memorable lunches in the nearby Bute Street cafés. It was the period during which the analysis of animal remains from archaeological sites was developing into a multidisciplinary science, and arguments and discussions on nomenclature prevailed in many international conferences. At the seminal conference on 'Equids in the Ancient World' held in Tübingen University in 1984 Colin's knowledge of equid taxonomy led the discussions (Groves, 1986).

In the context of archaeozoology and proposals for standardising the nomenclature of domestic animals that I have participated in with Colin, I give below a summary of the ways in which organisms in the animal kingdom have been named. This begins with the first written records in the prehistoric period, and progresses through the methods of Aristotle to the naturalists of the eighteenth century, Linnaeus, and modern taxonomy. The summary is followed with an outline of the suggestions and arguments for and against the formal naming of domestic animals that surrounded the development of archaeozoology in the twentieth century.

The beginnings

Since early hominins first began to use words to describe their surroundings they must have had names for the plants and animals around them, and these names would have been distinctive in the myriad languages that evolved around the world. However, it is only since the invention of written records that names could live on and be transcribed from ancient languages that have become extinct.

It is in the third millennium BC that the first written records of animal names begin to appear, and it is evident that elaborate systems of nomenclature already existed. To take an example from the 1984 Tübingen conference, the names for the different species and hybrids of equids, which were known to the Sumerians and Akkadians in ancient Mesopotamia, were decoded, as follows, from cuneiform texts by Nicholas Postgate (1986):

anse	= *generic term for equid, or E. asinus*
anse-DUN.GI or anse-LIBIR	= *E. asinus*
anse-eden-na	= *E. hemionus*
anse-BARxAN	= *E. asinus x E. hemionus*
anse-zi-zi or anse-kur-ra	= *E. caballus*

By the time that Genesis is believed to have been first recorded in writing, probably during the first millennium BC there must have been fully developed nomenclatures for every living and non-living thing in a people's environment. So it is not surprising that the Hebrew legend of creation included an explanation for the origin of animal names: 'and out of the ground the Lord God formed every beast of the field, and every fowl of the air; and brought them unto Adam to see what he would call them: and whatsoever Adam called every living creature, that was the name thereof' (Genesis 2, 19).

Furthermore, after the Flood had retreated, the Hebrew God

> [B]lessed Noah and his sons and said unto them, Be fruitful and multiply, and replenish the earth.
>
> And the fear of you and the dread of you shall be upon every beast of the earth, and upon every fowl of the air, upon all that moveth upon the earth, and upon all the fishes of the sea, into your hand are they delivered.
>
> Every moving thing that liveth shall be meat for you; even as the green herb have I given you all things (Genesis 9, 1–3).

This belief that everything in the world had been created by God for the benefit of humans seems not to have been inherited so rigidly by the ancient Egyptians but it certainly was by the ancient Greeks and its spread to Christianity was due in great part to the enormous influence of Aristotle's great works. This lasted in Western Europe until the time of Darwin, although doubts set in after the spread of new animals and plants (unknown to the classical world) from the Americas in the sixteenth century.

Aristotle was born in 384 BC and he died, aged 63 in 322 BC.[1] His approach to the natural world was teleological, that is, he believed that everything in Nature

---

[1] The sections on Aristotle and Linnaeus, in this chapter, were previously given as part of a paper presented to a conference, In the Company of Animals, held at the New School for Social Research, New York, in April

had a purpose, and this purpose was for the benefit of mankind (Clutton-Brock, 1999a). He wrote, 'plants are evidently for the sake of animals, and animals for the sake of Man; thus Nature, which does nothing in vain, has made all things for the sake of Man' (Peck, 1970: xli).

Aristotle's investigations into zoology are compiled into a series of books (authoritatively translated in the Loeb Library), known as the *History of Animals* (Peck, 1965, 1970; Balme, 1991) the *Generation of Animals* (Peck, 1990), and the *Parts of Animals,* the *Movement of Animals,* and the *Progression of Animals* (Peck and Forster, 1983). He wrote about more than 500 species including shellfish, insects, birds, reptiles, and quadrupeds, with humans being treated in the same way as all other animals. Aristotle's descriptions of animals were much quoted in the later classics, such as Pliny's *Natural History* (c. AD 77–79) and Aelian's (AD 175–235) *On Animals,* and it is from the classical writers on natural history that the long tradition of naming animals and plants in Latin was inherited.

Division of the animal kingdom is older than Aristotle; in Plato's philosophy the highest genus was divided by means of differentiae into subsidiary genera and each of these was then divided and subdivided by dichotomy, until the ultimate species was reached. At the upper end of Aristotle's scale he had main groups such as birds and fish, which were his genera, and at the lower end the commonly named animals such as dog, cat, eagle, etc., which were his species, but normally the intermediate stages are missing.

Aristotle did recognise a Scale of Nature but the rungs of his ladder were not the stages of a taxonomic scheme, and there is no evidence that he felt they should be. His purpose was not to construct a taxonomic system, but to collect data for ascertaining the Causes of observed phenomena; and this was to be done by looking to see whether certain characteristics were regularly found in combination: this was how the clues to the Causes would be brought to light. Aristotle believed that human beings were animals but at the same time he was certain that all other animals existed for the sake of Man. He asserted that it was impossible to produce a neat hierarchical order on the basis of obvious physical differences because these cut across each other.

Like the ancient civilisations of Mesopotamia and Egypt, that of classical Greece was a stratified society ruled by powerful hierarchies and in which all manual work was carried out by slaves. It is therefore only to be expected that the Greek philosophers would view the natural world as a gradation from the lowest to the highest, or as a scale of perfection, which was to become known as the Scale of Nature or the Great Chain of Being (Lovejoy, 1936).

1995. The proceedings were published in Clutton-Brock (1999a).

The Five Predicables

In the European-speaking world, until well into the eighteenth century, the method of classification of all organisms was based on the Five Predicables. This was a hierarchical system that had been adapted from Aristotle's classification of logic, as written in his work known as the *Topics*. The Five Predicables were genus, species, differentia, property, and accident. They were clearly defined by Simpson (1961: 24).

Linnaeus and binomial classification

Carl Linnaeus (1707–1778) was clearly an obsessional organiser who classified not only the plant and animal kingdoms but also the minerals and the kinds of diseases known in his day. Since the time of Aristotle, animals and plants had been named in Latin by using the genus and the differentia from the Five Predicables of classification. The two together made up the definition, which could be used as the name. However, with the classification and naming of more and more species over time, the differentia often became very long. The great innovation of Linnaeus was in creating the binomial or binary system by taking the old name for the genus and adding a single name from the many that had been used in the differentia, as the species.

Linnaeus's definitive tenth edition of his *Systema Naturae* (1758) was written in Latin and the long introduction has been seldom translated, although it is full of fascinating comments on eighteenth century attitudes to animals, as well as the first use of the term Mammalia. The translation of Robert Kerr (1792) has the title *The Animal Kingdom or Zoological System of the Celebrated Sir Charles Linnaeus*. After the short introduction there is a chapter translated as, 'The Empire of Nature', which begins with quotations from Aristotle on the Causes, and from the Roman writers Seneca (4 BC–AD 65) and Pliny the Elder (AD 23–79). Linnaeus followed Aristotle in believing that the three kingdoms of nature: minerals, vegetables, and animals met together in the Order of Zoophytes, and also in the belief from Genesis that everything in the world was created for Man, for he wrote: 'Hence one great employment of man, at the beginning of the world, must have been to examine created objects, and to impose on all the species names according to their kinds'.

Unlike his predecessors, Linnaeus saw that the unit of classification had to be the species, and he produced a strict hierarchical classification that ended at its summit with the Kingdom. Linnaeus summarised his ideas as follows (Kerr, 1792: 22–23):

Classes and Orders are the creatures of human invention, while the division of these into Genera and Species is the work of Nature. All true knowledge refers finally to the species of things, while at the same time, what regards the generic divisions is substantial in its Nature.

… God, beginning from the most simple terrestrial elements, advances through Minerals, Vegetables, and Animals, and finishes with Man. Man on the contrary, reversing this order, begins with himself, and proceeds downwards to the materials of the earth. The framer of a systematic arrangement begins his study by the investigation of particulars, from which he ascends to more universal proportions; while the teacher of this method, taking a contrary course, first explains the general propositions, and then gradually descends to particulars.

Vernacular names and early modern classifications

While the naturalists wrestled with trying to produce meaningful classifications of the natural world, the general population of each country of course had their own vernacular names for every living thing, and these names could be enormously complicated. This was especially so when the animal was part of a ritual such as the royal hunts of Medieval Europe. In modern English a male red deer is a stag, but in the Laws of Venery the red deer was a beast of the chase and the stag had many names, depending on its age. In its first year it was a calf, in its second, a brocket, in its third a spayard, in its fourth a staggard, in its fifth a hart of ten, and in its sixth a hart (Clutton-Brock, 1984). When the meat of an animal was to be eaten it also had a separate name and it was from the Normans that the English names, venison, beef, and pork were adopted.

Throughout the Medieval and early modern periods animals and plants were named according to their uses to humans and this applied to domestic animals as much as to wild ones. Thomas (1983: 55) quotes the sixteenth century book *Of English Dogges*, by Dr John Caius in which there were three categories of dogs: a 'generous' kind, used in hunting or by fine ladies; a 'rustic' kind used for necessary tasks, and a 'degenerate', currish kind, used as turnspits and for other menial purposes. This way of classifying dogs by their uses to humans was echoed by Linnaeus who 200 years later divided the dog (*Canis familiaris*) into 11 separate species, which included the sheep dog (*Canis domesticus*) and the turnspit (*Canis vertagus*).

After Linnaeus: Modern taxonomy and nomenclature

The fundamental unit of all classifications, including those of Aristotle and Linnaeus, is the species, which is composed of a population of interbreeding organisms. To Linnaeus, and to most biologists until the second half of the twentieth century, a species was considered to be a group of animals all of which were supposed to be identical with a type, officially recognised as such and preserved in a public institution. Following the growth of modern taxonomy, however, it was soon realised that a species comprises a population that is inherently variable in morphology and therefore the type specimen can have no special role in identifying other specimens. As explained in Simpson (1961: 31), 'A nomenclatural type is simply something to which a name is attached by purely legalistic convention'.

There is a commonly held view that the separation of two species can be determined by whether or not they will produce fertile offspring when interbred. However, on its own, the state of fertility of hybrid offspring is an inadequate means of defining a species. Many mammals that are normally considered to be good species will interbreed, although, because of a behavioural barrier, they may not usually do so in the wild, and their offspring will be fertile, for example the dog, wolf, jackal, coyote, and dingo will all interbreed and produce fertile offspring (Gray, 1972).

How then is the species to be defined? Since first proposed by Mayr (1940) an often-used definition has been the 'biological species concept'. This has gone through several revisions and expansions in the last 60 years, not least by Mayr himself, but remains essentially the same and is: *a species is a group of interbreeding natural populations that is genetically isolated from other such groups as a result of physiological or behavioural barriers*. However, Colin Groves argues (pers. comm., 27 July 2012) that:

> The Biological Species Concept gives no guidance in the case of allopatric forms; it does not satisfactorily cover cases where two parapatric taxa, which are homogeneous within their ranges, nonetheless interbreed where their ranges meet; and DNA studies show that there has been far, far more interbreeding between perfectly 'good' species, even sympatric ones, than we would have guessed.

Groves prefers the Phylogenetic Species Concept of Cracraft (1983) who described his views thus:

> As the 'biological species concept' really doesn't work, let us define species as being populations which are 'diagnosable', meaning that they

differ 100% from each other; you can always recognise individuals as belonging to a particular species (except in the case of demonstrable hybrids); they have fixed heritable differences between them; they are (in cladistic terms) the terminals on a cladogram – however one wishes to put it.[2]

Until well into the twentieth century there were no fixed concepts of what constituted or distinguished species, subspecies, or breeds. Linnaeus believed there were several species of domestic dogs and even Darwin was not sure of the distinctions. The subspecies is the lowest unit that may be included in zoological taxonomy and subspecies are designated with a trinomial, e.g. *Canis lupus arabs* (in botany, variations can also be given a Latin name). The status of the subspecies has been discussed at length by Simpson (1945: 16; 1961: 171). The modern definition that I consider most useful states: *a subspecies is a distinctive, geographical segment of a species, that is it comprises a group of wild animals that is geographically and morphologically separate from other such groups within a single species.*

Today, it is generally agreed that the end product of animal domestication is the breed and not the species or subspecies, and breeds are not given Latin names. My definition of a breed is: *a group of animals that has been bred by humans to possess uniform characters that are heritable and distinguish the group from other animals within the same species.* A breed parallels a subspecies, except that, whereas a subspecies is restricted to a geographical region a breed is not (Clutton-Brock, 1999b: 40).

But how should domestic forms be named? Up to the time of Linnaeus and beyond, there were no problems – domestic breeds were seen as species or subspecies in their own right, and the nineteenth and early twentieth century archaeozoologists were happy to allocate the bones they found associated with human settlements to taxa with Linnaean binomials and trinomials. Sheep remains were called *Ovis aries studeri* or *Ovis longipes egyptius*, while dog remains were *Canis poutiatini* or *Canis familiaris matris-optimae.*

The central difficulty for the naming of domestic species hangs on whether or not they should be considered as conspecific with their wild progenitors. To Linnaeus it was obvious that the dog was a separate species from the wolf, but to archaeozoologists who work on the identification of sub-fossil animal remains at the interface between the wild species and their earliest domesticated descendents there may be little or no evidence of an osteological and therefore a taxonomic distinction. In order to try to overcome this problem and with a widespread view that domesticates should be treated as conspecific with their

---

2 Their position as terminals on a cladogram is the origin of the 'phylogenetic' part of the name (Groves, pers. comm., 4 March 2013).

assumed progenitors, several different systems of nomenclature have been devised for domestic mammals, as reviewed by Gautier (1993). Although none has received international recognition, the most widely accepted system was that proposed by Bohlken (1961). Bohlken's solution was to call the domestic form by the first available name for the wild species, followed by the linking word 'forma' (f.) and then by the earliest name, according to the rule of priority, for the domestic animal. In this way we would have *Canis lupus* f. *familiaris* L. for the dog and *Capra aegagrus* f. *hircus* L. for the domestic goat. This arrangement is, however, clumsy and it has the disadvantage that it assumes certain identification of the wild progenitor, which for some domestic animals, for example the ferret, may never be established.

At one time, I also proposed that domestic species should be excluded from formal nomenclature but I have come to believe that domestication is an evolutionary process and if the domestic form of an animal is for all intents and purposes separated reproductively from the wild form then it should be classified as a separate species. It is then valid to use the Linnaean names, which have the great benefit of being widely known and in general usage.

If the Linnaean names are used for domestic mammals there has been a problem with certain wild species that were given the same names as the domestic by Linnaeus. When he was familiar with both the wild and the domestic form of a species and they looked alike, as with his native reindeer, Linnaeus gave them the same name, *Cervus tarandus*, now called *Rangifer tarandus*. On the other hand because he failed to see the relationship between the wolf and the dog he gave them the separate species names, *Canis lupus* and *Canis familiaris*. With yet others, for example the goats and sheep, he had no knowledge of the wild ancestor and so he named only the domestic form (see Clutton-Brock, 2012, Appendix, for the list of relevant species). In order to get over the difficulty of using say the Linnaean name *Equus asinus* for the African wild ass as well as for the donkey it has become usual to use the next available name according to the International Code of Zoological Nomenclature (ICZN) for the wild species, which is *Equus africanus*.

As a convention among zoologists and archaeozoologists this system of nomenclature worked well but it was not in accord with the rules of the ICZN and in the chapters on the Perissodactyla and Artiodactyla in the 1993 edition of the influential *Mammal Species of the World* edited by Wilson and Reeder, names were used for wild species irrespective of whether they were first described on a wild or a domestic form. Thus the Linnaean names *Equus asinus* and *Equus caballus* were used for both the wild and domestic forms of ass and horse respectively. It was clearly time to stabilise the nomenclature of the 15 wild species that Linnaeus had named on domestic forms. Accordingly, Anthea Gentry, Colin and I put a Case to the ICZN to conserve the usage of specific names for wild animals that are antedated by or are contemporary with names based on domestic animals (Gentry et al., 1996). Case 3010 was presented and for six years Comments for and against the Proposal went to the ICZN. Finally, the Commission voted on the Case – the names were to be conserved and the ruling was published (ICZN, 2003).

The controversy surrounding the ruling on these names may seem arcane to non-taxonomists, but in fact it is of considerable importance. For example the extinct aurochs would, strictly, have to be named *Bos taurus primigenius*, as indeed it still is in the latest edition of Wilson and Reeder (2005: 692–693), thus making it a subspecies of the domestic ox. It must be emphasised, however, that this ruling on the conservation of these names is for wild species and it does not affect the taxonomic status and nomenclature of domestic forms.

So what should be done to settle the continuing discussion about naming the domestic forms? After 40 years of consideration and several changes of view I now believe that if the domestic form of an animal is for all intents and purposes separated reproductively from the wild form then it should be classified as a separate species. It is for this reason that I, together with other mammalogists, including Colin, have argued that domestic animals should not be excluded from formal zoological nomenclature and that the traditional Latin binomial names such as *Capra hircus* and *Ovis aries* should hold (Clutton-Brock, 1999b, 2012; Gentry et al., 2004).

Nomenclature is the backbone of taxonomy, but I think it is important to remember its subjective element, and to support my view on this I will discuss the different formal names that have been given to the dingo over the past 50 years. Until the 1970s the dingo was generally known as the wild dog of Australia and it was not paid taxonomic attention, but then its depredations on livestock turned the sheep farmers against it and biologists and pest controllers were called in to study its behaviour. In 1973 Alan Newsome and colleagues published an account of the dingo in the *Australian Meat Research Committee Review* in which they named it *Canis familiaris dingo*. The aim of the work was, 'to provide the basic biological data to devise rational and effective control programs'. And by naming the dingo as a subspecies of domestic dog (*C. familiaris*) the biologists were justifying its control.

This justification was increased by the work of Laurie Corbett who with his morphological study of dingoes and Thai dogs claimed that 'dingo-like canids' were widespread throughout Southeast Asia (Corbett, 1985). The dingoes of Australia thereby lost their unique status and they could be classified as pests. Then in the 1990s, and after the notoriety of the Azaria Chamberlain Case (1980), the dingo began to be named *Canis lupus dingo*, and this trinomial remains today in numerous online publications and on the latest *IUCN Red List of Vulnerable*

Species. As a subspecies of wolf and a wild carnivore, it may be controlled when considered necessary but also conserved. However, genetic studies have now shown that the dingo was introduced to Australia possibly from South China and possibly at a single occasion, before the Neolithic expansion from Taiwan (Savolainen et al., 2004; Oskarsson et al., 2012). Since humans brought these first dogs to Australia, they have lived, bred and undergone natural selection in the wild, isolated from other canids until the arrival of Europeans. The case is clear to me that this unique dog should be recognised as part of the living heritage of Australia and it should revert to its first Latin name of *Canis dingo* Meyer, 1793, as argued by Crowther and others (2014). And finally I am pleased to learn online that the Merigal Dingo Sanctuary in Bargo (New South Wales), which I visited with Colin in 1987, is still active and promoting the conservation of dingoes.

References

Balme DM. [Transl.] 1991. Aristotle *Historia Animalium*. Books VII–X. Loeb Classical Library. Cambridge, MA and London: Harvard University Press.

Bohlken H. 1961. Haustiere und zoologische Systematik. *Zeitschrift für Tierzüchtung und Züchtungsbiologie* 76:107–113.

Clutton-Brock J. 1984. The master of game: The animals and rituals of Medieval venery. *Biologist* 31(3):167–171.

Clutton-Brock J. 1999a. Aristotle, the scale of nature, and modern attitudes to animals. In: Mack A, editor. *Humans and other animals*. Columbus: Ohio State University Press. pp. 5–24.

Clutton-Brock J. 1999b. *A natural history of domesticated mammals*. 2nd edition. Cambridge: Cambridge University Press & The Natural History Museum.

Clutton-Brock J. 2012. *Animals as domesticates a world view through history*. Michigan: Michigan State University Press.

Corbett LK. 1985. Morphological comparisons of Australian and Thai dingoes: A reappraisal of dingo status, distribution and ancestry. *Proc Ecol Soc Australia* 13:277–291.

Cracraft J. 1983. Species concepts and speciation analysis. *Current Ornithology* 1:159–187.

Crowther MS, Fillios M, Colman N, Letnic M. 2014. An updated description of the Australian dingo (*Canis dingo* Meyer, 1793). *J Zool* 293(3):192–203.

Gautier A. 1993. 'What's in a name?' A short history of the Latin and other labels proposed for domestic animals. In: Clason A, Payne S, Uerpmann H-P, editors. *Skeletons in her cupboard Festschrift for Juliet Clutton-Brock*. Oxbow Monograph 34, Oxford: Oxbow Books.

Gentry A, Clutton-Brock J, Groves CP. 1996. Case 3010. Proposed conservation of usage of 15 mammal specific names based on wild species which are antedated by or contemporary with those based on domestic animals. *Bull Zool Nomencl* 53(1):28–37.

Gentry A, Clutton-Brock J, Groves CP. 2004. The naming of wild animal species and their domestic derivatives. *J Archaeol Sci* 31:645–651.

Gray AP. 1972. *Mammalian hybrids*. Commonwealth Agricultural Bureau.

Groves CP. 1986. The taxonomy, distribution, and adaptations of Recent equids. In: Meadow RH, Uerpmann H-P, editors. *Equids in the Ancient World* I. Wiesbaden: Beihefte zum Tübinger Atlas des Vorderen Orients. pp. 11–65.

International Commission on Zoological Nomenclature (ICZN). 2003. Opinion 2027 (Case 3010). Usage of 17 specific names based on wild species which are predated by or contemporary with those based on domestic animals: conserved. *Bull Zool Nomencl* 60(1):81–84.

Kerr R. 1792. *The animal kingdom or zoological system of the celebrated Sir Charles Linnaeus; Class I Mammalia*. London: J. Murray.

Linnaei C. 1758 [1956]. *Systema Naturae*. A photographic facsimile of the first volume of the tenth edition (1758) *Regnum Animale*. London: British Museum (Natural History).

Lovejoy AO. 1936. *The great chain of being*. Cambridge, MA: Harvard University Press.

Mack A, editor. 1999. *Humans and other animals*. Columbus: Ohio State University Press.

Mayr E. 1940. Speciation Phenomena in Birds. *The American Naturalist* 74:249–278.

Newsome AE, Corbett LK, Best LW, Green B. 1973. *Australian Meat Research Committee Review*, No. 14.

Oskarsson MCR, Klütsch CFC, Boonyaprakob U, Wilton A, Tanabe Y, Savolainen P. 2012. Mitochondrial DNA data indicate an introduction through mainland Southeast Asia for Australian dingoes and Polynesian domestic dogs. *Proc Royal Soc Lond B* 279(1730): 967–974.

Peck AL. [Transl.] 1965. Aristotle *Historia Animalium*. Books I–III. Loeb Classical Library. Cambridge, Mass.: Harvard University Press & London: Heinemann.

Peck AL. [Transl.] 1970. Aristotle *Historia Animalium*. Books IV–VI. Loeb Classical Library. Cambridge, MA and London: Harvard University Press.

Peck AL. [Transl.] 1990. Aristotle *Generation of animals*. Loeb Classical Library. Cambridge, MA and London: Harvard University Press.

Peck AL, Forster ES. [Transl.] 1983. Aristotle *Parts of animals, movement of animals, progression of animals*. Loeb Classical Library. Cambridge, MA: Harvard University Press & London: Heinemann.

Postgate, JN. 1986. The equids of Sumer, again. In: Meadow RH, Uerpmann H-P, editors. *Equids in the Ancient World* I. Wiesbaden: Beihefte zum Tübinger Atlas des Vorderen Orients. pp. 194–206.

Savolainen P, Leitner T, Wilton AN, Matisoo-Smith E, Lundeberg J. 2004. A detailed picture of the origin of the Australian dingo, obtained from the study of mitochondrial DNA. *Nat Proc Acad Sci* 101:12837–12890.

Simpson GG. 1945. The principles of classification and a classification of mammals. *Bull Am Mus Nat Hist* 85: vii–xvi, 1–350.

Simpson GG. 1961. *Principles of animal taxonomy*. New York and London: Columbia University Press.

Thomas K. 1983. *Man and the natural world: Changing attitudes in England 1500–1800*. London: Allen Lane.

Wilson DE, Reeder DAM, editors. 2005. *Mammal species of the world: A taxonomic and geographic reference*. 2 vols. 3rd edition. Baltimore.

10. Changes in human tooth-size and shape with the Neolithic transition in Indo-Malaysia

David Bulbeck

Introduction

During my Master of Arts studies on Holocene human remains from Indonesia and Malaysia, I was the beneficiary of excellent supervision from Colin Groves and the late Alan Thorne. At the time, the general view was that the 'Mongoloid' features of most Southeast Asians reflect the late Holocene immigration of their ancestors from Northeast Asia into a region previously inhabited by large-toothed 'Australoids' (e.g. Jacob, 1967a; Howells, 1973; Bellwood, 1978). At the same time, Christy Turner was developing an alternative perspective of long-term continuity in Southeast Asia of a 'Sundadont' dental morphology complex, distinguished from the 'Sinodont' complex of Northeast Asia and the New World by features such as less marked incisor shovelling (e.g. Turner, 1983). Also, some biological anthropologists were developing a model that explained post-Neolithic craniodental changes in terms of biological adaptation to changed selection pressures (e.g. Carlson and van Gerven, 1977). The driving force, according to this model, was the reduced demand on the masticatory apparatus associated with the Neolithic transition, as people now grew crops low in fibrous content, and often cooked their food in pots to soften it further. Hence, my MA thesis proposed that Southeast Asia's late Holocene transition to smaller teeth and jaws, and broader and less robust crania, reflected local adaptation to the reduced need for large tooth mass (Bulbeck, 1981, 1982).

Colin Groves was very supportive of this 'Neolithic tooth-size reduction' model, and indeed highlighted it in his contribution at a major symposium (Groves, 1989). On the other hand, as I belatedly discovered, Loring Brace (1976) had already rejected the model's applicability for Southeast Asia. Brace accepted it for China, but argued that the retention of larger teeth in Southeast Asia pointed to a later onset of the Neolithic there, attributable, moreover, to immigration from South China (see also Brace and Hinton, 1981). Subsequently, Hirofumi Matsumura produced a series of studies that emphasised morphological similarities between Southwest Pacific and early Southeast Asian cranial remains. A critical aspect was Matsumura's removal of sheer tooth-size from the comparisons, allowing him to argue that the relative sizes of the different tooth diameters ('tooth shape') also pointed to Neolithic immigration from China into Southeast Asia (e.g. Matsumura and Hudson, 2004). These views accorded with a growing consensus amongst historical linguists that the Austroasiatic and Austronesian languages, which dominate Southeast Asia ethnographically, trace their origins to the north.

Other biological anthropologists have presented analyses that support long-term population continuity in Southeast Asia (e.g. Hanihara, 1994; Storm, 1995; Manser, 2005; Demeter, 2006; Pietrusewsky, 2006). However, only Manser's study found the Neolithic tooth-size reduction model useful. Indeed, Storm instead preferred an alternative explanation of body-size reduction related to post-Pleistocene warming.

The task of this contribution is to rigorously test whether biological adaptations to agriculture, and the use of pottery for cooking, could explain late Holocene craniodental change in Indo-Malaysia – the part of Southeast Asia where my specialisation lies. There are now enough well-dated burial series from Sulawesi, Borneo, Java and Malaya to test two main predictions of the Neolithic tooth-size reduction model:

1. Indo-Malaysian tooth-size should show continual reduction over time, not only between the pre-Neolithic and the Neolithic, but also continuing into the Early Metal Phase (EMP) and modern times.

2. Pre-Neolithic and late Holocene Indo-Malaysians should have similar tooth shape.

Depending on the obtained results, the discussion will also briefly examine the efficacy of tooth-size reduction as a driver for late Holocene change in Indo-Malaysian cranial shape, and review recent insights from human genetic and osteological comparisons.

Materials and methods

The dental metrical analyses presented here include male and bisexual samples. The male analyses cater for the critical Late Pleistocene Java sample, which is exclusively male. The bisexual analyses enable the inclusion of archaeological teeth that are difficult to sex – for instance, loose teeth, and teeth from sub-adults – and also cater for series where both males and females, on their own, are sparsely documented. In these analyses, tooth-size comparisons (but less so shape comparisons – Bulbeck, 1981; Bulbeck et al., 2005) are prone to distortion due to the samples' variable sex composition, given that male teeth are on average larger than female teeth. However, many of the prehistoric Indo-Malaysian samples are dominated by specimens that cannot be sexed, and so they can be analysed only as bisexual samples.

The present coverage of recent Indo-Malaysian populations focuses on Indonesia and the *Orang Asli* ('aboriginal people') of Malaya, including the Semang 'Negritos'. Several Southwest Pacific and Northeast Asian samples are also included so as to provide a regional context (Table 10.1). Most of the samples included here are based on dental casts from living subjects or anatomical collections of skulls from persons of known sex. Two exceptions are the ethnohistorical Motu cemetery on Motupore Island, Papua New Guinea (PNG), and the Euston cemetery on the Murray Valley, with an estimated age between 2,000 and 6,000 years ago (Pardoe, 1988). Postcranial material was available to assist sexing the Motupore skulls (Brown, 1978) but sexing of the Euston skulls relied on cranial size and robustness (Brown, 1981). A third exception is the 'historical Sulawesi' sample, which mainly comprises geographically dispersed archaeological finds (Table 10.2). Many of these remains cannot be reliably sexed and so the sample is best treated as bisexual.

Table 10.1: Recent/historical samples used in the comparisons.

Sample	Male sample size	Bisexual composition	Source
Shanghai Chinese	14–104	♂ and ♀ about equal	Brace et al. 1984
Historical Sulawesi	Not applicable	More ♂ than ♀ (probably)	See Table 10.2
Jahai Semang, Malaya	13–19	Pooled into Semang sample, more ♂ than ♀	Bulbeck et al. 2005
Batek Semang, Malaya	8–12		Bulbeck et al. 2005
Temiar Senoi, Malaya	6–30	♂ and ♀ about equal	Bulbeck et al. 2005
Temuan, Malaya	9–16	Pooled into Aboriginal Malay sample, ♂ and ♀ about equal	Bulbeck et al. 2005
Semelai, Malaya	14–22		Bulbeck et al. 2005
Batawi, Java	96–139	More ♂ than ♀	Snell 1938
Surabaya Javanese	35–63	Not available	Snell 1938
Motupore Island, PNG	9–11	♂ and ♀ about equal	Brown 1978
Eastern Highlands, PNG	32–53	Not available	Doran and Freedman 1974
Walbiri, Central Australia	29–136	Not used	Barrett et al. 1963, 1964
Euston, Murray Valley	14–27	♂ and ♀ about equal	Brown 1978

Source: All sources listed in the table and cited fully in the references.

Table 10.2: Historical Sulawesi dental metrics.[A]

Tooth	Mesiodistal (MD) diameters			Buccolingual (BL) diameters		
	Sample size	Mean	Standard deviation	Sample size	Mean	Standard deviation
Upper medial incisor (I^1)	11	8.3	0.90	14	7.2	0.45
Upper lateral incisor (I^2)	10	7.3	0.66	13	6.7	0.34
Upper canine (C)	20	7.9	0.54	21	8.2	0.66
Upper first premolar (P^3)	28	7.5	0.68	27	9.5	0.90
Upper second premolar (P^4)	21	7.2	0.45	21	9.5	0.60
Upper first molar (M^1)	23	10.7	0.72	23	11.7	0.59
Upper second molar (M^2)	25	9.8	0.69	24	11.8	0.80
Upper third molar (M^3)	19	9.4	0.64	19	11.9	0.75
Lower medial incisor (I$_1$)	11	5.8	0.57	12	6.3	0.49
Lower lateral incisor (I$_2$)	16	6.0	0.49	20	6.2	0.42
Lower canine (C)	14	7.3	0.47	16	8.0	0.40
Lower first premolar (P$_3$)	20	7.1	0.46	20	8.1	0.55
Lower second premolar (P$_4$)	24	7.4	0.64	24	8.3	0.61
Lower first molar (M$_1$)	31	11.5	0.57	33	10.8	0.48
Lower second molar (M$_2$)	34	11.3	0.71	35	10.4	0.54
Lower third molar (M$_3$)	25	11.3	0.94	25	10.7	0.75

Note: A. The sample includes teeth dated to the second millennium CE from the Talaud Islands in North Sulawesi (Bulbeck, 1981), teeth from ethnohistorical burials near Lake Towuti in central Sulawesi (laboratory notes), colonial period Bugis skulls (museum notes), seventeenth to twentieth-century Makassar teeth from Batu Ejaya in southwest Sulawesi (Bulbeck, 2004), and teeth from seventeenth to nineteenth century 'Macassan' skulls in the Northern Territory (museum notes).

Source: All sources listed in the notes section.

Numerous prehistoric samples from Indo-Malaysia are also available, dating between the Late Pleistocene and the EMP (Table 10.3). Where the sex composition of the bisexual samples could be assessed, they may have more males than females (the Gua Cha samples), approximately equal numbers of males and females (Mesolithic Java, Gilimanuk) or more females than males (the Niah samples). Some of the samples are composite, especially the 'Early Sulawesi' sample (Table 10.4). Early Sulawesi, along with Melanta Tutup in Sabah, and Neolithic and EMP Java, lack observations (as placed in the public domain) on some of their tooth diameters. Included for comparison are Khok Phanom Di, Thailand, with burial goods similar to those from Neolithic Malaya (Bellwood, 1993), and the terminal Pleistocene cemetery from Coobool Creek, in Australia's Murray Valley, for which only buccolingual diameters are available (Brown, 1989).

Table 10.3: Prehistoric samples used in the comparisons.

Site(s)	Location	Age	Comparisons	Data source	Dating source
Khok Phanom Di	Southern Thailand	Neolithic, 4–3.5 ka	Bisexual (♂, ♀ + children)	Tayles 1999	Tayles 1999
See Table 10.4	Early Sulawesi	mid-Holocene	Bisexual	See Table 10.4	See Table 10.4
Leang Buidane	Talauds, North Sulawesi	EMP, 700–1200 CE	Bisexual	Bulbeck 1981	Bellwood 1976
Leang Codong	Southwest Sulawesi	EMP, ~2-1 ka	Bisexual	Jacob 1967a	Bulbeck 1996–97
Pre-Neolithic Niah	Sarawak	20–8 ka	Bisexual	Manser 2005	Manser 2005
Neolithic Niah	Sarawak	Neolithic, 3.5–2 ka	Males, bisexual	Manser 2005	Manser 2005
Melanta Tutup	Sabah	Neolithic/EMP, 3.5-1 ka	Bisexual	Chia et al. 2005	Chia 2008
Gua Cha Hoabinhian	Malaya	mid-Holocene, 7–3 ka	Males, bisexual	Bulbeck 2005a	Bulbeck 2005a
Gua Cha Neolithic	Malaya	Late Neolithic, 3–2 ka	Males, bisexual	Bulbeck 2005a	Bulbeck 2005a
Guar Kepah	Malaya	Transitional Neolithic ~6 ka	Bisexual	Jacob 1967a	Tieng 2010[A]
Gua Harimau	Malaya	Neolithic/EMP, 3-2 ka	Bisexual	Bulbeck 2001	Bulbeck 2001
Wajak, Gua Lawa 1	Java	Pleistocene, ~30-20 ka	Males	Storm 1995; Détroit 2002	Storm et al. in press; Détroit 2002
Mesolithic Java[B]	Java	Pleistocene to mid-Holocene, 13–4 ka	Males, bisexual	Jacob 1967a; Détroit 2002; Noerwidi 2011–12	Jacob 1967a; Détroit 2002
Neolithic Java[C]	Java	Neolithic, 3.5–2.5 ka	Bisexual	Storm 1995; Détroit 2002; Noerwidi 2011–12	Storm 1995; Détroit 2002; Noerwidi 2011-12
EMP Java[D]	Java	~2-1 ka	Bisexual	Snell 1938; Jacob 1964; Noerwidi 2011–12	Jacob 1967a; Noerwidi 2011–12
Gilimanuk	Bali	EMP, 2–1 ka	Bisexual	Jacob 1967b	Anggraeni 1999
Coobool Creek	Murray Valley, Australia	Pleistocene, ~15 ka	Only males used	Brown 1989	Brown 1989

Notes: A. Tieng reports a radiocarbon date on marine shell of 5700 ± 50 BP for the Guar Kepah shell middens. The calibrated date (Intcal 09), allowing a delta R correction of 15 ± 38 (Singapore), would be 5967-6269 BP (2 sigma). The age of the shell middens serves as a maximum age for the burials. Pot sherds from all levels in the middens, and betelnut staining of the burials' teeth (Bulbeck, 2005b), indicate the burials are Neolithic.

B. Sampung; Pawon; Song Keplek 1 and 4; Song Terus 1; Gua Braholo 1, 4, 5, 7 and /H8.

C. Hoekgroet ♀; Gua Jimbe; Gua Kecil; Song Keplek 5 ♀; Gua Braholo loose teeth; Song Tritis.

D. Puger ♂; Anjar Lor ♀; Batujaya and Plawangan for lower premolars, M_1 and M_2.

Source: Includes data from all sources listed in the table and cited fully in reference list.

Table 10.4: Early Sulawesi dental metrical data.

Diameter	Leang Buidane Pre-ceramic, Talaud Islands[A]	Gua Mo'o Hono, Lake Towuti[B]	Leang Burung 1 Trench B, southwest Sulawesi[C]	Bola Batu, southwest Sulawesi[D]	Overall average
I¹ MD			8.4		8.4
I¹ BL			6.8		6.8
C MD	8.2				8.2
C BL	8.9				8.9
P³ MD	8.0				8.0
P³ BL	10.0			8.6	9.3
P⁴ MD	7.4	8.5		6.5	7.5
P⁴ BL	9.9	9.3		7.8	9.0
M¹ MD	11.95			9.9	11.3
M¹ BL	12.3			10.6	11.45
M² MD	11.3		9.5	10.4	10.6
M² BL	12.5		11.6	10.0	11.7
I₂ MD				6.8	6.8
I₂ BL				6.1	6.1
C BL				7.0	7.0
P₃ MD				7.0	7.0
P₃ BL				7.4	7.4
P₄ BL				7.7	7.7
M₁ MD				12.3	12.3
M₁ BL				11.2	11.2
M₂ MD		11.0			11.0
M₂ BL		10.15		9.5	9.9
M₃ BL				8.4	8.4

Notes: A. Unsexed adolescent stratified beneath the EMP cemetery (Bulbeck, 1981).

B. Teeth from spits 19 to 26, perhaps female, associated with 6–8 ka radiocarbon dates (Bulbeck et al., 2013). Data exclude teeth too worn for even their buccolingual diameters to be recorded.

C. Primary burial, perhaps male, directly radiocarbon dated to 4610 ± 220 BP (Bulbeck, 2004).

D. Slightly mineralised remains, some probably male, from pre-ceramic levels (Bulbeck, 2004).

Source: Includes data from all sources listed in notes and fully cited in reference section.

The statistical application employed in this study is Penrose's (1954) size and shape statistic, which divides Pearson's 'Coefficient of Racial Likeness' (CRL) into size and shape components. Like the CRL, Penrose's size and shape statistics are based on calculating a grand standard deviation for all the samples entered for analysis, dividing the samples' means by this grand standard deviation, and calculating the differences between the standardised means. I have developed an Approach database template that manages these steps, but only for up to 17 samples, which places a limit on how many samples with a reasonable sample size can be included in any analysis. However, I can freely add samples of very small sample size, because their variance would have minimal effect on the grand standard deviation, and so samples like these can be simply entered as mean values to be standardised. (More sophisticated statistical techniques that require individual specimens, with original measurements as observed or estimated for every analysed variable, are inappropriate for this study. This is because the dental metrical data are publicly available mainly in the form of means and standard deviations, and because most samples are dominated by incomplete dentitions or even loose teeth.)

Penrose's size component has the advantage that, when the calculated statistic is expressed as its square root, it is additive along a single dimension. For instance, if A is x larger than B, and B is y larger than C, then A is (x + y) larger than C. A second advantage is that the size difference tracks the average difference between samples in terms of grand standard deviations. So, for instance, if we assign C a size value of 0 (being smallest), and we then compute B's size as 0.5 and A's size as 1.0, we can state that A is on average one standard deviation larger than C, while B is half a standard deviation from both A and C.

Penrose's shape component essentially captures the variance that cannot be attributed to size. To make the shape distances more intuitive, two transformations are performed here. The first transformation is to express the calculated shape distances as their square roots, to convert them to Euclidean distances. The second transformation involves dividing each inter-sample shape distance by the square root of the product (or geometric average) of the average shape distances of the two samples being compared. For instance, if A has an average shape distance of 0.4 from the other samples, and B an average shape distance of 1.6, and their shape distance from each other is 0.8, their transformed (calibrated) shape distance would be 0.8/0.8 (the geometric average of 0.4 and 1.6), or 1. A value of 1 can be thought of as the 'expected' shape distance between any two samples, while values less than 1 (greater than 1) reflect cases of samples that are more similar (less similar) in shape than would be expected. In addition to relating shape differences to a benchmark of 1, this calibration process enables relatively small shape distances between a pair of samples to stand out, whether these samples' shape distances are on average large or small. (This calibration process also accommodates shape distances computed from different selections of variables in the same analysis, as later described for the analysis concerned.)

To present an overall view of the obtained shape distances, the samples are clustered into dendrograms using average linkage. In addition, two refinements are included, to the degree permitted by the dendrogram structure. (The calculations, performed using Excel spreadsheets, are available from the author on request.)

The first refinement involves seriating the samples along a single dimension. The samples most unlike each other are placed at the two extremes, and the other samples are positioned to the degree that they approach one or the other extreme. The success of the seriation can be calculated as the coefficient of variation between the seriated distances and the most similar, perfect seriation of those distances (see footnotes to Tables 10.5 to 10.8).

The second refinement is to vary the dendrogram's branch lengths according to the represented distances. A long branch in the dendrogram reflects a considerable shape disjunction, and a sample that accumulates great branch length with respect to the analysed samples' final joining distance (represented as 0 in the dendrogram) stands out as generally different from the other samples. The distance between any two samples is represented by the minimum horizontal distance that has to be traversed in tracing a path, through the dendrogram, that connects the two samples. How successful the traversed horizontal distances are in representing the shape distances can also be calculated in terms of their coefficient of variation. (The algorithm to calculate branch lengths is based on the average within-distance compared with the average outside-distance. For instance, if A and B cluster together at a distance of 0.5, but A is 0.1 farther from the other samples than B is, then the stem length of A is calculated as $0.5/2 + 0.1/2 = 0.3$, while the stem length of B is calculated as $0.5/2 - 0.1/2 = 0.2$.)

Five Penrose size and shape analyses are presented here. The first includes all of the male samples for all tooth diameters. The second analysis also focuses on male samples but is restricted to buccolingual diameters. Buccolingual diameters are not susceptible to reduction through interstitial wear, whereas, when mesiodistal diameters are included, there is a risk that the calculated size and shape distances mainly reflect differences in interstitial wear rates (but see Results). The third analysis includes the bisexual samples for all tooth diameters. The fourth comparison focuses on the same bisexual samples, but only on their buccolingual diameters, for the same reason as with the second analysis. Finally, the fifth analysis includes the four prehistoric Indo-Malaysian samples that are lacking data for some of the tooth diameters (see Results section).

Principal Components Analysis (PCA) was also undertaken of all of the analyses presented here, with the sample means submitted to PCA. In each case, the first principle component captured size, as would be expected of biological data (Joliffe 2002). The implications barely differ from the implications of the Penrose size analyses, and so it would be redundant to also present the PCA size results. After the 45–84% of variance accounted for by size was removed, the second, third and other principle components captured a maximum of 15% of variance. In most cases, this was too low to allow for ready interpretation. Accordingly, the PCA results are excluded from this contribution.

Results

First analysis: 17 male samples, all 32 diameters

The Penrose size comparisons are presented at the top of Figure 10.1 (Figure 10.1a). The results would be consistent with a scenario of pronounced tooth-size reduction in Indo-Malaysia between the Pleistocene and early Holocene, with continuing tooth-size reduction during the Neolithic and recent times. Surabaya males from Java show the smallest teeth, whereas the Late Pleistocene Java sample has the largest teeth, on average 2.3 standard deviations larger than Surabaya Javanese. The two pre-Neolithic Indo-Malaysian samples have larger teeth than the two Neolithic Indo-Malaysian samples, whose teeth are of above average size by recent Indo-Malaysian standards. Also of interest, Shanghai Chinese, and the Motupore males from Papua New Guinea, both resemble recent Indo-Malaysians in their tooth-size, whereas the other three Southwest Pacific samples have teeth that are much larger.

The Penrose shape distances (square roots) are presented at Table 10.5, both before and after calibration. The order of the samples reflects their order after seriation of the average-linkage dendrogram, which is illustrated in Figure 10.2. As shown there, the recent Southwest Pacific samples are all placed at one half of the seriated order and the recent Indo-Malaysian (and Chinese) samples toward the other half. Thus, seriation of the dendrogram appears to identify a distinction between 'Australoids' and 'Mongoloids' in their tooth shape, with the two Semang Negrito samples intermediate between the Australoids and the Mongoloids. The only Indo-Malaysian samples that cluster with the Australoids are the pre-Neolithic samples, including Pleistocene Java. The Neolithic samples, for their part, fall at the polar opposite from the Australoids. This result is consistent with the conventional wisdom (e.g. Bellwood, 1997) of a pre-Neolithic occupation of Indo-Malaysia by Australoid foragers, prior to the immigration of newcomers who introduced the Neolithic to the region.

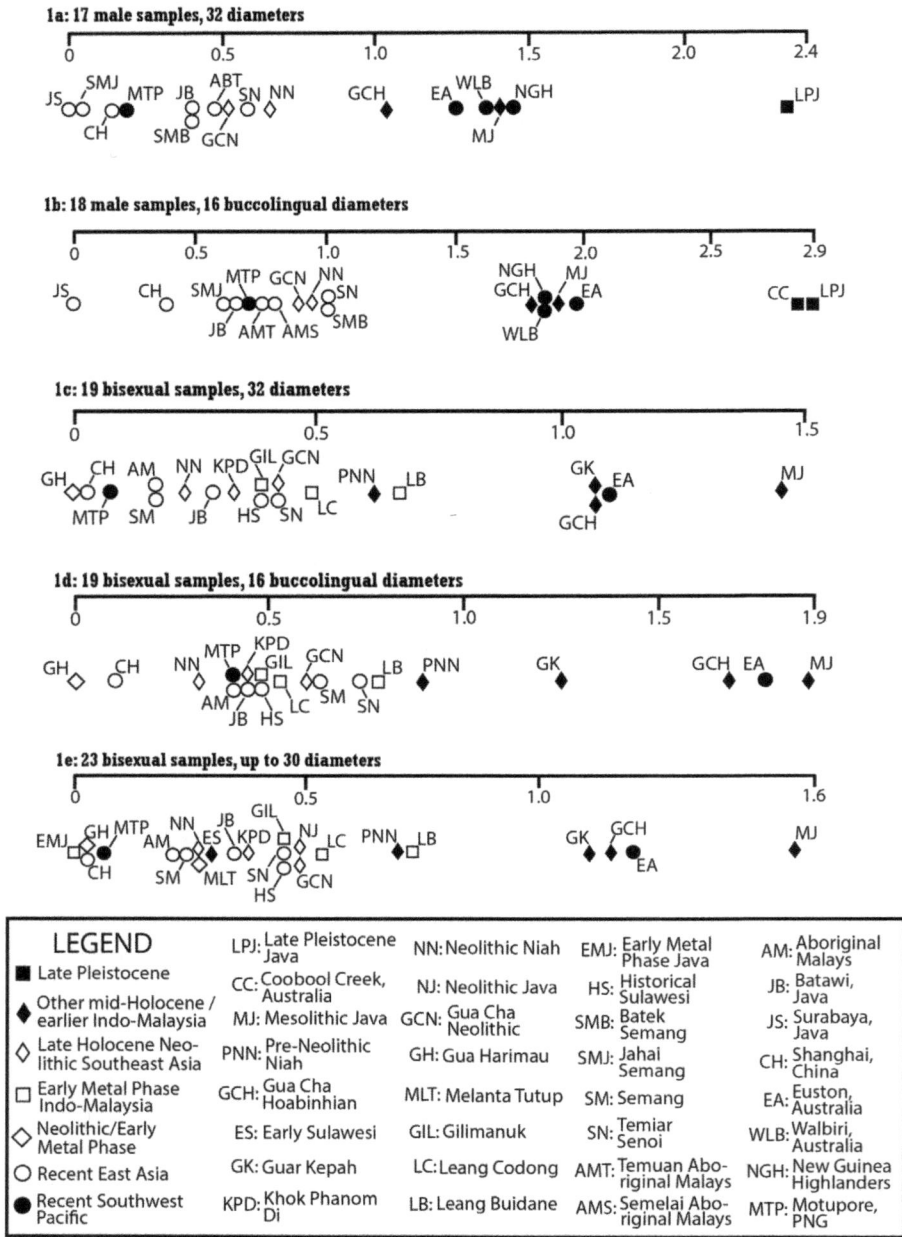

Figure 10.1: Dental metrics, Penrose size statistics comparisons.

Sources: Snell 1938; Barrett et al. 1963, 1964; Jacob 1964; Jacob 1967a, 1967b; Doran and Freedman 1974; Brown 1978; Bulbeck 1981; Brace et al. 1984; Brown 1989; Tayles 1989; Storm 1995; Bulbeck 2001; Détroit 2002; Bulbeck 2004; Bulbeck 2005a; Bulbeck et al. 2005; Chia et al. 2005; Manser 2005; Noerwidi 2011–12; Bulbeck et al. 2013; this paper.

Table 10.5: First analysis: Square roots of Penrose shape distances (top right) and after calibration and seriation (bottom left).[A]

Sample(B)	LPJ	EA	GCH	MJ	NGH	WLB	MTP	SMJ	SMB	SN	AMS	AMT	CH	JB	JS	GCN	NN
LPJ		0.832	0.681	0.807	0.611	0.667	0.568	0.730	0.773	0.757	0.782	0.860	0.688	0.722	0.850	0.937	0.976
EA	1.073		0.565	0.586	0.744	0.585	0.726	0.911	0.854	0.911	0.744	0.837	0.768	0.747	0.977	0.842	0.925
GCH	0.945	0.771		0.657	0.628	0.528	0.544	0.785	0.749	0.738	0.655	0.700	0.629	0.617	0.850	0.694	0.835
MJ	1.113	0.819	1.014		0.590	0.530	0.674	0.766	0.715	0.731	0.586	0.625	0.671	0.585	0.811	0.677	0.899
NGH	0.956	1.144	1.039	1.010		0.332	0.410	0.537	0.575	0.499	0.388	0.482	0.413	0.417	0.560	0.621	0.756
WLB	1.064	0.917	0.892	0.933	0.631		0.432	0.566	0.563	0.491	0.326	0.432	0.373	0.384	0.632	0.621	0.761
MTP	0.887	1.113	0.898	1.106	0.762	0.820		0.479	0.501	0.446	0.461	0.530	0.420	0.431	0.591	0.583	0.815
SMJ	1.046	1.282	1.190	1.133	0.917	0.986	0.818		0.297	0.411	0.563	0.608	0.605	0.599	0.758	0.635	0.982
SMB	1.121	1.218	1.151	1.075	0.995	0.993	0.866	0.470		0.472	0.539	0.564	0.592	0.561	0.743	0.569	0.914
SN	1.147	1.356	1.183	1.145	0.900	0.904	0.799	0.681	0.791		0.372	0.392	0.521	0.441	0.567	0.554	0.860
AMS	1.269	1.187	1.124	1.005	0.751	0.643	0.890	0.998	0.967	0.698		0.276	0.366	0.294	0.483	0.446	0.692
AMT	1.348	1.289	1.162	1.049	0.900	0.823	0.989	1.042	0.978	0.709	0.534		0.380	0.273	0.417	0.447	0.711
CH	1.095	1.201	1.060	1.147	0.783	0.721	0.795	1.051	1.041	0.956	0.719	0.722		0.252	0.402	0.530	0.673
JB	1.186	1.206	1.071	1.019	0.815	0.767	0.843	1.074	1.019	0.837	0.596	0.536	0.502		0.317	0.473	0.673
JS	1.218	1.375	1.288	1.220	0.956	1.101	1.008	1.185	1.176	0.937	0.854	0.713	0.697	0.568		0.592	0.693
GCN	1.368	1.209	1.073	1.041	1.080	1.103	1.013	1.013	0.918	0.934	0.806	0.779	0.939	0.864	0.943		0.896
NN	1.251	1.166	1.133	1.200	1.155	1.187	1.243	1.374	1.294	1.272	1.097	1.089	1.045	1.078	0.969	1.078	

Notes: A. The seriated calibrated distances, in the bottom-left half-matrix, show a tendency to be smallest along the diagonal and to increase with each step away from the diagonal, moving horizontally or vertically. In a perfect seriation, with each horizontal or vertical step away from the diagonal, the distances would increase (or at least stay the same). Therefore, the distances shown here were rearranged into the closest perfect seriation that could be found, and the coefficient of variation (68.8%) calculated between the distances shown here and the rearranged distances.

B. LPJ=Late Pleistocene Java; EA=Euston, Australia; GCH=Gua Cha Hoabinhian; MJ=Mesolithic Java; NGH=New Guinea Highlanders; WLB=Walbiri, Australia; MTP=Motupore Island, PNG; SMJ=Jahai Semang; SMB=Batek Semang; SN=Temiar Senoi; AMS=Semelai Aboriginal Malays; AMT=Temuan Aboriginal Malays; CH=Shanghai, China; JB=Batawi, Java; JS=Surabaya, Java; GCN=Gua Cha Neolithic; NN=Neolithic Niah.

Sources: Snell 1938; Barrett et al. 1963, 1964; Jacob 1967a; Doran and Freedman 1974; Brown 1978; Brace et al. 1984; Storm 1995; Détroit 2002; Bulbeck 2005a; Bulbeck et al. 2005; Manser 2005; Noerwidi 2011–12.

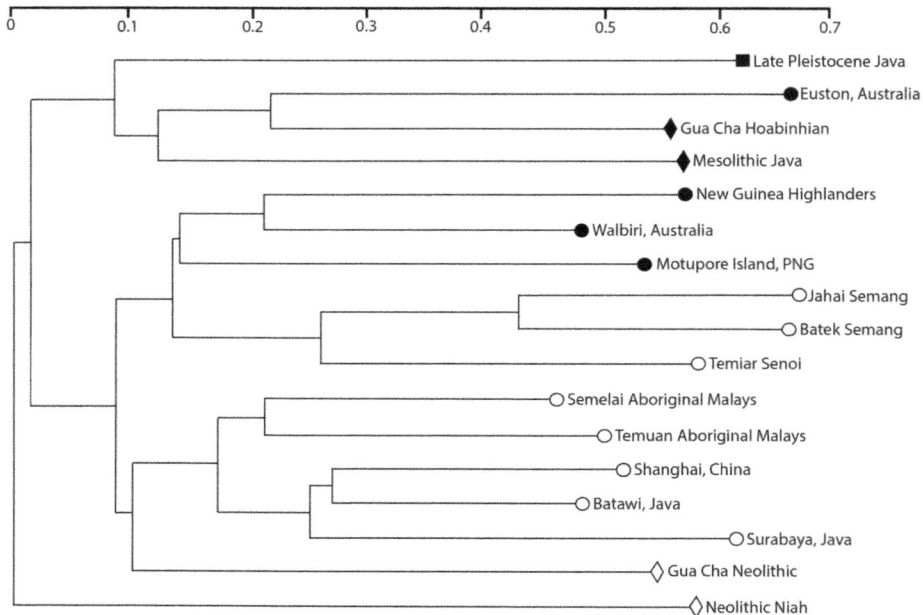

Coefficient of variation with a perfect seriation 68.8%. Branch lengths' coefficient of variation with distances 74.4%. For explanation of symbols, see Figure 1.

Figure 10.2: Dental metrics, Penrose shape distances, 17 male samples, 32 diameters, seriated average-linkage dendrogram.

Sources: Snell 1938; Barrett et al. 1963, 1964; Jacob 1967a; Doran and Freedman 1974; Brown 1978; Brace et al. 1984; Storm 1995; Détroit 2002; Bulbeck 2005a; Bulbeck et al. 2005; Manser 2005; Noerwidi 2011–12.

Second analysis: 18 male samples, 16 buccolingual diameters

The Penrose size comparisons are presented at Figure 10.1b. The results essentially echo those obtained for males using all diameters, with a clear indication of continual tooth-size reduction in Indo-Malaysia from the Pleistocene through the early Holocene, into the Neolithic and recent times. However, the inclusion of Coobool Creek (Late Pleistocene Australia) in the comparison offers two additional insights. First, the Coobool Creek teeth are larger than those of Holocene Australian Aborigines, as emphasised by Brown (1989). Secondly, Pleistocene Java and Australian teeth appear very similar in size, just as Mesolithic Java tooth-size appears very similar to Holocene Australian tooth-size.

Table 10.6: Second analysis: Square roots of Penrose shape distances (top right) and after calibration and seriation (bottom left).[A]

Sample(B)	CC	EA	MJ	GCH	WLB	AMS	AMT	SN	CH	JB	MTP	JS	NN	GCN	NGH	SMB	SMJ	LPJ
CC		0.501	0.624	0.696	0.582	0.623	0.709	0.857	0.673	0.708	0.833	0.856	0.842	0.904	0.850	0.917	0.940	0.828
EA	0.686		0.561	0.656	0.464	0.591	0.651	0.805	0.655	0.603	0.756	0.746	0.825	0.869	0.723	0.877	0.880	0.767
MJ	0.935	0.875		0.690	0.464	0.467	0.522	0.575	0.558	0.487	0.604	0.608	0.748	0.669	0.536	0.643	0.614	0.570
GCH	1.058	1.038	1.196		0.516	0.526	0.438	0.577	0.450	0.360	0.479	0.423	0.535	0.650	0.630	0.747	0.755	0.551
WLB	0.976	0.811	0.888	1.002		0.274	0.290	0.433	0.359	0.348	0.457	0.483	0.584	0.611	0.402	0.532	0.552	0.577
AMS	1.064	1.052	0.911	1.038	0.597		0.267	0.333	0.379	0.341	0.404	0.425	0.506	0.447	0.392	0.534	0.542	0.605
AMT	1.228	1.174	1.031	0.877	0.642	0.600		0.302	0.318	0.284	0.318	0.360	0.483	0.465	0.421	0.503	0.533	0.586
SN	1.386	1.356	1.062	1.079	0.896	0.700	0.644		0.443	0.368	0.406	0.384	0.551	0.466	0.432	0.544	0.494	0.557
CH	1.171	1.186	1.107	0.906	0.797	0.858	0.730	0.949		0.252	0.255	0.334	0.450	0.441	0.448	0.426	0.446	0.494
JB	1.276	1.131	1.001	0.750	0.801	0.799	0.675	0.818	0.600		0.239	0.192	0.433	0.450	0.395	0.506	0.497	0.413
MTP	1.429	1.350	1.181	0.949	1.002	0.901	0.718	0.859	0.580	0.562		0.177	0.427	0.408	0.382	0.463	0.466	0.518
JS	1.457	1.323	1.182	0.834	1.050	0.940	0.807	0.805	0.752	0.449	0.394		0.391	0.426	0.391	0.537	0.513	0.457
NN	1.268	1.294	1.284	0.931	1.124	0.990	0.958	1.023	0.898	0.894	0.839	0.763		0.447	0.584	0.642	0.716	0.687
GCN	1.383	1.385	1.167	1.149	1.194	0.890	0.938	0.878	0.893	0.944	0.816	0.845	0.784		0.509	0.501	0.581	0.698
NGH	1.371	1.216	0.987	1.175	0.828	0.822	0.895	0.858	0.959	0.875	0.804	0.819	1.081	1.091		0.450	0.446	0.506
SMB	1.394	1.388	1.115	1.312	1.033	1.056	1.007	1.019	0.858	1.055	0.919	1.057	1.119	0.901	0.841		0.280	0.558
SMJ	1.426	1.390	1.063	1.325	1.070	1.069	1.066	0.924	0.895	1.035	0.923	1.008	1.245	0.885	0.832	0.491		0.530
LPJ	1.245	1.201	0.979	0.958	1.108	1.183	1.162	1.032	0.983	0.852	1.017	0.892	1.184	1.223	0.935	0.971	0.921	

Notes: A. The seriated calibrated distances, in the bottom-left half-matrix, show a tendency to be smallest along the diagonal and to increase with each step away from the diagonal, moving horizontally or vertically. In a perfect seriation, with each horizontal or vertical step away from the diagonal, the distances would increase (or at least stay the same). Therefore, the distances shown here were rearranged into the closest perfect seriation that could be found, and the coefficient of variation (73.2%) calculated between the distances shown here and the rearranged distances.

B. CC=Coobool Creek; EA=Euston, Australia; MJ=Mesolithic Java; GCH=Gua Cha Hoabinhian; WLB=Walbiri, Australia; AMS=Semelai Aboriginal Malays; AMT=Temuan Aboriginal Malays; SN=Temiar Senoi; CH=Shanghai, China; JB=Batawi, Java; MTP=Motupore Island, PNG; JS=Surabaya, Java; NN=Neolithic Niah; GCN=Gua Cha Neolithic; NGH=New Guinea Highlanders; SMB=Batek Semang; SMJ=Jahai Semang; LPJ=Late Pleistocene Java.

Sources: Snell 1938; Barrett et al. 1964; Jacob 1967a; Doran and Freedman 1974; Brown 1978; Brace et al. 1984; Brown 1989; Storm 1995; Détroit 2002; Bulbeck 2005a; Bulbeck et al. 2005; Manser 2005; Noerwidi 2011–12.

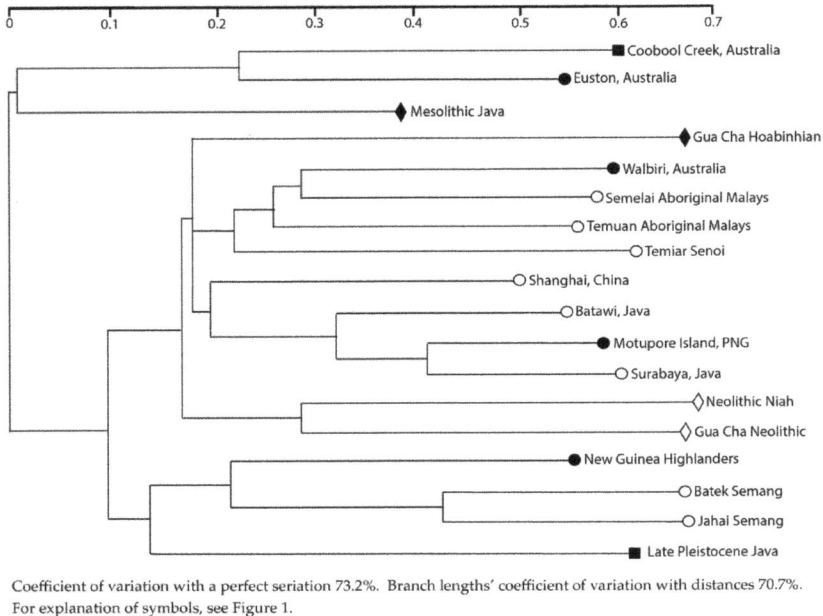

Coefficient of variation with a perfect seriation 73.2%. Branch lengths' coefficient of variation with distances 70.7%. For explanation of symbols, see Figure 1.

Figure 10.3: Dental metrics, Penrose shape distances, 18 male samples, 16 buccolingual diameters, seriated average-linkage dendrogram.

Sources: Snell 1938; Barrett et al. 1964; Jacob 1967a; Doran and Freedman 1974; Brown 1978; Brace et al. 1984; Brown 1989; Storm 1995; Détroit 2002; Bulbeck 2005a; Bulbeck et al. 2005; Manser 2005; Noerwidi 2011–12.

However, when the calibrated shape distances (Table 10.6) are clustered and seriated, the results differ from Figure 10.2 (the first analysis). The Southwest Pacific samples now split between Australian samples (including Coobool Creek) at one half of the seriation, and New Guinea samples at the other half of the seriation (Figure 10.3). Mesolithic Java and Gua Cha Hoabinhians align with the Australian samples, whereas Late Pleistocene Java takes up a polar position away from Australians. There is however one concordance between Figures 10.2 and 10.3: the two Neolithic samples fall closer to recent Indo-Malaysians than to the pre-Neolithic Indo-Malaysian samples.

The similarity between Australian and New Guinea samples in Figure 10.2, lacking from Figure 10.3, suggests the existence of a 'Southwest Pacific tooth shape' based on mesiodistal diameters and their relation to buccolingual diameters. This similarity cannot be attributed to interstitial wear. The Euston Aboriginal teeth were affected by much greater interstitial wear than the Motupore teeth (personal observation). Hence, if interstitial wear were at stake, any Euston-Motupore similarity should be evident in Figure 10.3, not Figure 10.2 – the reverse of what we find. Accordingly, the dentition as a whole apparently reflects a genetically based difference between 'Australoids' and 'Mongoloids', detectable notwithstanding differences between populations in their interstitial wear rates.

Third analysis: 19 bisexual samples, all 32 diameters

In the Penrose size comparisons (Figure 10.1c), the Neolithic/EMP sample from Gua Harimau shows the smallest teeth. The Niah samples also appear relatively small-toothed, with Neolithic Niah smaller than Neolithic Gua Cha, and pre-Neolithic Niah smaller than Gua Cha Hoabinhians and Mesolithic Java. This however may be affected by the greater representation of females than males in the Niah samples. Another complication is that the Guar Kepah sample, while qualifying as Neolithic, is of similar mid-Holocene antiquity to the Gua Cha Hoabinhians, and indeed their teeth are similarly large. The late Holocene Neolithic samples, for their part, tend to have slightly smaller teeth than the EMP samples. For all that, we can safely conclude that mid-Holocene and earlier Indo-Malaysian teeth appear to have been larger than their late Holocene counterparts.

The shape distances (Table 10.7), upon analysis, produce a pattern similar to that observed for the male samples with all 32 diameters included. The top half of the dendrogram (Figure 10.4) features Southwest Pacific samples, along with mid-Holocene and earlier Indo-Malaysian samples, as well as the Semang and Temiar Senoi from Malaya. The bottom half of the dendrogram includes recent Indo-Malaysians (except the Semang and Senoi), Shanghai Chinese and the late Holocene prehistoric samples. The extreme examples are Gua Harimau and Neolithic Niah, whereas Neolithic Gua Cha now tends towards the middle of the dendrogram. The simplest interpretation of Figure 10.4 may be that it points to broadly 'Australoid' (Euston Aborigines to Motupore) and 'Mongoloid' (Gua Harimau to Aboriginal Malay) groupings, with pre-Neolithic Niah, Guar Kepah and Gua Cha Neolithic intermediate.

Table 10.7: Third analysis: Square roots of Penrose shape distances (top right) and after calibration and seriation (bottom left).[A]

Sample[B]	EA	GCH	MJ	SM	SN	MTP	PNN	GK	GCN	AM	JB	CH	KPD	LB	LC	GIL	HS	NN	GH
EA		0.558	0.627	0.903	0.872	0.795	0.705	0.896	0.821	0.768	0.791	0.824	0.886	0.836	0.909	0.845	0.856	0.934	1.015
GCH	0.757		0.591	0.730	0.672	0.615	0.465	0.675	0.657	0.605	0.603	0.673	0.703	0.651	0.663	0.660	0.670	0.776	0.875
MJ	0.875	0.924		0.614	0.586	0.602	0.580	0.633	0.518	0.529	0.535	0.616	0.628	0.578	0.641	0.609	0.669	0.806	0.837
SM	1.248	1.130	0.977		0.389	0.473	0.637	0.606	0.577	0.464	0.545	0.550	0.593	0.575	0.699	0.661	0.702	0.758	0.947
SN	1.315	1.134	1.017	0.669		0.463	0.544	0.589	0.454	0.291	0.388	0.502	0.476	0.463	0.509	0.459	0.545	0.627	0.765
MTP	1.174	1.017	1.024	0.797	0.851		0.550	0.543	0.526	0.423	0.422	0.412	0.482	0.512	0.581	0.529	0.609	0.642	0.826
PNN	1.091	0.805	1.032	1.122	1.046	1.037		0.539	0.469	0.389	0.360	0.449	0.504	0.460	0.402	0.414	0.481	0.543	0.640
GK	1.333	1.123	1.084	1.027	1.089	0.983	1.022		0.483	0.482	0.448	0.371	0.360	0.370	0.449	0.464	0.535	0.557	0.872
GCN	1.303	1.168	0.947	1.044	0.896	1.017	0.949	0.940		0.362	0.322	0.422	0.454	0.339	0.412	0.325	0.446	0.460	0.615
AM	1.276	1.126	1.013	0.879	0.602	0.855	0.824	0.982	0.788		0.246	0.332	0.326	0.353	0.382	0.313	0.464	0.515	0.661
JB	1.360	1.162	1.060	1.069	0.829	0.884	0.789	0.946	0.725	0.579		0.260	0.295	0.256	0.273	0.264	0.331	0.450	0.587
CH	1.337	1.223	1.151	1.016	1.012	0.814	0.929	0.738	0.896	0.739	0.599		0.237	0.294	0.361	0.352	0.403	0.514	0.724
KPD	1.410	1.252	1.150	1.076	0.943	0.934	1.022	0.702	0.946	0.710	0.665	0.504		0.323	0.345	0.359	0.413	0.515	0.725
LB	1.372	1.195	1.092	1.076	0.946	1.023	0.962	0.743	0.727	0.794	0.596	0.644	0.694		0.298	0.346	0.327	0.455	0.679
LC	1.363	1.192	1.131	1.215	0.921	1.040	0.851	0.918	0.687	0.692	0.605	0.759	0.760	0.754		0.281	0.354	0.438	0.574
GIL	1.451	1.185	1.178	1.273	1.010	1.130	0.818	0.879	0.861	0.834	0.619	0.770	0.722	0.642	0.596		0.419	0.459	0.632
HS	1.423	1.247	1.281	1.330	1.126	1.232	1.020	1.091	0.970	0.959	0.720	0.827	0.900	0.735	0.925	0.774		0.478	0.594
NN	1.350	1.256	1.341	1.249	1.128	1.131	1.001	0.988	0.871	1.021	0.923	0.993	0.976	0.890	0.883	0.833	0.947		0.517
GH	1.311	1.265	1.244	1.394	1.229	1.299	1.053	1.382	1.040	1.170	1.075	1.251	1.229	1.186	1.086	0.976	1.052	0.797	

Notes: A. The seriated calibrated distances, in the bottom-left half-matrix, show a tendency to be smallest along the diagonal and to increase with each step away from the diagonal, moving horizontally or vertically. In a perfect seriation, with each horizontal or vertical step away from the diagonal, the distances would increase (or at least stay the same). Therefore, the distances shown here were rearranged into the closest perfect seriation that could be found, and the coefficient of variation (76.0%) calculated between the distances shown here and the rearranged distances.

B. EA=Euston, Australia; GCH=Gua Cha Hoabinhian; MJ=Mesolithic Java; SM=Semang; SN=Temiar Senoi; MTP=Motupore Island, PNG; PNN=pre-Neolithic Niah; GK=Guar Kepah; GCN=Gua Cha Neolithic; AM=Aboriginal Malays; JB=Batawi, Java; CH=Shanghai, China; KPD=Khok Phanom Di; LB=Leang Buidane; LC=Leang Codong; GIL=Gilimanuk; HS=Historical Sulawesi; NN=Neolithic Niah; GH=Gua Harimau.

Sources: Snell 1938; Barrett et al. 1963, 1964; Jacob 1964; Jacob 1967a, 1967b; Doran and Freedman 1974; Brown 1978; Bulbeck 1981; Brace et al. 1984; Tayles 1989; Storm 1995; Bulbeck 2001; Détroit 2002; Bulbeck 2004; Bulbeck 2005a; Bulbeck et al. 2005; Chia et al. 2005; Manser 2005; Noerwidi 2011–12; Bulbeck et al. 2013; this paper.

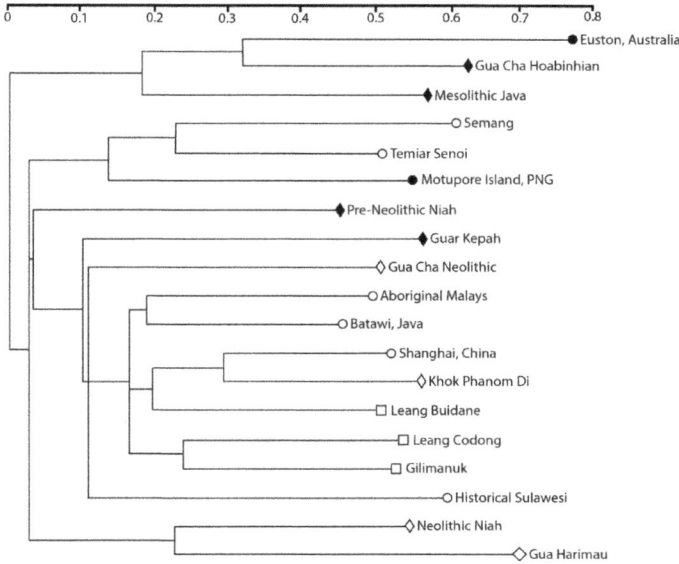

Coefficient of variation with a perfect seriation 76.0%. Branch lengths' coefficient of variation with distances 79.2%.
For explanation of symbols, see Figure 1.

Figure 10.4: Dental metrics, Penrose shape distances, 19 bisexual samples, 32 diameters, seriated average-linkage dendrogram.

Sources: Snell 1938; Barrett et al. 1963, 1964; Jacob 1964; Jacob 1967a, 1967b; Doran and Freedman 1974; Brown 1978; Bulbeck 1981; Brace et al. 1984; Tayles 1989; Storm 1995; Bulbeck 2001; Détroit 2002; Bulbeck 2004; Bulbeck 2005a; Bulbeck et al. 2005; Chia et al. 2005; Manser 2005; Noerwidi 2011–12; Bulbeck et al. 2013; this paper.

Fourth analysis: 19 bisexual samples, 16 buccolingual diameters

When the Penrose size distances for bisexual samples are limited to buccolingual diameters, the resulting graph (Figure 10.1d) can be viewed as a clarification of Figure 10.1c. There is now no overlap in tooth-size between the late Holocene and the mid-Holocene and earlier samples. Moreover, the lack of a systematic size distinction, comparing the Neolithic, EMP and recent/historical samples with each other, is very apparent. (The small size of the Gua Harimau teeth is even more apparent than in Figure 10.1c.)

The structure of the shape distances (Table 10.8; Figure 10.5) is difficult to interpret. The extreme positions of the seriated dendrogram are taken up by Euston Aborigines and Gua Harimau, as in the third analysis. However, the three pre-Neolithic Indo-Malaysian samples now split between Mesolithic Java, which clusters with Euston Aborigines, and pre-Neolithic Niah and Gua Cha Hoabinhians, which seriate adjacently to Gua Harimau. The Neolithic samples (including Guar Kepah) lie in the same half of the dendrogram as Mesolithic Java.

199

Table 10.8: Fourth analysis: Square roots of Penrose shape distances (top right) and after calibration and seriation (bottom left).[A]

Sample[B]	EA	MJ	SM	GK	CH	KPD	MTP	NN	GCN	LB	SN	AM	GIL	JB	LC	HS	PNN	GCH	GH
EA		0.596	0.820	0.792	0.612	0.589	0.753	0.738	0.721	0.632	0.684	0.509	0.620	0.553	0.636	0.635	0.594	0.613	0.855
MJ	0.992		0.460	0.552	0.447	0.424	0.561	0.688	0.506	0.470	0.496	0.421	0.584	0.454	0.507	0.591	0.581	0.670	0.780
SM	1.370	0.849		0.334	0.357	0.445	0.436	0.607	0.437	0.398	0.479	0.469	0.573	0.500	0.535	0.644	0.595	0.709	0.911
GK	1.374	1.058	0.644		0.289	0.329	0.351	0.493	0.432	0.389	0.494	0.455	0.477	0.429	0.489	0.580	0.589	0.618	0.908
CH	1.193	0.963	0.772	0.650		0.161	0.298	0.442	0.370	0.284	0.403	0.314	0.386	0.271	0.351	0.421	0.468	0.501	0.751
KPD	1.153	0.918	0.967	0.743	0.409		0.326	0.435	0.389	0.310	0.370	0.287	0.332	0.249	0.313	0.397	0.499	0.502	0.714
MTP	1.382	1.138	0.887	0.742	0.708	0.778		0.282	0.365	0.371	0.441	0.370	0.419	0.320	0.414	0.547	0.525	0.590	0.680
NN	1.304	1.344	1.189	1.004	1.011	0.998	0.607		0.419	0.396	0.477	0.418	0.404	0.331	0.398	0.506	0.523	0.556	0.567
GCN	1.334	1.036	0.897	0.921	0.887	0.936	0.823	0.911		0.419	0.703	0.788	0.858	0.852	0.989	0.994	1.157	1.174	1.222
LB	1.232	1.013	0.862	0.874	0.717	0.786	0.883	0.906	0.652		0.349	0.834	0.836	0.948	0.636	0.799	0.802	0.470	0.678
SN	1.264	1.012	0.981	1.051	0.964	0.888	0.991	1.033	0.703	0.349		0.272	0.310	0.239	0.317	0.350	0.497	0.574	0.689
AM	0.998	0.911	1.020	1.026	0.797	0.730	0.884	0.961	0.788	0.834	0.272		0.263	0.267	0.326	0.460	0.406	0.496	0.672
GIL	1.160	1.058	1.190	1.028	0.936	0.809	0.956	0.887	0.858	0.836	0.310	0.263		0.313	0.296	0.317	0.415	0.420	0.706
JB	1.139	1.032	1.142	1.018	0.723	0.667	0.803	0.800	0.852	0.948	0.239	0.267	0.313		0.191	0.326	0.357	0.374	0.567
LC	1.240	1.092	1.156	1.098	0.887	0.794	0.983	0.911	0.989	0.636	0.317	0.326	0.296	0.508		0.326	0.349	0.377	0.577
HS	1.122	1.154	1.264	1.182	0.963	0.912	1.179	1.050	0.994	0.799	0.350	0.460	0.317	0.766	0.746		0.504	0.415	0.604
PNN	1.037	1.121	1.152	1.185	1.057	1.132	1.117	1.071	1.157	1.085	1.064	1.062	1.010	0.900	0.789	1.031		0.297	0.683
GCH	1.037	1.252	1.332	1.205	1.098	1.105	1.217	1.104	1.174	1.085	1.191	1.092	0.883	0.866	0.825	0.824	0.582		0.717
GH	1.247	1.257	1.474	1.527	1.418	1.354	1.209	0.971	1.222	1.280	1.231	1.274	1.279	1.130	1.089	1.034	1.154	1.175	

Notes: A. The seriated calibrated distances, in the bottom-left half-matrix, show a tendency to be smallest along the diagonal and to increase with each step away from the diagonal, moving horizontally or vertically. In a perfect seriation, with each horizontal or vertical step away from the diagonal, the distances would increase (or at least stay the same). Therefore, the distances shown here were rearranged into the closest perfect seriation that could be found, and the coefficient of variation (66.1%) calculated between the distances shown here and the rearranged distances.

B. EA=Euston, Australia; MJ=Mesolithic Java; SM=Semang; GK=Guar Kepah; CH=Shanghai, China; KPD=Khok Phanom Di; MTP=Motupore Island, PNG; NN=Neolithic Niah; GCN=Gua Cha Neolithic; LB=Leang Buidane; SN=Temiar Senoi; AM=Aboriginal Malays; GIL.=Gilimanuk; JB=Batawi, Java; LC=Leang Codong; HS=Historical Sulawesi; PNN=pre-Neolithic Niah; GCH=Gua Cha Hoabinhian; GH=Gua Harimau.

Sources: Snell 1938; Barrett et al. 1964; Jacob 1967a, 1967b; Doran and Freedman 1974; Brown 1978; Bulbeck 1981; Brace et al. 1984; Tayles 1989; Storm 1995; Bulbeck 2001; Détroit 2002; Bulbeck 2004; Bulbeck 2005a; Bulbeck et al. 2005; Manser 2005; Noerwidi 2011–12.

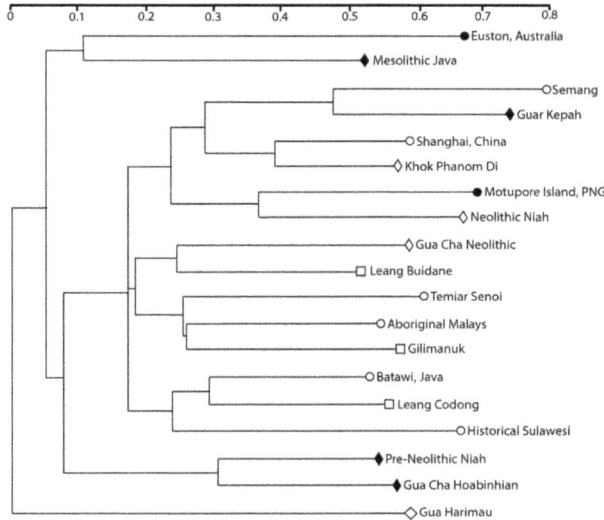

Coefficient of variation with a perfect seriation 66.1%. Branch lengths' coefficient of variation with distances 66.6%.
For explanation of symbols, see Figure 1.

Figure 10.5: Dental metrics, Penrose shape distances, 19 bisexual samples, 16 buccolingual diameters, seriated average-linkage dendrogram.

Sources: Snell 1938; Barrett et al. 1964; Jacob 1967a, 1967b; Doran and Freedman 1974; Brown 1978; Bulbeck 1981; Brace et al. 1984; Tayles 1989; Storm 1995; Bulbeck 2001; Détroit 2002; Bulbeck 2004; Bulbeck 2005a; Bulbeck et al. 2005; Manser 2005; Noerwidi 2011–12.

Fifth analysis: 23 bisexual samples, up to 32 diameters

The fifth analysis has the complication that it includes four samples lacking data for some of the diameters. The available data for these samples cover 30 diameters (EMP Java), 24 diameters (Neolithic Java and Melanta Tutup) and 23 diameters (Early Sulawesi). Further, when these samples are compared with each other, the number of diameters they have in common may be further reduced, to as few as 16 (Neolithic Java compared with Melanta Tutup). The approach adopted here to missing data is to base the pair-wise comparisons on as many diameters as there are data available.

For the size comparisons, each of the additional four samples was individually compared with the 19 samples (those in the third analysis) on all of the diameters for which the individual sample has data. Of these, EMP Java was found to have smaller teeth than Gua Harimau. Therefore, with EMP Java established as the new 'ground zero' for the size comparisons, the tooth-size of the other samples is represented by their distance from EMP Java for as many diameters as they have in common with EMP Java, up to 30 (Figure 10.1e).

Table 10.9: Fifth analysis, four additional samples: Square roots of Penrose shape distances (left) and after calibration and seriation (right).[A]

Sample[B]	NJ	MLT	EMJ	ES	ES	EMJ	MLT	NJ
EA	1.140	1.517	1.240	1.563	1.694	1.464	1.630	1.230
GCH	0.927	1.414	1.119	1.312	1.733	1.552	1.637	1.166
MJ	1.049	1.191	0.943	1.223	1.663	1.274	1.423	1.301
MTP	0.948	1.168	0.967	1.324	1.832	1.368	1.472	1.250
SM	1.037	1.158	1.024	1.318	1.620	1.355	1.436	1.251
SN	0.887	1.059	0.851	1.280	1.733	1.231	1.444	1.182
PNN	0.894	1.121	0.852	1.042	1.472	1.274	1.462	1.239
GK	1.118	1.005	0.898	1.156	1.594	1.285	1.318	1.503
GCN	0.992	0.951	0.637	1.162	1.692	1.141	1.287	1.411
AM	0.915	0.988	0.752	1.163	1.796	1.201	1.451	1.347
JB	0.840	0.941	0.698	1.044	1.694	1.153	1.386	1.276
KPD	1.019	0.911	0.748	1.109	1.659	1.177	1.231	1.472
CH	0.999	0.938	0.745	1.060	1.613	1.169	1.328	1.425
LB	0.936	0.965	0.735	1.057	1.622	1.160	1.364	1.367
GIL	1.017	0.875	0.808	1.057	1.568	1.254	1.235	1.460
LC	0.933	0.879	0.686	0.989	1.466	1.061	1.201	1.312
HS	0.807	1.014	0.748	1.105	1.623	1.117	1.377	1.113
NN	1.031	0.989	0.840	1.214	1.663	1.167	1.233	1.296
GH	0.920	1.079	0.854	1.234	1.411	1.066	1.184	1.079
NJ		1.650	1.230	1.570	1.357	1.401	1.475	
MLT			0.934	1.163	1.112	0.979		
EMJ				0.897	0.837			
ES	Square roots of shape distances				Calibrated shape distances			

Notes: A. The calibrated distances in the top 19 rows were calculated from four separate 20 by 20 half-matrices of shape distances based on all of the tooth diameters recorded for the sample named in the column heading. For instance, to compare EMJ (EMP Java) with the 19 samples from the third analysis (EA to GH), a 20 by 20 half-matrix of shape distances was calculated based on the 30 diameters recorded for EMJ. The calibrated distance of EMJ from EA (for instance) is 1.240 divided by the geometric average of the average distances obtained for EMJ and for EA. However, the calibrated distances comparing NJ (Neolithic Java), Melanta Tutup (MLT), EMJ and Early Sulawesi (ES) with each other were calculated from six separate, 21 by 21 half-matrices. For instance, to compare NJ and MLT, the 21 compared samples included the 19 samples from the third analysis (EA to GH) as well as NJ and MLT. The shape distances in the 21 by 21 half-matrix were calculated from the 16 diameters NJ and MLT have in common. Their calibrated distance is 1.650 divided by the geometric average of the average distances now obtained for NJ and MLT.

B. EA=Euston, Australia; GCH=Gua Cha Hoabinhian; MJ=Mesolithic Java; MTP=Motupore Island, PNG; SM=Semang; SN=Temiar Senoi; PNN=pre-Neolithic Niah; GK=Guar Kepah; GCN=Gua Cha Neolithic; AM=Aboriginal Malays; JB=Batawi, Java; KPD=Khok Phanom Di; CH=Shanghai, China; LB=Leang Buidane; GIL=Gilimanuk; LC=Leang Codong; HS=Historical Sulawesi; NN=Neolithic Niah; GH=Gua Harimau; NJ=Neolithic Java; MLT=Melanta Tutup; EMJ=Early Metal Phase Java; ES=Early Sulawesi.

Sources: Snell 1938; Barrett et al. 1963, 1964; Jacob 1964; Jacob 1967a, 1967b; Doran and Freedman 1974; Brown 1978; Bulbeck 1981; Brace et al. 1984; Tayles 1989; Storm 1995; Bulbeck 2001; Détroit 2002; Bulbeck 2004; Bulbeck 2005a; Bulbeck et al. 2005; Chia et al. 2005; Manser 2005; Noerwidi 2011–12; Bulbeck et al. 2013; this paper.

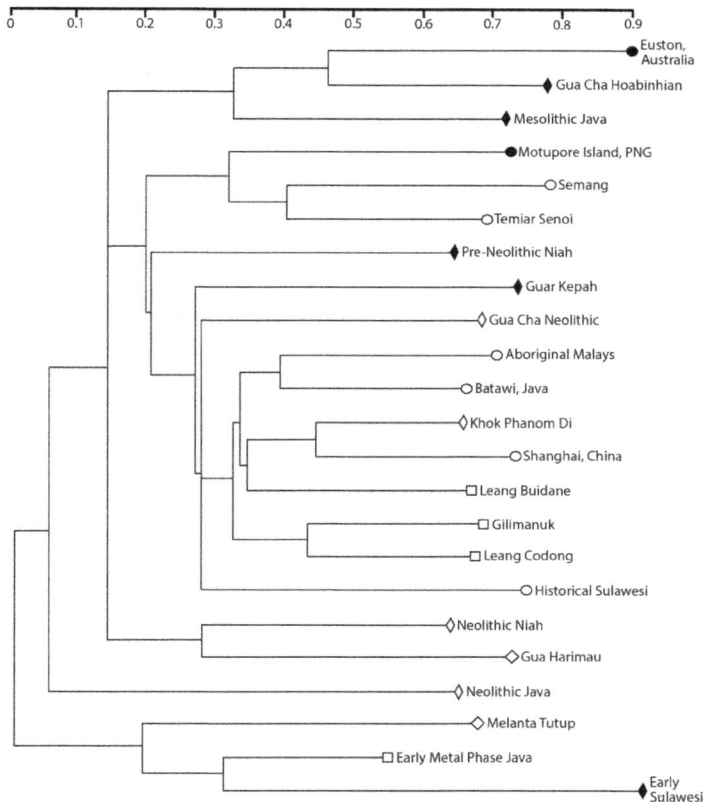

Coefficient of variation with a perfect seriation 79.5%.

Branch lengths' coefficient of variation with distances 82.7%.

For explanation of symbols see Figure 1.

Figure 10.6: Dental metrics, Penrose shape distances, 23 bisexual samples, up to 32 diameters, seriated average-linkage dendrogram.

Sources: Snell 1938; Barrett et al. 1963, 1964; Jacob 1964; Jacob 1967a, 1967b; Doran and Freedman 1974; Brown 1978; Bulbeck 1981; Brace et al. 1984; Tayles 1989; Storm 1995; Bulbeck 2001; Détroit 2002; Bulbeck 2004; Bulbeck 2005a; Bulbeck et al. 2005; Chia et al. 2005; Manser 2005; Noerwidi 2011–12; Bulbeck et al. 2013; this paper.

As shown there, the Early Sulawesi teeth are small (as previously noted by Bulbeck, 2004), smaller than EMP (Leang Codong, Leang Buidane) and historical Sulawesi teeth. The Neolithic Java and Melanta Tutup teeth are small, and the EMP Java teeth are very small. Caution should be exercised in the use of these results, given that they are based on very small sample sizes and incomplete coverage of the dentition. Nonetheless, it is interesting that the late Holocene tooth-size reduction indicated for Java, Borneo and Malaya appears to be reversed for Sulawesi.

Turning to the shape comparisons, we first note that the calibrated shape distances from the third analysis (Table 10.7) were migrated wholesale to the fifth analysis, supplemented here by the calibrated shape distances that involve the four additional samples. As explained in footnote A to Table 10.9, the calibrated shape distances in the fifth analysis vary in terms of the tooth diameters used to generate them, but they are comparable in that the value of 1 serves as the benchmark for the 'expected' shape distance in each cell.

Figure 10.6 shows the seriated dendrogram that results from the calibrated shape distances in Tables 10.7 and 10.9. The same general structure emerges as for Figure 10.4. Euston Aborigines, Gua Cha Hoabinhians, Mesolithic Java, Motupore and pre-Neolithic Niah fall towards one pole, while Gua Harimau, Niah Neolithic and historical Sulawesi fall towards the other pole. The four additional samples take up the extreme position at the latter pole. In other words, Neolithic Java, Melanta Tutup, EMP Java and especially Early Sulawesi (but not the Chinese sample) appear particularly non-Australoid. Thus, the Early Sulawesi sample differs from the other mid-Holocene and earlier Indo-Malaysian samples not just by its smaller teeth but also its distinct dental metrical shape.

Discussion

The first assignment for this study was to test the hypothesis that the advent of agriculture and pottery selected for smaller teeth. The expectation required to confirm this hypothesis was evidence for continual tooth-size reduction in Indo-Malaysia throughout the late Holocene. The male comparisons are broadly consistent with this expectation. However, only two late Holocene, prehistoric samples from Indo-Malaysia qualified for inclusion in these comparisons. Further, one of them, Gua Cha Neolithic males, had slightly smaller teeth than Temiar males, even though Temiar ancestry may have included the population represented by the Gua Cha Neolithic burials (Bulbeck, 2011: 237). Certainly, when analysis included the bisexual samples, evidence for continual tooth-size reduction throughout that late Holocene was hard to find. For instance, in the fifth analysis, the Neolithic and recent samples overlap comprehensively in tooth-size, both falling centrally within the EMP range of variation (Figure 10.1e). Similarly, Brace and others (1984) could find no evidence for late Holocene tooth-size reduction in north China, while in mainland Southeast Asia, the Neolithic teeth from Ban Kao in Thailand are smaller than those of recent Thais, and the Iron Age Dong Son teeth smaller than those of recent Vietnamese (Matsumura and Hudson, 2005: Figure 3). Therefore, detailed comparisons appear to falsify the 'Neolithic tooth-size reduction' model.

Keeping to a local evolutionary paradigm, we turn to an alternative possible explanation for the tooth-size reduction which characterises Java, Malaya and Borneo, when late Holocene samples are compared with older samples. This explanation, general body-size reduction, is the one preferred by Brown (1989) for the smaller size of Holocene Murray Valley teeth compared to Coobool Creek. It is also Storm's (1995) proposed explanation for the differences between the Wajak and recent Javanese skulls in their general morphology. However, the available data, sparse as they are, would suggest very large body size in Java right up to the Neolithic (Table 10.10), despite the small teeth as currently documented for Neolithic Java. As for Malaya, Bulbeck (2011) finds evidence for around 10% reduction in stature between the early Holocene and recent times (with reference to the *Orang Asli*). However, according to Bulbeck's data, this stature reduction continued throughout the late Holocene, whereas, as shown here, the Gua Cha Neolithic and Gua Harimau teeth appear no larger than those of the *Orang Asli*. Finally, Zuraina and Pfister (2005) find evidence for a circa 10 cm *increase* in stature of the Niah Neolithic burials compared to the pre-Neolithic Niah burials. While the small stature estimated for the pre-Neolithic Niah burials (154 cm for males, 145 cm for females) is consistent with the modest size of their teeth, Neolithic Niah appears to combine stature increase with tooth-size reduction. In summary, while a broad correlation between tooth-size and general body size would be expected, general body-size reduction fails as an explanation for late Holocene tooth-size reduction in Borneo, Malaya and Java.

With the falsification of the two mooted, local evolutionary explanations for tooth-size changes in late Holocene Indo-Malaysia, the incursion of 'new people' with smaller teeth is left as the default explanation. Although this study was not designed to investigate the source of any such newcomers, the fifth analysis did identify a potential candidate – Early Sulawesi. On the available data, the circa mid-Holocene teeth from Sulawesi are of very typical size by the standards of late Holocene teeth across Indo-Malaysia (Figure 10.1e). The Early Sulawesi teeth also stand at the polar extreme from Australoids in their shape (Figure 10.6), and so would serve as an appropriate precursor for the markedly non-Australoid tooth shape recorded for late Holocene samples from other locations (notably Neolithic and EMP Java, Melanta Tutup, Gua Harimau and Neolithic Niah). The status of Sulawesi as a donor region is reasonable in view of mitochondrial DNA evidence that it was a centre for the dispersal of the E1a, E2a, M7c1c and D5 haplogroups, which together constitute a substantial proportion of recent Indo-Malaysians' mitochondrial DNA (Hill et al., 2007; Soares et al., 2008).

Table 10.10: Comparative data on major limb-bone lengths[A] in Java.

Specimen/series	Femur length (mm)	Tibia length (mm)	Humerus length (mm)	Source
Wajak (male)	Larger than Hoekgroet	Too fragmentary to assess	Too fragmentary to assess	Storm 1995
Song Terus 1 (Mesolithic male)	470 mm (right)	Not reported	Not reported	Détroit 2002:224
Song Keplek 5 (Neolithic female)	Too fragmentary to assess	390 mm (right)	346 mm (left)	Noerwidi 2011-12
Hoekgroet (Neolithic, female?)	~450 mm	~390 mm	Too fragmentary to assess	Storm 1995
Recent Surabaya males	414.3 + 23.0 (right)	348.0 + 21.0 (right)	296.6 + 16.2 (left)	Bergman and The 1955

Note: A. Femur length in natural position (Martin 2), total tibia length (Martin 1) and maximum humerus length (Martin 1) are the definitions employed by Bergman and The (1955), and also by Noerwidi (2011–12). Détroit (2002) and Storm (1995) do not define their measurements.

Source: All listed in notes section.

At the same time, the evidence relating tooth-size reduction to broader cranial shape is ambiguous, at best. Table 10.11 focuses on the 'cranial index' (cranial breadth as a percentage of cranial length) as this is one of the most widely reported observations for prehistoric remains from Indo-Malaysia. The comparison between Wajak and early to mid-Holocene Indo-Malaysians suggests a transition to smaller teeth associated with more elongated, not broader, crania. Late Holocene Indo-Malaysians, for their part, can generally be distinguished from early to mid-Holocene Indo-Malaysians both by their smaller teeth and their broader crania. However, there are exceptions, such as the Puger skull (EMP Java), which combines a narrow braincase with very small teeth.

The dental metrical results outlined here are at variance with recently presented results on the same burial series. Manser (2005) was unable to distinguish between the pre-Neolithic and Neolithic Niah remains on their facial shape, cranial non-metric traits and dental morphology, finding that both resemble recent Southeast Asians, Polynesians and Australian Aborigines. Here, however, we find that the Neolithic teeth from Niah are not only smaller than the pre-Neolithic teeth, as noted by Manser (2005), but also different in shape. As for Gua Cha, Bulbeck (2005a) supported earlier findings of population continuity between the Hoabinhian and Neolithic series, consistent with the decision by Matsumura and Hudson (2004) to pool them into a single 'Gua Cha' sample. However, tooth shape analysis would be compatible with a scenario in which the Gua Cha Neolithic burials reflect admixture between the Gua Cha Hoabinhians and Neolithic immigrants from the north, as represented by Khok Phanom Di (Figures 10.4 to 10.6; cf. Bellwood, 1993). Finally, the Java Mesolithic burials appear to be more homogeneous in their cranial shape (Table 10.11) and their dental metrics, and more distinct from the Java Neolithic burials, than Détroit (2002) inferred – although this discrepancy may reflect Détroit's reliance on a now superseded, pre-Neolithic dating for the critical Song Keplek 5 burial.

Table 10.11: Comparison of Indo-Malaysian tooth-size and cranial index.

Location/Age	Series/specimen	Tooth size[A]	Cranial index[B]	Cranial index data source
Java, Late Pleistocene	Wajak	Very large	Mesocranic (75.5)	Storm 1995
Java, Pleistocene/ Holocene junction	Gua Braholo	Large	Dolichocranic (70.3, 73.9)	Détroit 2002
Malaya, Early Holocene	Gua Peraling 4	Small	Dolichocranic (74.7)	Bulbeck and Adi 2005
Java, mid-Holocene	Song Keplek 4	Large	Mesocranic (77.8)	Détroit 2002
Java, mid-Holocene	Sampung	Large	Dolichocranic	Jacob 1967a
Malaya, mid-Holocene	Gua Cha Hoabinhian	Large	Dolichocranic (72.8, 73.1)	Bulbeck 2005a
Malaya, mid-Holocene	Guar Kepah	Large	Dolichocranic (63.6)	Jacob 1967a
Sulawesi, mid-Holocene	Leang Buidane Pre-ceramic	Medium (Table 10.4)	Dolichocranic (66.8)	Bulbeck 1981
Java, Neolithic	Song Keplek 5	Medium	Brachycranic	Noerwidi 2011–12
Java, Neolithic	Hoekgroet	Small	Brachycranic (80.5)	Storm 1995
Malaya, Neolithic	Gua Cha Neolithic	Small	Mesocranic (76.0), brachycranic (>80)	Bulbeck 2005a
Sulawesi, EMP	Leang Buidane	Medium	Mesocranic	Bulbeck 1981
Java, EMP	Puger	Very small	Dolichocranic (73.9)	Snell 1938
Sulawesi, Recent	Historical Sulawesi	Small	Brachycranic (80.6)	Pietrusewsky 1981[C]
Malaya, Recent	Semang	Small	Mesocranic	Bulbeck and Lauer 2006
Malaya, Recent	Senoi	Small	Mesocranic	Bulbeck and Lauer 2006
Malaya, Recent	Aboriginal Malays	Small	Mesocranic	Bulbeck and Lauer 2006
Java, Recent	Surabaya, Batawi	Small to very small	Brachycranic (81.0)	Pietrusewsky 1981[C]

Notes: A. Tooth-size from this paper except for Gua Peraling 4, from Bulbeck and Adi (2005), and Song Keplek 5, from Noerwidi (2011–12).

B. A cranial index below 75 corresponds to narrow braincases (dolichocrany), while a cranial index above 80 corresponds to broad braincases (brachycrany). Mesocrany (cranial index between 75 and 80) is intermediate.

C. Calculated from the male means for maximum cranial length and cranial breadth.

Sources: Snell 1938; Jacob 1964; Jacob 1967a; Bulbeck 1981; Pietrusewsky 1981; Storm 1995; Détroit 2002; Bulbeck 2004; Bulbeck 2005a; Bulbeck and Adi 2005; Bulbeck et al. 2005; Bulbeck and Lauer 2006; Noerwidi 2011–12; Bulbeck et al. 2013; this paper.

Détroit (2002) hypothesised that Indo-Malaysia was a region of population movements, from the terminal Pleistocene onwards, with links to what is now mainland East Asia to the north and New Guinea (and the Northern Territory) to the southeast. The population history of Indo-Malaysians was more of a swirl in all directions of the compass than a two-layer sequence involving indigenous Australoid foragers and immigrant Neolithic farmers. This hypothesis is consistent with Sulawesi's possible status as a source for the small teeth and non-Australoid tooth shape found widely across late Holocene Indo-Malaysia. Détroit's hypothesis is also consistent with the mtDNA findings of Martin Richards and his associates. These findings include significant population movements from Taiwan into Indonesia and from Mainland Southeast Asia into Malaya associated with the introduction of the Neolithic, but also numerous other dispersals into, within and out from Indo-Malaysia's 'entangled bank' (Hill et al., 2007; Soares et al., 2008; Soares et al., 2011; Bulbeck, 2011).

Colin Groves (1976, 1980) developed a strong interest in the macaques and other mammals of Sulawesi, which he explored in the context of both the natural (Groves, 1984) and human-mediated (Groves, 1985, 1995) dispersal of mammals across Indo-Malaysia. Of particular relevance to this paper are his proposals for a sea-borne dispersal of early dogs through Indonesia into New Guinea, and the eastward transport of the Sulawesi warty pig (*Sus celebensis*) as a domesticate or game animal, prior to the introduction of the dog and pig breeds that could be reasonably associated with the late Holocene migration of Austronesian speakers from Taiwan to Indo-Malaysia. Without reviewing the considerable information now available for human-mediated dispersals of mammals across Indo-Malaysia, we can note that it broadly justifies and indeed extends Colin's open-minded, exploratory perspective on this topic. Colin taught his students to follow the evidence wherever it leads, an approach which this contribution hopefully illustrates.

Conclusion

Analysis of dental metrical data from Indo-Malaysia suggests a three-stage sequence: very large teeth in the Late Pleistocene; large teeth (except for Sulawesi) between the terminal Pleistocene and mid-Holocene; and small teeth during the late Holocene. Evidence for tooth-size reduction between the Neolithic and recent times was hard to find. This falsifies the model that attributes the small size of late Holocene teeth in Indo-Malaysia to changed selection pressures associated with agriculture and cooking in pots. Also, body-size reduction fails as an explanation for late Holocene tooth-size reduction in Borneo, Java and Malaya. Thus, the change appears to reflect the influx of newcomers, whose teeth were not only smaller than those of their mid-Holocene counterparts but also different in 'shape'. Although the study was not designed to investigate the origins of these newcomers, mid-Holocene Sulawesi emerged as a possibility, admittedly based on a small sample with incomplete coverage of the dentition.

Acknowledgements

The author's data collection from Sulawesi skulls in museums was funded by a large Australian Research Council grant to the author and Colin Groves for the 'Contribution of South Asia to the Peopling of Australasia' project. The author's study of teeth from Lake Towuti in Sulawesi was funded by a large Australian Research Council grant (DP 110101357) to Sue O'Connor, Jack Fenner, Janelle Stevenson (Australian National University) and Ben Marwick (University of Washington). The author thanks Marc Oxenham and an anonymous reviewer for their comments on this paper.

References

Anggraeni. 1999. The introduction of metallurgy to Indonesia. MA thesis, The Australian National University. Available from: ANU Library.

Barrett MJ, Brown T, Arato G, Ozols IV. 1964. Dental observations on Australian Aborigines: Buccolingual crown diameters of deciduous and permanent teeth. *Aust Dent J* 9:280–285.

Barrett MJ, Brown T, MacDonald MR. 1963. Dental observations on Australian Aborigines: Mesiodistal crown diameters of permanent teeth. *Aust Dent J* 8:150–155.

Bellwood P. 1976. Archaeological research in Minahasa and the Talaud Islands, northeastern Indonesia. *AP* 19:240–288.

Bellwood P. 1978. *Man's conquest of the Pacific*. Sydney: Collins.

Bellwood P. 1993. Cultural and biological differentiation in Peninsular Malaysia: The last 10,000 years. *AP* 32:37–60.

Bellwood P. 1997. *Prehistory of the Indo-Malaysian Archipelago*. Revised edition. Honolulu: University of Hawai'i Press.

Bergman RAM, The TH. 1955. The length of the body and long bones of the Javanese. *Doc Med Geogr Trop* 7:197–214.

Brace CL. 1976. Tooth reduction in the Orient. *AP* 19:203–219.

Brace CL, Hinton RJ. 1981. Oceanic tooth-size variation as a reflection of biological and cultural mixing. *Curr Anthropol* 22:549–659.

Brace CL, Shao X-q, Zhang Z-b. 1984. Prehistoric and modern tooth-size in China. In: Smith FH, Spencer, F, editors. *The origins of modern humans: A world survey of the fossil evidence.* New York, NY: Alan R. Liss. pp. 485–516.

Brown PJ. 1978. The ultrastructure of dental abrasion: Its relationship to diet. BA Hons thesis, The Australian National University.

Brown P. 1981. Sex determination of Aboriginal crania from the Murray River Valley: A reassessment of the Larnach and Freedman technique. *Archaeol Ocean* 16:53–63.

Brown P. 1989. *Coobool Creek.* Terra Australis 13. Canberra: The Australian National University.

Bulbeck FD. 1981. Continuities in Southeast Asian evolution since the Late Pleistocene. Some new material described and some old questions reviewed. MA thesis, The Australian National University. Available from: ANU Library.

Bulbeck FD. 1982. A re-evaluation of possible evolutionary processes in Southeast Asia since the late Pleistocene. *Bull Indo Pac Pre hi* 3:1–21.

Bulbeck D. 1996–97. The Bronze-Iron Age of South Sulawesi, Indonesia: <ortuary traditions, metallurgy and trade. In: Bulbeck FD, Barnard N, editors. *Ancient Chinese and Southeast Asian Bronze Age cultures.* Taipei: Southern Materials Center Inc. Vol II, pp. 1007–1076.

Bulbeck D. 2001. Human remains from Gua Harimau, West Malaysia. Report to the Department of Museums and Antiquity, Kuala Lumpur, Malaysia. Available from: the author on request.

Bulbeck D. 2004. South Sulawesi in the corridor of island populations along East Asia's Pacific rim. In: Keates SG, Pasveer JM, editors. Quaternary research in Indonesia. *Mod Quat Re* 18. Leiden: A.A. Balkema. pp. 221–258.

Bulbeck FD. 2005a. The Gua Cha burials. In: Zuraina M, editor. *The Perak Man and other prehistoric skeletons of Malaysia.* Penang: Penerbit Universiti Sains Malaysia. pp. 253–309.

Bulbeck D. 2005b. The Guar Kepah human remains. In: Zuraina M, editor. *The Perak Man and other prehistoric skeletons of Malaysia.* Penang: Penerbit Universiti Sains Malaysia. pp. 383–423.

Bulbeck D. 2011. Biological and cultural evolution in the population and culture history of Malaya's anatomically modern inhabitants. In: Enfield N, editor. *Dynamics of human diversity: The case of Mainland Southeast Asia*. Pacific Linguistics 627. Canberra: The Australian National University. pp. 207–255.

Bulbeck D, Adi T. 2005. A description and analysis of the Gua Peraling human remains. In: Zuraina M, editor. *The Perak Man and other prehistoric skeletons of Malaysia*. Penang: Penerbit Universiti Sains Malaysia. pp. 311–343.

Bulbeck D, Kadir RA, Lauer A, Zamri R, Rayner D. 2005. Tooth-sizes in the Malay Peninsula past and present: Insights into the time depth of the indigenous inhabitants' adaptations. *Int J Indigen Res* 1:41–50.

Bulbeck D, Lauer A. 2006. Human variation and evolution in Holocene Peninsular Malaysia. In: Oxenham M, Tayles N, editors. *Bioarchaeology of Southeast Asia*. Cambridge: Cambridge University Press. pp. 133–171.

Bulbeck D, Marwick B, O'Connor S et al. 2013. Archaeology of Lake Towuti: Survey and excavation in South Sulawesi. Poster presented at the Society of American Archaeology Annual Meeting, 3–7 April, Honolulu, Hawaii.

Carlson DS, Van Gerven DP. 1977. Masticatory function and post-Pleistocene evolution in Nubia. *Am J Phys Anthropol* 46:495–506.

Chia S. 2008. Prehistoric sites and research in Semporna, Sabah, Malaysia. *Bulletin of the Society for East Asian Archaeology* 2, www.seaa-web.org/bulletin2008/bul-essay-08-01.htm.

Chia S, Arif J, Matsumura H. 2005. Dental characteristics of prehistoric human teeth from Melanta Tutup, Sabah. In: Zuraina M, editor. *The Perak Man and other prehistoric skeletons of Malaysia*. Penang: Penerbit Universiti Sains Malaysia. pp. 239–251.

Demeter F. 2006. New perspectives on the peopling of Southeast and East Asia during the late Upper Pleistocene. In: Oxenham M, Tayles N, editors. *Bioarchaeology of Southeast Asia*. Cambridge: Cambridge University Press. pp. 112–132.

Détroit F. 2002. Origine et evolution des *Homo sapiens* in Asie du Sud-Est: Descriptions et analyses morphométriques de nouveaux fossils. PhD thesis, Museum National d'Histoire Naturelle. Available from: www.dsifilex.mnhn.fr/get?k=5ZIu9om0ReQJidRUlY9.

Doran GA, Freedman L. 1974. Metrical features of the dentition and arches of populations from Goroka and Lufa, Papua New Guinea. *Hum Biol* 3:583–594.

Groves CP. 1976. The origin of mammalian fauna of Sulawesi (Celebes). *Z Säugetierk* 41:201–216.

Groves CP. 1980. Speciation in *Macaca*: The view from Celebes. In: Lindberg D, editor. *The macaques: Studies in ecology, behavior and evolution.* New York: Van Nostrand Reinhold. pp. 84–124.

Groves CP. 1984. Mammal faunas and palaeogeography of the Indo-Australian region. *Cour Forsch Inst Senckenberg* 69:267–273.

Groves CP. 1985. On the agriotypes of domestic cattle and pigs in the Indo-Pacific region. In: Misra VN, Bellwood P, editors. *Recent advances in Indo-Pacific prehistory.* Leiden: EJ Brill. pp. 429–438.

Groves CP. 1989. A regional approach to the problem of the origin of modern humans in Australasia. In: Mellars P, Stringer C, editors. *The human revolution: Behavioural and biological perspectives on the origins of modern humans.* Princeton: Princeton University Press. pp. 274–285.

Groves CP. 1995. Domesticated and commensal mammals of Austronesia and their histories. In: Bellwood P, Fox JJ, Tryon D, editors. *The Austronesians: Historical and comparative perspectives.* Canberra: The Australian National University. pp. 161–173.

Hanihara T. 1994. Craniofacial continuity and discontinuity of Far Easterners in the late Pleistocene and Holocene. *J Hum Evol* 27:417–441.

Hill C, Soares P, Mormina M, et al. 2007. A mitochondrial stratigraphy for Island Southeast Asia. *Am J Hum Genet* 80:29–43.

Howells WW. 1973. *The Pacific Islanders.* London: Weidenfeld & Nicolson.

Jacob T. 1964. A human mandible from the Anjar urn field, Indonesia. *J Natl Med Assoc* 56:421–426.

Jacob T. 1967a. *Some problems pertaining to the racial history of the Indonesian region.* Utrecht: Netherlands Bureau for Technical Assistance.

Jacob T. 1967b. Racial identification of the Bronze Age human dentitions from Bali, Indonesia. *J Dent Res* 46:903–910.

Joliffe IT. 2002. *Principal components analysis.* Second edition. New York, NY: Springer.

Manser JM. 2005. Morphological analysis of the human burial series at Niah Cave. PhD thesis. Available from: UMI Dissertation Services, Ann Arbor, MI.

Matsumura H, Hudson MJ. 2005. Dental perspectives on the population history of Southeast Asia. *Am J Phys Anthropol* 127:182–209.

Noerwidi S. 2011–12. The significance of the Holocene human skeleton Song Keplek in the history of human colonization of Java: A comprehensive morphological and morphometric study. MA thesis, Museum National d'Histoire Naturelle. Available from: www.hopsea.mnhn.fr/pc/thesis/M2Noerwidi_5.pdf.

Pardoe C. 1988. The cemetery as symbol: The distribution of prehistoric Aboriginal burial grounds in southeastern Australia. *Archaeol Ocean* 23:1–16.

Penrose LS. 1954. Distance, size and shape. *Ann Eugen* 18:337–343.

Pietrusewsky M. 1981. Cranial variation in Early Metal Age Thailand and Southeast Asia studied by multivariate procedures. *Homo* 32:1–26.

Pietrusewsky M. 2006. A multivariate craniometric study of the prehistoric and modern inhabitants of Southeast Asia, East Asia and surrounding regions: A human kaleidoscope? In: Oxenham M, Tayles N, editors. *Bioarchaeology of Southeast Asia*. Cambridge: Cambridge University Press. pp. 59–90.

Snell CARD. 1938. *Meschelijke skeletresten uit de duin formatie van Java's zudkust nabij Poeger* (Z. Banjoewangi). Surabaya: G. Kolff.

Soares P et al. 2008. Climate change and post-glacial human dispersals in Southeast Asia. *Mol Biol Evol* 25:1209–1218.

Soares P et al. 2011. Ancient voyaging and Polynesian origins. *Am J Hum Genet* 88:239–247.

Storm P. 1995. The evolutionary significance of the Wajak skulls. *Scripta Geol* 110.

Storm P et al. In press. U-series and radiocarbon analyses of human and faunal remains from Wajak. *J Hum Evol*.

Tayles NG. 1999. *The excavation of Khok Phanom Di: A prehistoric Site in Central Thailand. Volume V: The people*. London: The Society of Antiquaries.

Tieng FS. 2010. Hoabinhian rocks: An examination of Guar Kepah artifacts from the Heritage Conservation Centre in Jurong. MA thesis, National University of Singapore. Available from: www.scholarbank.nus.edu.sg/bitstream/handle/10635/20404/FooST.pdf?sequence=1.

Turner CG II. 1983. Sundadonty and Sinodonty: A dental anthropological view of Mongoloid microevolution, origin, and dispersal into the Pacific basin,

Siberia and the Americas. In: Vasilievsky RS, editor. *Late Pleistocene and early Holocene cultural connections of Asia and America.* Novosibirsk: USSR Academy of Sciences, Siberian Branch. pp. 147–157.

Zuraina M, Pfister L-A. 2005. The Niah collection of 122 skeletons at the University of Nevada. In: Zuraina M, editor. *The Perak Man and other prehistoric skeletons of Malaysia.* Penang: Penerbit Universiti Sains Malaysia. pp. 155–173.

11. Variation in the Early and Middle Pleistocene: The phylogenetic relationships of Ceprano, Bodo, Daka, Kabwe and Buia

Debbie Argue

Introduction

Despite the increased number of hominin fossils available for the period from one million years ago to c.600 ka clarity about their phylogenetic relationships has not emerged. This is because, while studies of each of these hominins typically include comparative analyses with similar fossil material, in most cases this has resulted in controversy as to their affinities and phylogenetic relationships. Variation during this period is explained by some as representing a single species, *H. erectus*. For others it represents multiple taxa among which *H. erectus* is an exclusively Asian species; or a modified version of this in which observed variation is viewed as continual remodelling of the vault and face that does not involve speciation events (Mbua and Bräuer, 2012). As well, the morphological boundaries of *H. erectus* continue to be stretched. Asfaw and others (2002) referred the Daka cranium (Middle Awash, Ethiopia dated c.800 ka years ago) to *H. erectus* despite clear differences in cranial characters from *H. erectus s.s.*, and concluded from cladistic analyses that its morphology is consistent with the hypothesis of a widespread polymorphic and polytypic species existing one million years ago representing a single evolving lineage series of *Homo erectus* fossils in Africa. Ascenzi and others (1996) referred the Ceprano cranium (Italy), at the time thought to be c.800 ka but recently dated to c.450 ka (Muttoni et al., 2009), to *H. erectus* while acknowledging differences from *H. erectus s.s.* such as a larger endocranial volume, no sagittal keel or parasagittal depression. Bodo (600 ka) and Kabwe (date unknown) from Africa are viewed as more derived than Early Pleistocene *Homo*. Kabwe has been placed in a separate species, *H. rhodesiensis*, or as a subspecies of *H. sapiens*, *H. sapiens rhodesiensis*, in recognition of some perceived relatively modern characters; and Bodo has also been placed in *H. sapiens rhodesiensis*.

There are two other hominins known from this period. The Buia cranium from Eritrea (Abbate et al., 1998) dated to 992 ka (Albianelli and Napoleone, 2004) has not been fully described and was unavailable for study at the time this research was undertaken. There are, however, morphological descriptions from which similarities and differences with its contemporaries, Ceprano, Daka, Kabwe and Bodo, may be observed. Secondly, ATD6-69 (partial face) and the frontal (ATD6-15) from the same individual, a child, from Gran Dolina, Spain dating to 780–857 ka (Falguères et al., 1999) has been placed in a new species, *Homo antecessor* (Bermúdez de Castro et al., 1997). Few of the characters used in the present study are available for this hominin. It proved quite unstable in the phylogenetic analyses I performed earlier (Argue, 2010) and its relationships indeterminable. I, therefore, do not include it in these analyses.

I use cladistic analyses incorporating Early and Middle Pleistocene fossil crania from Africa and Asia to test existing hypotheses about the phylogenetic relationships of Ceprano, Daka, Kabwe and Bodo to propose an alternative hypothesis for human evolution during the period c.1 Ma–600 ka.

Background

Daka

The Daka cranium, BOU-VP-2/66, (Ethiopia), was found *in situ* in sediments with a basal 40Ar/39Ar age of 1.042 ± 0.009 Ma; the sediments are reverse polarity and their minimum age is therefore estimated to be c.0.8 Ma (Asfaw et al., 2002). The cranium is well preserved although it has some distortion. Its discoverers concluded from cladistic analyses that it is *Homo erectus* (Asfaw et al., 2002). Manzi and others (2003), however, using a phenetic approach which quantifies overall similarity of single specimens, found that Daka shares the greatest affinities with two fossil specimens from the Koobi Fora region in Africa, KNM-ER 3733 and KNM-ER 3883, which they attributed to *H. ergaster*, and that Daka is very different from *H. erectus*. They proposed that Daka is best viewed as part of a local African evolutionary lineage spanning 1.8 Ma–c.1 Ma.

Ceprano

Many fragments of a cranium were found near the town of Ceprano, Italy, in 1994 (Ascenzi et al., 1996). Originally dated to >700 ka (Ascenzi et al., 1996), further studies have proposed two younger dates. Muttoni and others (2009) and Manzi and others (2010) report that the level that yielded the hominin cranium has an age of c.0.45 Ma; while a date of 0.35 ± 4 Ma (Nomade et al., 2011) based on 40Ar/39Ar dating on K-feldspars retrieved from the sediments that hosted the skull has also been proposed.

Ceprano was at first referred to *H. erectus*, particularly to late *H. erectus*, by which Ascenzi and others (1996) were referring to the Middle Pleistocene fossils Arago, Petralona, and contemporaries which, as they acknowledge, some would attribute to *H. heidelbergensis*. Clarke (2000) undertook a reconstruction of the cranium during 1997 which resulted in a revision of its reported metric values. Although these changes altered a number of characteristics of the cranium, Clarke (2000) retained it in *H. erectus*; this was supported by Ascenzi and others (2000) based upon their comparison of the character states observed on Ceprano with the distinctive features of *H. erectus* listed by Wood (1991: Table 2.11, p.37). Manzi and others (2001), after declaring their confidence in the new reconstruction, calculated phenetic distances using two methods (Unweighted pair group method (UPGMA); Neighbour Joining (NJ)) to generate unrooted phylogenetic trees. The UPGMA method yielded a tree in which Ceprano is grouped with the African Middle Pleistocene sample (Kabwe, Bodo, Saldhana) and the NJ tree shows Ceprano in an isolated position but nevertheless closer to the African Middle Pleistocene group than to Sangiran (Indonesian) and Zhoukoudian (China) *H. erectus*, Dmanisi, and other Early Pleistocene African crania. Mallegni and others (2003) performed a cladistic analysis using 30 cranial characters yielding eight equally parsimonious trees from which a strict-consensus tree showed Ceprano and Daka in a monophyletic (sister taxa) group with 84% bootstrap support although they found only one unambiguous synapomorphy (short cranial vault) and one ambiguous synapomorphy (presence of sharply angulated occipital profile) for the clade. They proposed a new species for Ceprano, *H. cepranensis sp. nov.*, based upon their assessment that it possesses a unique suite of characters, Beyond a brief discussion of shared characters, they did not engage in a discussion of the apparent close relationship of Ceprano and Daka.

Ceprano, then, has been referred to 'late *H. erectus*' (= *H. heidelbergensis*) (Ascenzi et al., 1996); *H. erectus* (Clarke, 2000; Ascenzi et al., 2000), and specifically *not H. heidelbergensis* (Clarke, 2000); a new species *H. cepranensis sp. nov.* (Mallegni et al., 2003); and, possibly, *H. heidelbergensis* (Manzi et al., 2001).

Bodo

Bodo is a partial cranium (Bodo d'Ar, Ethiopia) estimated to be 600 ka based on biostratigraphic and archaeological considerations (Clarke et al., 1994). In their original announcement, Conroy and others (1976) refrained from making a taxonomic determination. Later, Kalb and others (1982) assigned Bodo to *H. sapiens rhodesiensis,* including it in a taxon with Kabwe (*H. rhodesiensis*; Smith Woodward, 1921); Stringer (1984) conditionally compared Bodo to *H. erectus s.s.*, although he recognised the possible phylogenetic significance of some *H. sapiens* features of this cranium; Groves (1989) attributed Bodo to a subspecies of *H. sapiens*, as did Adefris (1992) in his dissertation on this fossil (although he preferred the term 'archaic *Homo sapiens*'); Rightmire (1996) undertook a detailed description and comparative analysis of the Bodo cranium, concluding that it seems most reasonable to group it with Kabwe and similar specimens from the Middle Pleistocene sites in Africa and Europe.

Kabwe

Kabwe 1 was found during mining operations in the basal wall of a steeply sloping cleft emanating from a cave within a small hillock at Broken Hill, Zimbabwe (then Rhodesia). It has not been dated, and, as it seems to have rolled down the cleft at an unknown time, and was annually inundated by a high water table (Hrdlicka, 1930), so attempts to date it, particularly by using the Electron Spin Resonance method, would be compromised. Further, the hillock no longer exists, having been completely mined. It has not, then, been possible to reliably estimate Kabwe's age. It was originally attributed to a new species, *H. rhodesiensis* by Smith Woodward (1921) who viewed it as quite different from *H. neanderthalensis* and *H. erectus*. Having compared the Broken Hill skull to those of *H. neanderthalensis*, Omo I and II, Hopefield, and OH 9, Rightmire (1976) gave Kabwe a subspecific designation *H. sapiens rhodesiensis* that probably evolved from local groups of *H. erectus*.

Bräuer (1984) assigned Kabwe to 'early archaic *H. sapiens*'[1] in a group that included Bodo, Hopefield, Eyasi, Ndutu and other African crania, as the cranial vaults are more expanded than *H. erectus*. Groves (1989) placed Kabwe in a subspecies of *H. sapiens*, *H. sapiens heidelbergensis*, that includes the African and European Middle Pleistocene fossils; Kabwe, Bodo, Tighenif and later fossils from Europe and Africa.

Daka, Ceprano, Bodo and Kabwe are, then, each variously attributed to a range of species. The objective is to resolve the phylogenetic position of each so that we may generate hypotheses concerning human evolution during his period.

Materials and methods

Information about cranial characteristics was obtained from original fossil material and casts of Early and Middle Pleistocene *Homo* so that cladistic analyses could be performed.

---

1   The term 'early archaic H. sapiens' is now considered unsatisfactory, being a descriptive category rather than a taxonomic term, and has been generally replaced with H. heidelbergensis; the taxon is considered to comprise similar hominins from Africa and Europe.

The comparative sample comprises:

- Sangiran 2, 4, 9, 17 and Trinil (*H. erectus*);
- KNM-ER 1813, OH 24 (*Homo habilis*; casts);
- KNM-ER 3733;
- KNM-ER 3883;
- Daka;
- Ceprano;
- Bodo; and
- Kabwe (*H. rhodesiensis*).

I compiled 89 cranial character states (Appendix 1) of which 62 are phylogenetically informative for this set of OTUs (Operational Taxonomic Units). Character states are derived from Lahr (1996), and Schwartz and Tattersall (2002); and Zeitoun (2000) who referred to Weidenreich (1943), MacIntosh and Larnach (1972), Sartono and Grimaud (1983), Grimaud (1982), and Hublin (1978).

Cladistic analysis is widely used in the biological sciences as a methodological approach to phylogenetic reconstruction and has been applied to hominin taxa since the 1970s. It assumes that shared features observed among taxa can be explained by hypotheses of common ancestry that are represented by sets of characters in a hierarchical pattern of taxa (Faith and Cranston, 1991) and is based upon Hennig's (1966) approach to systematics, specifically his approach to descent with modification. Descendants acquire traits transmitted genetically from their ancestors and these are passed on to subsequent descendants (Humphries, 2002). The aim of cladistic analysis is to identify taxa that share a common ancestor by finding, or distinguishing, shared derived character states (synapomorphies) from among the characters in the data set. Cladistic analysis produces possible phylogenetic trees, called cladograms, which are branching diagrams that depict sister group relationships. The cladogram groups OTUs into clusters called clades, and these represent hypotheses about relationships among OTUs. Cladistic analysis is based upon the total number of character changes necessary to support the relationship of OTUs in a tree. The shortest trees are those that account for the observed differences among taxa in the smallest number of evolutionary steps. They are the most parsimonious trees and are generally considered to present the best working hypotheses (after Argue et al., 2009). The phylogenetic trees presented here are derived from my data and analyses.

Before performing a cladistic analysis it seems useful to assess the probability that the phylogenetic trees derived from the data could have arisen by chance alone (Faith, 1991). To test for this I performed a Permutation Tail Probability Test (PTP) in which a 'p' value of 0.05 or less would indicate that the null hypothesis, that the data are random, is rejected while noting that the meaning of a 'p' value of >0.05 is controversial (Carpenter et al., 1998; Källersjö et al., 1992; Trueman, 1993; Fu and Murphy, 1999). For this data and taxon set, the PTP value (excluding the outgroup) is $p = 0.01$ and I consider that these data therefore represent a non-random tree structure.

I perform a heuristic search of the cranial data using PAUP* (Phylogenetic Analysis Using Parsimony) Version 4.0b10 for Macintosh (Swofford, 2002), to find the most parsimonious tree or trees. I examine the relative strength of the clades identified by using 1,000 replications of the PAUP bootstrap test. The bootstrap analysis begins by creating a number of pseudo replicate data matrices from the original dataset by resampling the original data matrix and creating a matrix of the same size as the original. Any data series might be represented in any pseudo matrix once, twice, many times, or none at all. The frequency at which a set of clades appear in these reiterations constitutes the bootstrap score. If a particular clade has a high number of characters supporting it, and few characters refuting its monophyly, the chances are that at least some of these will appear in the resampling process and the group will appear in many of the trees generated. If evidence for the monophyly is weak or there is a high level of homoplasy (similar character state in more than one taxon derived from mechanisms other than immediate shared ancestry) in the original matrix, the group might not appear at all, or at low frequency; and the probabilities obtained are not the probability of the reality of clades as such, but reflect the relative support of the clades, given the assumptions in the technique (Wiley and Liebermann, 2011).

The most parsimonious trees found in the initial PAUP search are transposed into MacClade (Maddison and Maddison, 1982). MacClade provides an interactive environment for exploring phylogeny and was developed to help biologists explore relationships between OTUs (Maddison and Maddison, 1982). In MacClade's tree window, phylogenetic trees (cladograms) can be manipulated and alternative hypotheses for an individual taxon, or groups of taxa, may be explored. I, therefore, reproduce the shortest tree produced in the initial analysis (that used PAUP*) in MacClade so that I can test hypotheses that have been presented for each OTU by constraining relevant OTUs and comparing the ensuing tree lengths with the length of the shortest tree. I can test, for example, the likelihood that Daka shared a common ancestor with Ceprano or *H. erectus*; or if Kabwe and Bodo likely shared a common ancestor.

Clades are further tested using a topology-dependent permutation tail probability test (T-PTP); this tests the support for clades, or sister taxa, shown in the cladogram (Faith and Cranston, 1991; Faith, 1991). The test is defined as the estimate of the proportion of times that a given clade can be found, generated from permuted data that are as short as, or shorter than, the original tree. That is, it compares the degree of corroboration for the observed data to that expected by chance alone, so is a test of monophyly of selected nodes. I reject the null hypothesis, that the data have no cladistic structure beyond that produced by chance, at the 0.05 level if fewer than five out of 100 of the trees have a length as short, or shorter, than the cladogram, that is, if the T-PTP result is ≥0.05.

Results

Shortest trees

There are three shortest trees 245 steps long (Figure 11.1). In each case, Daka, Ceprano and Bodo form a branch but its configuration varies; in one tree these OTUs form a branch with *H. rhodesiensis* (Tree 3).

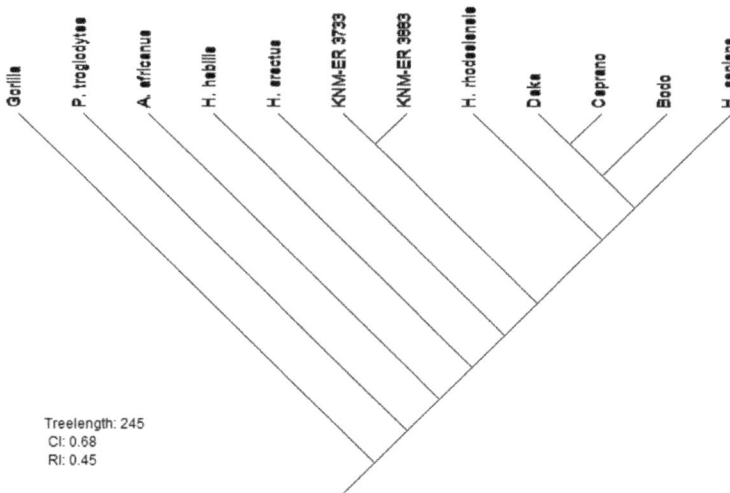

Treelength: 245
CI: 0.68
RI: 0.45

Figure 11.1a: Three shortest trees.

Source: Author's calculations.

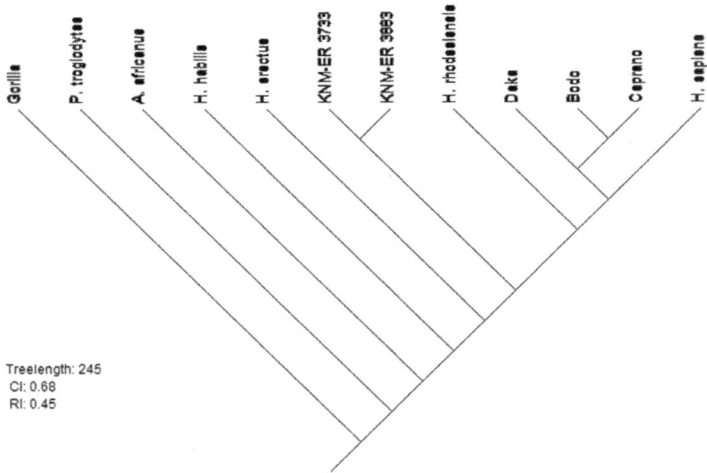

Figure 11.1b: Three shortest trees.

Source: Author's calculations.

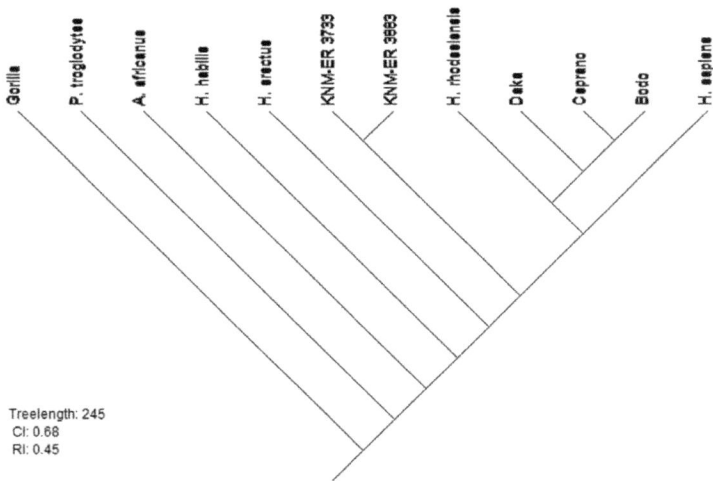

Figure 11.1c: Three shortest trees.

Source: Author's calculations.

Bootstrap analysis

The bootstrap analysis (Figure 11.2) shows two clades with >50% support. The value for the clade comprising Bodo and Ceprano is 54%; for the Koobi Fora group it is 61%. Neither of these results reaches the threshold of 70% (Hillis and Bull, 1993). This would suggest that either there is a high level of homoplasy in the data or evidence for the clades is not strong (or both). The Consistency Index (CI) for these data is 0.68 where a value of 1.0 indicates that there is no homoplasy. This CI suggests that there is indeed likely to be homoplasy in these data but one might also suppose that the support for the clades is not particularly strong.

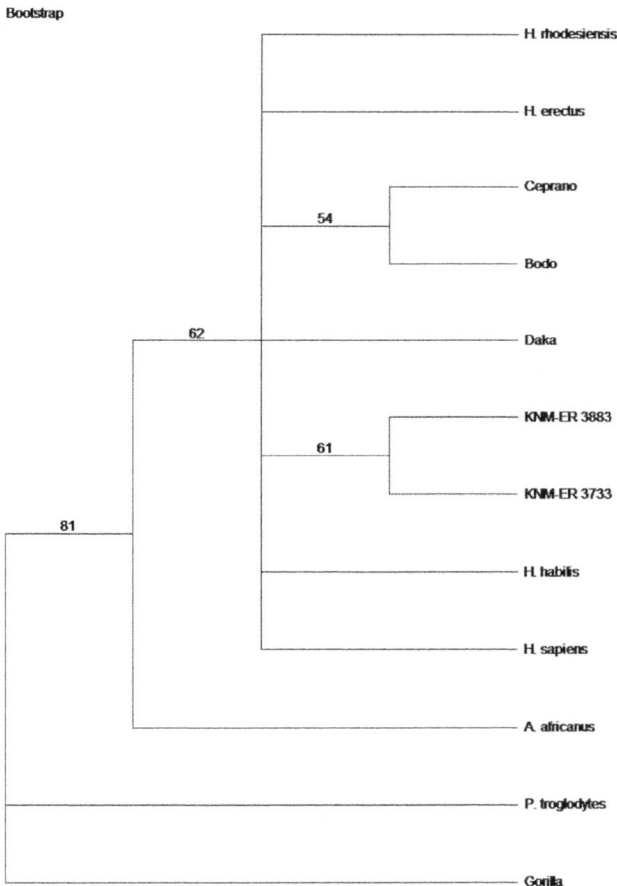

Figure 11.2: Bootstrap analysis.

Source: Author's calculations.

I had earlier tested our data for randomness using the PTP test (above), which led me to conclude that the data is not random and represents phylogenetic structure. I, therefore, am satisfied that the results provide a valid basis upon which to test hypotheses and I treat the clades as representing phylogenetic relationships. Nevertheless I test for the likelihood that each clade on the shortest tree is monophyletic as opposed to occurring by chance alone.

Testing clades and branches

Bodo, Daka and Ceprano

As Bodo and Ceprano form a clade in two of the shortest trees I test the likelihood that this apparent sister taxon relationship occurred by chance alone. The T-PTP result is $p = 0.04$ indicating that Bodo and Ceprano do not group by chance alone.

The clade shares four possible[2] synapomorphies:

- Post-orbital lateral depression (a depression on the lateral supraorbital region bounded by the temporal line)
- Weak metopic keeling
- Anteroposterior width of mandibular fossa is narrow
- The posterior edge of the tuberculum articular in *norma basilaris* is arched.

In another of the shortest trees, Ceprano and Daka form sister taxa. The T-PTP for the clade is, however, $p = 0.19$. This would suggest that this clade would come together by chance alone. In fact, the clade shares only two possible synapomorphies: the articular eminence is higher relative to posterior wall of glenoid fossa; and each has an angular torus. In comparison, Mallegni and others' (2003) analysis shows a strongly supported (84%) Daka-Ceprano clade, notwithstanding that the OTUs share only two synapomorphies: a short cranial vault; and the presence of a sharply angled occipital profile. Based upon our T-PTP result, we would reject Tree 1 (Figure 11.1) as the most parsimonious solution to the data.

The monophyly for Ceprano, Bodo and Daka, however, is not rejected; the T-PTP is $p = 0.04$; that is, it is unlikely that this clade would form by chance alone. If we accept the T-PTP result above for Ceprano and Daka, rejecting it as a viable clade within the branch, the most parsimonious solution is that Bodo and Ceprano are sister taxa; and Daka shares a common ancestor with each of them. Ceprano, Bodo and Daka share the following five possible synapomorphies:

---

2 A particular character state might occur in taxa that are not included in this analysis, so we cannot say categorically that a given state is uniquely synapomorphic for the sister taxa in this study.

- The frontal edge is convex anteriorly in *norma verticalis*
- The supraorbital torus is interrupted in the medial zone, forming two 'mono-tori'
- Temporal squama low in relation to vault
- Posterior part of tympanic joins anterior part of mastoid process
- Postglenoid process does not extend laterally beyond extent of tympanic.

Daka, Ceprano, Bodo, *H. rhodesiensis*

The T-PTP for the branch comprising Bodo, Ceprano, Daka and *H. rhodesiensis* is also p = 0.04. It shares five synapomorphies:

- Depression at glabella
- Angulation between the pre-glenoid planum and the posterior slope of the articular tuberculum (homoplasy with *A. africanus*)
- The supraorbital margin is thick, rounded and not demarcated from the roof of the orbit
- A pre-glenoid planum precedes the glenoid cavity (homoplasy with *A. africanus* and *H. erectus*)
- A very prominent temporal band on the frontal (homoplasy with the two Koobi Fora hominins).

Before presenting hypotheses for human evolution during this period I test hypotheses previously presented for Daka, Bodo, and Ceprano. In this case, I use the phylogeny in Tree 3 (Figure 11.1) in which Bodo, Ceprano and Daka are on a branch with *H. rhodesiensis*, as the analyses so far would suggest that this is the most parsimonious solution to the data.

Testing other hypotheses for OTUs

Bodo

Bodo has been named *H. sapiens rhodesiensis* (Kalb et al., 1982); a subspecies of *H. sapiens* (Groves, 1989; Adefris, 1992) and, conditionally, *H. erectus* (Stringer, 1984). The shortest trees in this analysis show Bodo on a branch with, amongst others, *H. rhodesiensis*.

Bodo and *H. rhodesiensis*

To argue *H. rhodesiensis* subspecies status for Bodo we would expect the tree in which these OTUs are constrained to be as short, or nearly so, as the shortest tree for this set of OTUs. When Bodo is constrained to form a clade with *H. rhodesiensis*, however, the shortest trees length is 253 steps (Figure 11.3). This is eight steps longer than the shortest tree (L = 245); and the CI has decreased to 0.66. Such a tree, then, is a less parsimonious explanation of the data. Bodo is unlikely to be a subspecies of *H. rhodesiensis*.

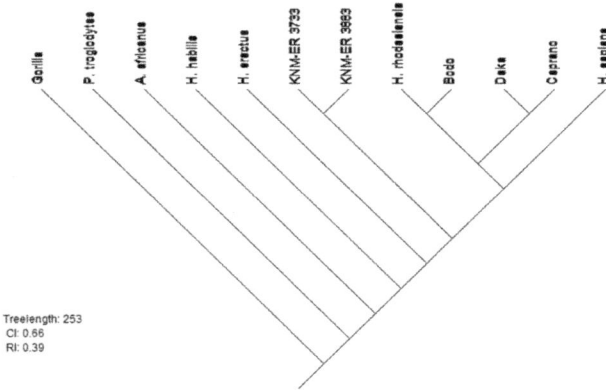

Figure 11.3: Bodo and *H. rhodesiensis*.

Source: Author's calculations.

Bodo and *H. erectus*

The shortest tree in which Bodo and *H. erectus* are constrained to form a clade is L = 249 (Figure 11.4), four steps longer than the shortest tree (L = 245) for Bodo. The T-PTP for the constrained clade is p = 0.18. This is, therefore, also an unlikely solution for Bodo.

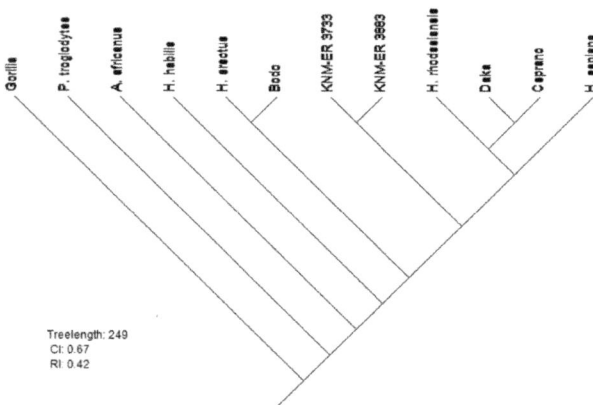

Figure 11.4: Bodo and *H. erectus*.

Source: Author's calculations.

Daka

Daka and *H. erectus*

Asfaw and others (2002) proposed that Daka is *H. erectus*. When this hypothesis is tested in the present study by constraining Daka to form a clade with *H. erectus* the shortest tree length is 251 (Figure 11.5); this is six steps longer than the most parsimonious tree (L = 245), and the T-PTP is p = 0.27; a phylogeny in which Daka and *H. erectus* form sister taxa is a less parsimonious solution for Daka than that identified in the shortest trees.

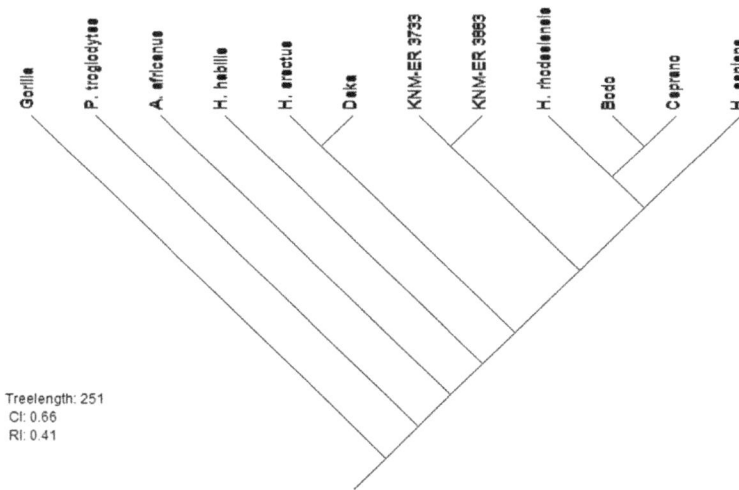

Treelength: 251
CI: 0.66
RI: 0.41

Figure 11.5: Daka and *H. erectus*.

Source: Author's calculations.

Daka and KNM-ER 3733 and KNM-ER 3883

I explore the possibility for a chronological and anatomical morphocline from the Koobi Fora hominins KNM-ER 3733 and KNM-ER 3883 to Daka/Buia as proposed by Asfaw and others (2002) by testing if it is possible that Daka shared an immediate common ancestor with KNM-ER 3733 and KNM-ER 3883. The outcome (Figure 11.6) is a tree nine steps longer than the most parsimonious; it is most unlikely Daka and the Koobi Fora hominins are phylogenetically closely related.

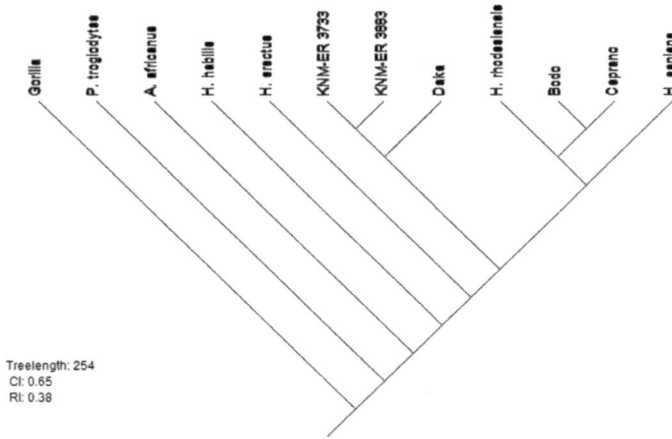

Figure 11.6: Daka and KNMs.

Source: Author's calculations.

Ceprano

Ceprano and *H. erectus*

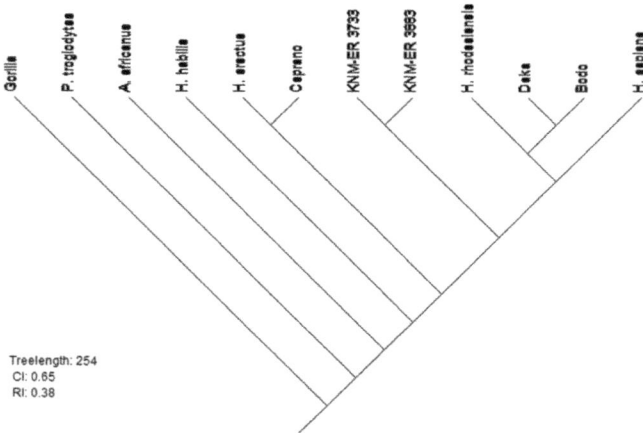

Figure 11.7: Ceprano and *H. erectus*.

Source: Author's calculations.

Ascenzi and others (2000) attributed Ceprano to *H. erectus* based upon a comparison of Ceprano's characteristics with a list of *H. erectus* characteristics compiled by Wood (1991). I tested for a possible phylogenetic relationship between Ceprano and *H. erectus s.s.*, The shortest tree that includes the clade {Ceprano, *H. erectus*} is 254 steps (Figure 11.7), nine steps longer than the shortest tree (L = 245); it is difficult to argue, then, that Ceprano shared an immediate common ancestor with Sangiran *H. erectus*.

Buia

Although Buia is unavailable for study, there is considerable published morphological information (Abbate et al., 1998; Macchiarelli et al., 2004). The cranium is estimated to date from 992 ka (Albianelli and Napoleone, 2004) and is thus close in geological age to Daka (1.042 ± 0.009 Ma; Asfaw et al., 2002). Buia comprises a cranium, a large part of the facial skeleton, the base (Macchiarelli et al., 2004); and a left symphysis (Bondioli et al., 2006). Preliminary descriptions (Abbate et al., 1998; Macchiarelli et al., 2004) indicate that the braincase is very long (204 mm) compared to its width (130 mm), and is relatively high (Abbate et al., 1998).

Buia and Daka share many similarities. While the Buia cranium is longer than Daka (180 mm; Asfaw et al., 2002), both have an endocranial volume (ECV) of 995 cc (Daka: Asfaw et al., 2002; Buia: Macchiarelli et al., 2004). In lateral view the frontal profiles are rounded and rise relatively steeply from the supraorbital sulcus; the occipital profiles are rounded with an incipient bun. In frontal profile, both crania are widest inferiorly and have relatively straight-sided parietal walls but Buia's lateral walls converge inferiorly (Macchiarelli et al., 2004); reminiscent of Ceprano. Both Buia and Daka have reduced post-orbital constriction. The only section of the supraorbital available for Buia, the right lateral half, closely matches the form of the same region on Daka. There are, then, a number of phenetic similarities between these almost contemporaneous *Homo* that lived 600 km apart in the Danakil Depression, making it difficult to argue that they are from separate populations. A more detailed comparative analysis may show otherwise, of course, when Buia is available for study.

Buia and Ceprano also share a number of similarities: the parietals converge slightly inferiorly (Buia: Macchiarelli et al., 2004; Ceprano: Ascenzi et al., 1996: 419; pers. obs.); they have a small depression on the same area laterally on the front of the supraorbitals (Ceprano: pers. obs.; Buia: Macchiarelli et al., 2004, Fig 1); on both the frontals rise steeply; supraorbitals are interrupted at glabella (Buia: Macchiarelli et al., 2004); they have reduced post-orbital constriction; mastoid processes are short and broad; there are only modest external occipital protrusions, and slight angular tori. Ceprano has a slightly greater ECV, of 1185 cc (72 cc larger than Buia and Daka). They differ, however, in that the temporal lines on Buia disappear early on the parietals, whereas Ceprano's temporal lines continue to asterion; Buia does not have an occipital torus (Macchiarelli et al., 2004), whereas Ceprano does; glabella is in a forward position on Buia (Macchiarelli et al., 2004), while this area is depressed on Ceprano; and Buia has an occipital 'bun' (Abbate et al., 1998, Figure 2b), which is absent on Ceprano.

Discussion

Identifying relationships between taxonomic units is of critical concern to the study of *Homo*. Cranial analyses were undertaken so that predictions could be made about the phylogenetic relationships between Early and Middle Pleistocene hominins Ceprano, Daka, Bodo and Kabwe.

Ceprano, Bodo, Daka and Kabwe form a supported branch in which Ceprano and Bodo appear to be the more derived taxa. Groves (1989) had also found Kabwe, Bodo and others fitted a common pattern and did not comprise a western variant of *H. erectus*; he attributed the group to *H. sapiens heidelbergensis*.

I hypothesise that Ceprano, Bodo, Daka and Kabwe form a species for which the prior available name is *H. rhodesiensis* (Smith Woodward, 1921). In this study *H. rhodesiensis* is characterised by an angulation between the pre-glenoid plane and the posterior slope of the articular tuberculum; a depression at glabella; a supraorbital margin that is thick, rounded and not demarcated from the roof of the orbit; a pre-glenoid plane anterior to the glenoid fossa; and a very prominent temporal band on the frontal.

While Buia was unavailable and could not be included in the cladistic analysis, a comparative assessment shows that it is very similar to Daka, and, to a slightly lesser extent, Ceprano. While I would propose from the discussion above that Buia is closely related to Daka at least, further analyses need to be undertaken when this cranium is fully described.

H. rhodesiensis thus composed is not closely related to *H. erectus*. *H. erectus* forms a separate lineage in the most parsimonious trees in which Daka, Ceprano and Bodo do not belong. Nor can Daka be shown to belong in the same species as the Koobi Fora hominins.

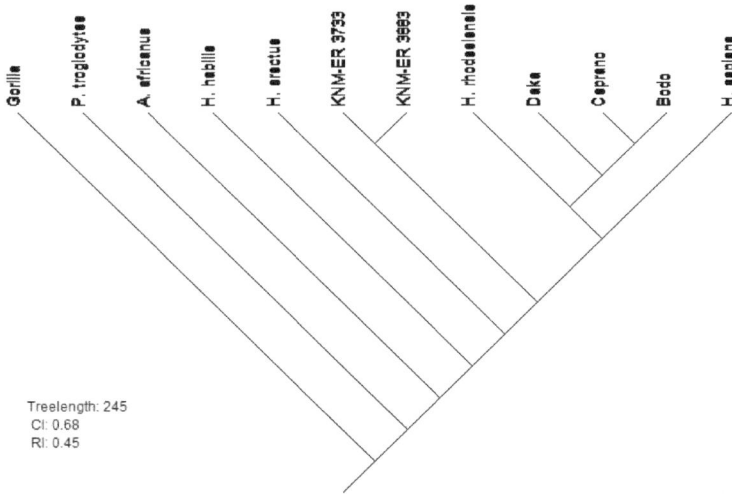

Figure 11.8: Preferred phylogeny.

Source: Author's calculations.

The key development in the evolution of the clade exemplified by Daka, Bodo, and Ceprano is a marked expansion of the vault, and, in Ceprano and Buia, the inferior contraction in the sides of the vault suggesting brain expansion during this period. Brain expansion in *Homo* is associated with a number of cranial characters: parietal bossing; high contour of the temporal squama; rounding of the occiput; and a relatively steeply sloping frontal (Rightmire, 1996). Kabwe has the first two of these characters, and a relatively large cranium with an ECV of 1300 cc (Holloway et al., 2004: 120). While Ceprano, Daka and Bodo differ in some aspects of cranial shape, they nevertheless represent an evolutionary shift in the overall form of the braincase, a shift we can also observe in Buia. Like Kabwe, vault walls are nearly parallel in Daka, but converge slightly inferiorly in Ceprano and Buia; and postorbital constriction is reduced in all. Ceprano and Daka are more derived than *H. rhodesiensis* in that their frontals rise more steeply, and they have rounded occipitals. Ceprano also has a relatively high temporal squama but does not have parietal bossing. Daka, too, lacks some of the characters associated with cranial expansion, such as a relatively high temporal squama. Although Bodo has a relatively large cranial capacity (1250 cc; Holloway et al., 2004) it does not show some of the other characters associated with an expanded braincase such as parietal bossing, greatest width at the parietals, or reduced post-orbital constriction, although its occipital seems to be rounded and it has a relatively high temporal squama contour. That is, although Bodo, Daka, and Ceprano all show marked cranial expansion in ECVs and related characters, none show all characters.

Buia shares many characteristics with Daka and Ceprano and I hypothesise that there is a close phylogenetic relationship between them. Further, Buia and Daka are the same age and, arguably, the populations from which they came lived in geographical proximity.

Conclusions

The paradigm for human evolution in the late Early Pleistocene–Middle Pleistocene comprised a model in which one species, *H. erectus*, is present. As more hominin crania were discovered, for the most part they were placed in this species, requiring an ongoing broadening of the definition of *H. erectus* to accommodate the increasing variation in cranial morphology the newer discoveries presented. This is particularly so for the larger-brained crania Daka, Ceprano, and Bodo.

When these are assessed using cladistic analyses, in which sister-taxon relationships are sought, this group of hominins formed a supported branch that included *H. rhodesiensis*. This species shares a number of synapomorphies, but of equal importance is the evidence for expansion of the brain; and, for the first time, two hominins, Ceprano and Buia, show expansion at the parietals, rather than the prevailing condition in which crania are widest in a lower plane.

H. rhodesiensis is well separate from *H. erectus* in the phylogeny. Further, none of the target OTUs could be shown to have shared an immediate common ancestor with *H. erectus*.

The morphological variation in hominins during this period, then, is better explained by speciation than stasis; and the tendency to place new hominins within a framework comprising one species, *H. erectus*, is not supportable.

References

Abbate E, Albianelli A, Azzaroli A, Benvenuti M, Tesfamariam B, Bruni P, Cipriani N, Clarke RJ, Ficcarelli G, Macchiarelli R, Napoleone G, Papini M, Rook L, Sagri M, Tecle TM, Torre D, Villa I. 1998. A one-million-year-old *Homo* cranium from the Danakil (Afar) Depression of Eritrea. *Nature* 393:458–460.

Adefris T. 1992. A Description of the Bodo cranium: An archaic *Homo sapiens* cranium from Ethiopia. A dissertation in the Department of Anthropology, York University. Michigan, USA: UMI dissertation services.

Albianelli A, Napoleone G. 2004. Magnetostratigraphy of the *Homo*-bearing Pleistocene Dandiero Basin (Danakil Depression, Eritrea). *Riv Ital Paleontol S* 110:35–44.

Argue D, Morwood M, Sutikna T, Jatmiko, Saptomo EW. 2009. *Homo floresiensis*: A cladistic analysis. *J Hum Evol* 57:623–629.

Argue D. 2010. The Genus *Homo* in the Early and Middle Pleistocene of Africa and Europe. PhD thesis, The Australian National University.

Ascenzi A, Biddittu I, Cassoli PF, Segre AG, Naldini E. 1996. A calvaria of late *Homoerectus* from Ceprano, Italy. *J Hum Evol* 31:409–423.

Ascenzi A, Mallegni F, Manzi G, Segre AG, Naldini ES. 2000. A re-appraisal of Ceprano calvaria affinities with *Homo erectus*, after the new reconstruction. *J Hum Evol* 39:443–450.

Asfaw B, Gilbert WH, Beyenne Y, Hart WK, Renne PR, Wolde G, Vrba ES, White TD. 2002. Remains of *Homo erectus* from Bouri, Middle Awash, Ethiopia. *Nature* 416:317–319.

Bermúdez de Castro JM, Arsuaga JL, Carbonell E, Rosas A, Martine I, Mosquera, M. 1997. A hominid from the Lower Pleistocene of Atapuerca, Spain: Possible ancestor to Neanderthals and modern humans. *Science* 276:1392–1395.

Bondioli L, Coppa A, Frayer DW, Libsekal Y, Rook L, Macchiarelli R. 2006. A one-million-year-old human pubic symphysis. *J Hum Evol* 50:479–483.

Bräuer, G. 1984. The Afro-European sapiens-hypothesis, and hominid evolution in Asia during the late Middle and Upper Pleistocene. In: Andrews, Franzen, JL. editors: *The early evolution of man with special emphasis on Southeast Asia and Africa*. Frankfurt am Main: Courier Forschungsinstitut Senckenberg. pp. 145–167.

Carpenter JM, Goloboff PA, Farris JS. 1998. PTP is meaningless, T-PTP is contradictory: A reply to Trueman. *Cladistics* 14:105–116.

Clarke JD, de Heinzelin J, Schlick KD, Hart WK, White D, Wolde Grabriel G, Walter RC, Suwa G, Asfaw B, Vrba E, H-Selassie Y. 1994. African *Homo erectus*: Old radiometric ages and young Oldowan assemblages in the Middle Awash Valley, Ethiopia. *Nature* 264:1907–1909.

Clarke RJ. 2000. A corrected reconstruction and interpretation of the *Homo erectus* calvaria from Ceprano, Italy. *J Hum Evol* 39:433–442.

Conroy GC, Jolly CJ, Cramer D, Kalb JE. 1976. Newly discovered fossil hominid skull from the Afar depression, Ethiopia. *Nature* 275:67–70.

Faith DP, 1991. Cladistic permutation tests for monophyly and nonmonophylly. *Syst Zool* 40:266–375.

Faith DP, Cranston PS. 1991. Could a cladogram this short have arisen by chance alone? On permutation tests for cladistic structure. *Cladistics* 7:1–28.

Falguères C, Bahain J-J, Yokoyama Y, Arsuaga JL, de Castro BJM, Carbonell E, Bischoff JL, Dolo J-M. 1999. Earliest humans in Europe: The age of TD6 Gran Dolina, Atapuerca, Spain. *J Hum Evol* 37:343–352.

Fu J, Murphy RW. 1999. Discriminating and locating character covariance: An application of Permutation Tail Probability (PTP) analysis. *Syst Biol* 48(2):380–385.

Grimaud, D. 1982. Évolution du parietal de l'Homme fossile, position de l'Homme de Tautavel parmi les hominidés. Thése de 3éme cycle. Muséum National d'Histoire Naturelle (indet).

Groves CP. 1989. *A theory of human and primate evolution*. Oxford: Clarendon Press.

Hennig W. 1966. *Phylogenetic systematics*. Urbana, Chicago: University of Illinois Press.

Hillis D, Bull JJ. 1993. An empirical test of bootstrapping as a method for assessing confidence in phylogenetic analysis. *Syst Biol* 42(2):182–192.

Holloway RL, Broadfield DC, Yuan MS. 2004. *The human fossil record, Volume 3. Brain endocasts – the paleoneurological evidence*. Somerset, NJ: John Wiley and Sons. pp. 129.

Hrdlicka A. 1930. *The skeletal remains of early man*. Washington: Smithsonian Institution.

Hublin JJ. 1978. Le *torus occipital transverse* et les structures associées: Evolution dans le genre *Homo*. Thése de 3éme cycle, Université P. and M. Curie, Paris IV (inédet).

Humphries CJ. 2002. Homology, characters and continuous variables. In: MacLeod N, Forey PL, editors. *Morphology, shape and phylogeny*. Systematics Association Special Volume Series 64. London and New York: Taylor and Francis. pp. 8–27.

Kalb J, Jolly CJ, Mebrate A, Tebedge S, Smart CM, Oswald EB, Cramer D, Whitehead P, Wood CB, Conroy GC, Adefris T, Sperling I, Kana B. 1982. Fossil mammals and artefacts from the Middle Awash Valley, Ethiopia. *Nature* 298:25–29.

Källersjö M, Farris JS, Kluge AG, Bult C. 1992. Cladistics: What's in a word? *Cladistics* 9:183–199.

Lahr MM. 1996. *The evolution of modern human diversity: A study on cranial variation.* Cambridge: Cambridge University Press.

Macchiarelli R, Bondioli M, Chech M, Coppa A, Fiore I, Russom R, Vecchi F, Libsekal Y, Rook L. 2004. The late Early Pleistocene human remains from Buia, Danakil Depression, Eritrea. In: Abbate E, Woldehaimanot B, Libsekal Y, Tecle TM, Rook L, editors. A step towards human origins. The Buia *Homo* one million years ago in the Eritrean Danakil Depression (East Africa). *Riv Ital Paleontol S* 110 Supplement:133–144.

MacIntosh NWG, Larnach SL. 1972. The persistence of *Homo erectus* traits in Australian Aboriginal crania. *Archaeology and Physical Anthropology in Oceania* 7:1–7.

Maddison WP, Maddison DR. 1992. *MacClade: Analysis of phylogeny and character evolution.* Version 3.0. Sunderland, Massachusetts: Sinauer Associates.

Mallegni F, Carnieri E, Bisconti M, Tartarelli G, Ricci S, Biddittu I, Segre A. 2003. *Homo cepranensis sp nov* and the evolution of African-European Middle Pleistocene hominids. *Comptes Rendus Palevol* 2:153–159.

Manzi G, Bruner E, Passarello P. 2003. The one-million-year-old *Homo* cranium from Bouri (Ethiopia): A reconsideration of its *H. erectus* affinities. *J Hum Evol* 44:731–736.

Manzi G, Magri DB, Salvatore M, Palombo MR, Margari V, Celiberti V, Barbieri M, Barbieri, Mu, Melis RT, Rubini M, Ruffo M, Saracino B, Tzedakis PC, Zarattin A, Biddittu I. 2010. The new chronology of the Ceprano calvarium (Italy). *J Hum Evol* 59:580–585.

Manzi G, Mallegni F, Ascenzi A. 2001. A cranium for the earliest Europeans: Phylogenetic position of the hominid from Ceprano, Italy. *Proc Natl Acad Sci* 98:10011–10016.

Mbua E, Bräuer G. 2012. Patterns of Middle Pleistocene hominin evolution in Africa and the emergence of modern humans. In: Reynolds S, Gallagher A, editors. *African genesis: Perspectives on hominin evolution.* Cambridge: Cambridge University Press. pp. 366–422.

Muttoni G, Scardia G, Kent DV, Swisher Carl C, Manzi G. 2009. Pleistocene magnetochronology of early hominin sites at Ceprano and Fontana Ranuccio, Italy. *Earth Planet Sci Lett* 286:255–268.

Nomade S, Muttoni G, Guillou G, Robin E, Scardia G. 2011. First 40Ar/39Ar age of the Ceprano man (central Italy). *Quaternary Geology* 6(5):453–457.

Rightmire GP. 1976. Relationships of Middle and Upper Pleistocene hominids from sub-Saharan Africa. *Nature* 260:238–240.

Rightmire GP. 1996. The human cranium from Bodo, Ethiopia: Evidence of speciation in the Middle Pleistocene? *J Hum Evol* 31:21–39.

Sartono S., Grimaud, D. 1983. Les parétaux des Pithécanthropines Sangiran 12 et Sangiran 17. *L'Anthropologie* 87:475–482.

Schwartz JH, Tattersall I. 2002. *The human fossil record: Craniodental morphology of Genus* Homo. Volume 2: *Africa and Asia*. New York: Wiley Liss.

Stringer CB. 1984. The definition of *Homo erectus* and the existence of the species in Africa and Europe. In: Andrews P, Franzen JL, editors. *The early evolution of man with special emphasis on Southeast Asia and Africa*. Frankfurt am Main: Courier Forschungsinstitut Senckenberg. pp. 131–143.

Smith Woodward A. 1921. A new cave man from Rhodesia, South Africa. *Nature* 17:371–372.

Swofford D. 2002. PAUP* Version 4.0b10. Distributed by Sinauer Associates, Inc. Smithsonian Institute.

Trueman JWH. 1993. Randomization confounded: A response to Carpenter. *Cladistics* 9:101–109.

Weidenreich F. 1943. The Skull of *Sinanthropus Pekinensus*. *Palaeontologia Sinica*. Pehpei, Chungking. Oxford: Geological Survey of China. pp. 270–272.

Wiley EO, Lieberman BS. 2011. *Phylogenetics. theory and practice of phylogenetic systematics*. Hoboken, NJ: John Wiley and Sons, Inc. pp. 90–193.

Wood B. 1991. *Koobi fora research project, vol. 4: Hominid cranial remains*. Clarendon Press, Oxford.

Zeitoun V. 2000. Revision of the species *Homo erectus* (Dubois, 1893). Use of morphologic and metric data in cladistic investigation of the case of *Homo erectus*. *B Mem Soc Anthro Par* 12:1–200.

Appendix 1. Character states

1. continuity of post orbital sulcus

0 = absent because of continuity of frontal and supraorbital

1 = present but incomplete, interrupted in the medial zone

2 = present – complete and with a distinct edge or border

2. postorbital lateral depression

(a depression on the lateral supraorbital region bounded by the temporal line)

0 = absent

1 = present

3. depression at glabella in *norma facialis*

0 = absent

1 = present

4. shape of frontal edge in *norma verticalis*

0 = linear

1 = convex frontwards

5. position of glabella in *norma verticalis*

0 = glabella zone is depressed

1 = glabella is neither depressed or protruding

2 = glabella projects beyond the frontal

6. continuity of the supraorbital torus

0 = no supraorbital torus

1 = incomplete, interrupted in the medial zone – there are 2 distict tori 'mono-orbitares'

2 = continuous torus

7. superior surface of orbit margins

0 = flow smoothly into frontal squama

1 = horizontal posttoral plane from which squama rises posteriorly

2 = there is a sulcus between posterior aspect of elevated supraorbital rim and frontal squama

8. type of orbital arcade – supraorbitals

Where 'a' is central, 'b' is middle and 'c' is lateral:

0 = a>b, b<c and a < c

1 = a>b, b<c and a>c

2 = a<b, b>c and a>c

3 = a>b, b>c and a>c

4 = no variation in form

The objective is to determine differences in superior-inferior height of supraorbital across the orbit. Measurements were used to determine 'a', 'b', 'c' for each specimen.

9. prominence of temporal line on the frontal

0 = weak

1 = very prominent

10. metopic keeling

0 = absent

1 = present but weak

2 = strong

11. development of the keeling

0 = parallel edges

1 = wider and flatter posteriorly

2 = absent (no keeling)

12. bregmatic eminence

0 = absent

1 = present

13. upper coronal reinforcement

0 = absent

1 = present

14. frontal bosse

0 = absent

1 = present

15. obelionic region

0 = keeling present

1 = no keeling

2 = presence of obelionic depression

16. pre-lambdaic depression

0 = keeling on 4th quarter

1 = no keeling on 4th quarter

2= present

17. presence of the temporal band after the coronal suture

0 = absent

1 = present

18. asterionic process

0 = absent

1 = present

19. parietal bosse

0 = absent

1 = present

20. angular tuber

0 = absent

1 = present

21. curvature of nuchal plane in *norma lateralis*

0 = convex posteriorly

1 = flat to lightly concave posteriorly

22. importance of the occipital torus

0 = weak

1 = strong

2 = no occipital torus

23. extension of external occipital protrusion

0 = absent

1 = present

24. extension of the tuberculum linearum

This refers to the degree of elevation, or relief, at the junction of the superior nuchal line and occipital crest.

0 = absent

1 = moderate

2 = strong

25. medial concavity of the occipital lip to the tuberculum linearum

Is there a depression above where nuchal lines meet?

0 = absent

1 = depression

26. external occipital crest, where present

0 = absent

1 = present for whole of nuchal

2 = present above inferior nuchal line

3 = present below inferior nuchal line

27. occipitomastoid crest

0 = absent

1 = present

28. height of temporal squama cf vault

0 = high

1 = low

29. shape of the temporal squama

0 = polygon to round

1 = triangular

30. strength of supramastoid crest in the region of porion

0 = weak

1 = strong

31. relation between the supramastoid crest and zygomatic process in lateral view

0 = zygomatica forms an angle with supramastoid crest

1 = zygomatica is continuous with supramastoid crest

32. continuity of the supramastoid crest with the inferior temporal line

0 = no direct link

1 = continuity

33. tuberculum supramastoid anterius

Is there a tubercle where supramastoid crest stops at squamous suture?

0 = absent

1 =present

34. strength of the mastoid crest

0 = weak

1 = strong

35. continuity between mastoid crest and superior temporal line

0 = no direct link

1 = continuity

36. supramastoid sulcus, where present

0 = closed posteriorly

1 = open posteriorly

37. importance of supramastoid sulcus

0 = absent

1 = narrow

2 = wide

38. convergence of mastoid crest and supramastoid crest

0 = divergent anteriorly

1 = parallel

39. suprameatum spine

0 = absent

1 = present

40. section of tympanal in *norma lateralis*

0 = rounded

1 = ellipsoid to ovoid

41. orientation of main axis of tympanal in *norma lateralis*

0 = orientated anteriorly

1= vertical

2 = orientated posteriorly

42. thickness of tympanic in *norma lateralis* (anterior edge of tympanic)

0 = weak

1 = strong (>2mm)

43. contribution of the tympanal to mandibular fossa

0 = postglenoid process is strongly involved in the wall

1 = the tympanal makes up most of the wall

2 = rudimentary or no postglenoid process

44. relative development of mastoid process in *norma lateralis*

(i.e. does it project below the base of the cranium?)

0 = does not project below the base of the cranium

1 = projects below base

45. extension of the pre-glenoid planum

(Is there a level surface of bone preceding the mandibular fossa from the articular eminence either for the whole, or at least half, of the width of the eminence?)

0 = no pre-glenoid planum precedes the glenoid cavity

1 = a pre-glenoid planum precedes the glenoid cavity

46. space between the tympanal and anterior of mastoid process

0 = posterior part of tympanal joins anterior part of mastoid process

1 = 'split'

2 = wide space

47. anteroposterior width of mandibular fossa

0 = narrow

1 = wide

48. deepness of glenoid fossa

0 = very shallow

1 = deep

49. height of articular eminence relative to posterior wall of glenoid fossa (basal view)

0 = slope is shorter

1= similar

2 = higher

50. orientation of mastoid process

0 = not orientated inwards

1 = orientated inwards

51. deepness of digastric fossa

1 = shallow

2 = deep

52. size of juxtamastoid eminence

0 = no eminence

1 = weak

2 = strong

53. importance of deepness between entoglenoid formation and tympanic plate

0 = fused

1 = groove

2 = space

54. anterior wall of glenoid fossa

0 = the anterior wall is horizontal

1 = oblique

2 = almost vertical

55. inferior projection of the entoglenoid process compared to that of the tuberculum zygomaticum anterior

1 = entoglenoid projects to a greater extent than the tuberculum zygomaticum anterior

2 = entoglenoid is similar to tuberculum zygomaticum anterior in degree of inferior projection

3 = entoglenoid is less projected than the tuberculum zygomaticum anterior

56. relative position of the entoglenoid formation to the tuberculum zygomaticum anterior

0 = the entoglenoid formation is at the same level as the tuberculum zygomaticum

1 = the entoglenoid formation is posterior to the tuberculum zygomaticum

2 = entoglenoid formation is very posterior to the tuberculum zygomaticum

57. inferior projection of the entoglenoid process and the tuberculum zygomaticum compared to the tuberculum articulare

0 = **very** large inferior projection relative to the tuberculum articulare

1 = large inferior projection relative to the tuberculum articulare

2 = small inferior projection relative to the tuberculum articulare

58. antero-posterior convexity of the tuberculum articular (articular eminence)

0 = the tuberculum articular is flat in *norma lateralis*

1 = the tuberculum articular forms a large round arc

2 = the tuberculum articular forms a small round arc

59. shape of posterior edge of the tuberculum articular in norma basilaris

0 = flat

1 = arched

2 = sigmoid

60. continuity between the pre-glenoid planum and the posterior slope of the articular tuberclum

0 = the two are continuous

1 = there is an angulation between them

61. crest on lateral edge of mandibular fossa

0 = absent

1 = present

62. inferior projection of entoglenoid process compared to the sphenoid border/edge

0 = the entoglenoid process projects inferiorly to a greater extent than sphenoid edge

1 = the entoglenoid process is equivalent in inferior projection to sphenoid edge

2 = the entoglenoid process is les projected than sphenoid edge

63. prominence of entoglenoid formation

0 = very prominent
1 = not prominent

64. lateral extension of entoglenoid process

0 = very extended posteriorly
1 = marginally extended backward
2 = not extended posteriorly
3 = tubercle
4 = not extended posteriorly or tubercle

65. does postglenoid process extend out beyond tympanic?

0 = doesn't overlap the tympana
1 = does overlap the tympana
2 = no postglenoid process or rudimentary process

66. profile of nasal saddle and nasal roof

1 = flat nasal bones
2 = slightly raised nasals, forming a curve
3 = nasals forming well-defined curve, ranging in size from medium to large
4 = deep angled nasal bones forming a 'pinched nose'

67. relationship of rhinion to nasospinale

1 = nasospinale lies in front of rhinion
2 = nasospinale is on same plane as rhinion
3 = nasospinale lies behind rhinion

68. condition of the margo limitans

1 = the *margo limitans* forms a sill
2 = *margo limitans* forms a smooth curve
3 = *margo limitans* includes a prenasal groove

69. the condition of the facies anterior of maxilla and alveolar process

1 = the *facies anterior* and alveolar process is inflated,
2 = the *facies anterior* and alveolar process is well filled out
3 = the *facies anterior* and alveolar process is sunken
4 = the *facies anterior* and alveolar process forms a flat surface

70. presence of jugum alveolar

1 = there is no *jugum alveolar*

2 = the *jugum alveolar* forms a narrow ridge

3 = the *jugum alveolar* forms a broad and prominent ridge (width of 1+ premolar)

71. presence of a sulcus infraorbitalis (i.e. under the infraorbital foramen)

1 =there is no *sulcus infraorbitalis*

2 = the *sulcus infraorbitalis is narrow*

3 = the *sulcus infraorbitalis is wide*

72. zygomaticoalveolar crest (ordered)

1= relatively straight

2 = curved

3 = forms an arc

4 = forms an arch

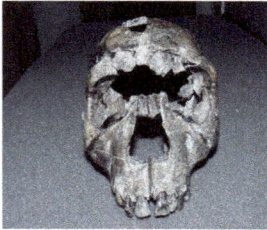 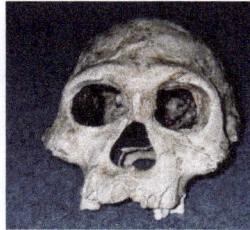 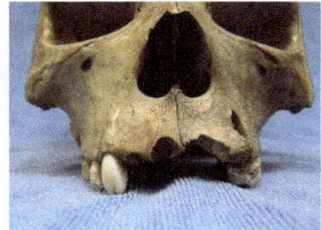

State 2 (curved)   State 3 (arc)   State 4 (arch)

Description: KNM-WT 15000 frontal view 102

Description: D2700 caste frontal view

Description: P1221728

73. shape of naso-alveolar clivus

1 = naso-alveolar clivus is convex

2 = naso-alveolar clivus is flat

3 = naso-alveolar clivus is concave

74. palate surface has low irregular crests or fine ridges arranged in more or less longitudinal direction

1 = present

2 = absent

75. location and direction of orifice of incisive canal

1 = orifice of incisive canal is immediately posterior to incisors
2 = orifice of incisive canal is on a plane with canines
3 = orifice of incisive canal is on a plane with 1st premolar
4 = orifice of incisive canal is on a plane with 2nd premolar

76. location of zygomatic arch

1 = the zygomatic arch runs below the Frankfurt horizontal
2 = the zygomatic arch runs at level of Frankfurt horizontal
3 = the zygomatic arch runs above the Frankfurt horizontal

77. condition of the supraorbital margin

1 = the supraorbital margin is thick, rounded and not demarcated from roof of orbit
2 = the margin is thick with an edged crest not demarcated from roof of orbit
3 = the supraorbital margin is an edged crest demarcated from the roof of orbit
4 = the supraorbital margin is thin with an edged crest and demarcated from the roof of orbit

78. condition of infraorbital margin of the orbits

1 = sharp high line dividing the floor of the orbit from the facial portion of the malar
2 = relatively rounded orbital margin but raised in relation to floor of the orbit
3 = pronounced rounding of the inferior *lateral* border which is leveled with the floor of the orbit (i.e. lower outside edge for half the lower edge of orbit is rounded but other half of lower orbit not rounded)

79. character of superior fissure

1 = the superior fissure is small and round
2 = superior fissure is a slit-like lateral prolongation
3 = there is a strut dividing the fissure into 2

80. styloid process

1 = present
2 = absent

81. tympanic trough

A coronally oriented long narrow trough along tympanic tube in basal view

0 = absent
1 = present

82. sagittal keeling on first half of parietal.

0 = absent

1= present

83. presence of external occipital crest

0 = absent

1 = present

84. presence of glasseri fissure

0 = absent

1 = present

85. supraorbital torus

0 = absent

1 = present

86. tuberculum linearum

0 = absent

1 = present

87. maximum cranial breadth

0 = at supramastoid region

1 = at parietal region

88. length of nuchal dominates over length of occipital

Determined by comparing measurement for lambda-inion (occipital length) and measurement for inion-opisthion (nuchal length)

0 = yes

1 = no

89. foramen magnum

0 = round

1 = oval

12. Human evolution in Sunda and Sahul and the continuing contributions of Professor Colin Groves

Michael C Westaway, Arthur Durband and David Lambert

Introduction

In terms of understanding evolutionary processes in the human origins story, much can be gained by employing an approach of splitting the known fossil record into numerous taxonomically recognised species. By focusing on variation in fossil specimens, as defined by differences in characters in both time and space, we can start to explore possible evolutionary relationships among the range of fossil specimens. The opposite approach, of lumping such taxonomically species will potentially obscure interesting differentiation, resulting in the possible loss of understanding for important phylogenetic relationships (Groves, 1989a). Whether such taxonomically recognised species are biologically real or not is exceedingly difficult to establish in the fossil record, however, splitting fossils into divisions based on character traits enables the effective testing of hypotheses regarding the validity of such species.

Approaches in Australian palaeoanthropology have largely been characterised by splitting and lumping. For more than 15 years, the charismatic lumper Alan Thorne had carried the day for the evolutionary story of the Australasian region. His interpretation of the evolution of *Homo sapiens* in our region evoked an argument for deep geological links with regional *Homo erectus*. It represented a record of regional continuity anchored to the evidence first discovered in Java some 120 years ago (Dubois, 1894). The model was unlike the earlier variants of the regional continuity brand of human evolution, which proposed a direct 'candelabra' treatment of the fossil record. Thorne proposed that there was some limited gene flow between populations from different regions, however, the crux of his argument was that there was general continuity of traits in the regions. There had not been much modification to this regional pattern from outside. For the colonisation of Sahul, Thorne envisaged that there had been two major migrations into Australia, one derived from Ancient China, the other from Ancient Java. His hypothesis was referred to as the dihybrid model of Aboriginal origins.

With a background in journalism and a multitude of connections, Thorne delivered a powerful ABC documentary to the nation on human origins in Sunda and Sahul in 1988, known as 'Man on the Rim'. Together with his American colleague Milford Wolpoff and links forged with Chinese and Indonesian colleagues, he was generally very successful in promoting his model of human evolution for the region on the international stage. His partnerships with Indonesian colleagues had helped him obtain an extensive cast collection (now at the Shellshear Museum, University of Sydney) representative of Sunda, which enabled him to draw his comparisons with the available record from Sahul. He added dramatically to our understanding of the Sahul record, and amassed, through an inspiring two decades of fieldwork, an assemblage of Pleistocene remains from Australia that numbered over 100 individuals, from the internationally significant sites of Kow Swamp and the Willandra Lakes (including Lake Mungo). He held a virtual monopoly over the fossil series, and developed strong links with the Willandra Elders to ensure that research could continue on the series, following the scientific disaster of the Kow Swamp reburial. Thorne for these years appeared to reign supreme, but his position was not to last.

In 1974, Colin Groves, fresh from the Duckworth Laboratory of Physical Anthropology at Cambridge, arrived in Australia. His arrival was in time for some quite significant events relating to research on Aboriginal origins. He was present at the time of the excavation of Mungo Man (WLH 3). This taste of Australian fieldwork was soon after followed by the 'Origins of the Australians' conference held in Canberra at the Australian Institute of Aboriginal Studies. At this conference the mandible of Mungo Man, with his extraordinary pattern of occlusal wear, was exhibited. Macintosh commented at the time that it was morphologically within the range of modern Australians (Macintosh and Larnach, 1976). The Canberra conference defined the parameters for much future debate on the origins of the Australians, a debate that we have moved only somewhat closer to answering today.

From these beginnings at The Australian National University, it was here that Groves was to base much of his interpretations of the fossil record for human origins. His contributions to our understanding of human origins have been of international significance. While there is much of importance in his work to discuss, such as his naming of the species *Homo ergaster* with Mazák (1975), which has helped to make sense of the confusing morphological diversity that is all too often lumped into *Homo erectus*, this chapter is restricted to a summary of his contributions to our understanding of human origins in Sunda and Sahul. His work on human origins in Sunda and Sahul did not really begin to emerge in print until 15 years after he first had his feet burnt at Lake Mungo, but many publications strongly (and always politely) argued for replacement over regional continuity. This chapter will attempt to tie his work into current research and understanding of human evolution in Sunda and Sahul, and in doing so we hope, shall demonstrate how well his science has held up. While the picture seems far more complicated today compared to the original sketch from 1974, encouragingly a great deal more is now known as a result of a series of new discoveries (some of which are nothing short of spectacular) and the development of improved analytical techniques.

In this chapter we shall consider five major themes that Groves has devoted some thought towards, including i) the question of the taxonomic affinity of the first hominins in Sunda, ii) the first crossing of the Wallace Line by archaic hominins, iii) the evolutionary trajectory of *Homo erectus* (with a focus on the meaning of late derived *erectus*), iv) the second crossing of the Wallace Line by *Homo sapiens* resulting in the subsequent colonisation of Sahul, and v) the important new insights that studies of ancient DNA (aDNA) are contributing to our rewriting of the human evolutionary narrative in Sunda and Sahul. We feel that these represent the key topics in human evolutionary studies on our genus within the region, all being topics that Groves has contributed to.

The first Javan hominins: Is there a case for a pre-*erectus* taxon in Sunda?

The possibility of an earlier species being present within the lower units of the site of Sangiran, Java, is a proposition that has quite a long history. Von Koenigswald argued that there were two separate species represented in the fossil record of Java, which he called *Pithecanthropus erectus* and an earlier species *Meganthropus palaeojavanicus* (von Koenigswald 1956). Indeed Robinson (1953 and 1955) had suggested that the specimens often associated with *Meganthropus* were best placed within the genus *Paranthropus* and identified them as a separate species, *Paranthropus palaeojavanicus*. Certainly when one considers the hyper robust corpus of Sangiran 6 it is reminiscent of the heavy masticatory apparatus of African *Paranthropus*, a dentition adapted for processing hardy open woodland vegetation.

A significant comparison of these fossils by Philip Tobias and Ralph von Koenigswald (1964) in Cambridge compared the fossils from the African record (primarily Olduvai) with the Javan specimens. They had a particular focus on those fossils that had been termed *Meganthropus*. The picture that emerged from the study was an identification of four grades of hominisation, with a suggestion that *Meganthropus* fitted within a grade similar to that of *Homo habilis*. Later Tyler (1995) in describing the Sangiran 31 calotte suggested that either the range of *Homo erectus* needed to be redefined or this fossil represented a different species, of which *Homo habilis* is one that he favoured. The extreme occipital torus is well beyond that of any of the Javan specimens, and a strange region of raised bone has been identified as a sagittal crest (Tyler et al., 1995), a description that has been dismissed by Grimaud-Hervé (2001) as being not possible due to the presence of an angular torus. She suggests that the character may instead be a post-mortem anomaly, the feature perhaps representing raised external compact bone.

A study by Kaifu and colleagues (2005) reviewed this proposition, and identified that the specimens that had in the past been recognised as *Meganthropus* (Weidenreich 1943) supported two evolutionary propositions. Either there had been a great deal more variability in *Homo erectus* prior to the Middle Pleistocene, or that there may indeed be another species present. One of the key points that Kaifu had made was that there were a number of characters in the jaw fragments that were more like *Homo habilis* than *Homo erectus*.

In an article in *Australasian Science* in 2008 in relation to the taxonomic affinities of *Homo floresiensis,* Groves discussed the possibility of a connection between the enigmatic fossil and the earliest known hominins from Java. Specifically he asked:

> The question is: can we find traces of its (<u>Homo floresiensis'</u>) passage? There are almost no fossils of the relevant time period anywhere between Africa and Java, but in Java itself there are (mostly rather scrappy) remains from levels somewhat earlier than those from which <u>Homo erectus</u> have come. Most authors have considered that these early Javanese fossils also represent <u>Homo erectus</u>, but recently a joint Japanese and Indonesian team, led by the noted palaeoanthropologist Kaifu, have found that they are actually rather different. This raises a question: if there is something different, something non-erectus, in these early levels, could <u>Homo floresiensis</u> have a hitherto unrecognized ancestor among them? (Groves 2008)

This idea was discussed further by Groves and one of his students (Westaway and Groves 2009), where it was suggested that a process of replacement in the evolution of hominins perhaps was a regular pattern in hominin evolution in Sunda. They suggested that the later *erectus* extinction event, following the arrival of *Homo sapiens,* was just the next stage of hominin replacement in Java. It was also suggested in this paper that the infant fossil from Mojokerto, dating to sometime around 1.49 Ma (Morwood et al., 2003), may in fact represent a juvenile of this species.

The presence of pre-*erectus*-like characters in the early Java series is intriguing. At a time when Sunda was not a chain of islands, a savannah corridor up to 150 km wide (Bird, Taylor and Hunt, 2005) supported a more open fauna. It is not unreasonable to imagine that there may have been pre-*erectus* hominins in the Early Pleistocene occupying the savannah and coastal plains at times when the sea levels were low. Provisionally such a species has been called *Homo modjokertensis* (Westaway and Groves 2009). The view that *Homo modjokertensis* is a real taxon requires the discovery of more complete fossils dating to the early to Middle Pleistocene. There is a lot of uncertainty relating to the earlier Sangiran fossils, and it is perhaps time to revisit Robinson's original proposal that *Paranthropus* may be present in the early Middle Pleistocene in Sunda. Certainly the dimensions of Sangiran 6 seem to fall closer within the range of *Paranthropus* than *Homo erectus* (Figure 12.1a–d). While Sunda is on the edge of the range of the genus *Homo*, there has been somewhere in the vicinity of 1.5 Mya of hominin evolution. The discovery of *Homo floresiensis* has exposed palaeoanthropologists to a view that there was much greater diversity in hominin species in Sunda than previously considered. Indeed it has been questioned as to whether *Homo floresiensis* should in fact continue to be included in the genus *Homo* (Collard and Wood, 2007), which is a point that we shall return to in the next section. Further studies are required to arrive at a clearer understanding of this diversity.

Figure 12.1a: In recent years the Sangiran 6 mandible has been considered by only a few researchers to be outside the range of *Homo erectus*. The extreme thickness of Sangiran 6 is compared to that of Sangiran 1b.

Source: Photographs taken by Michael Westaway.

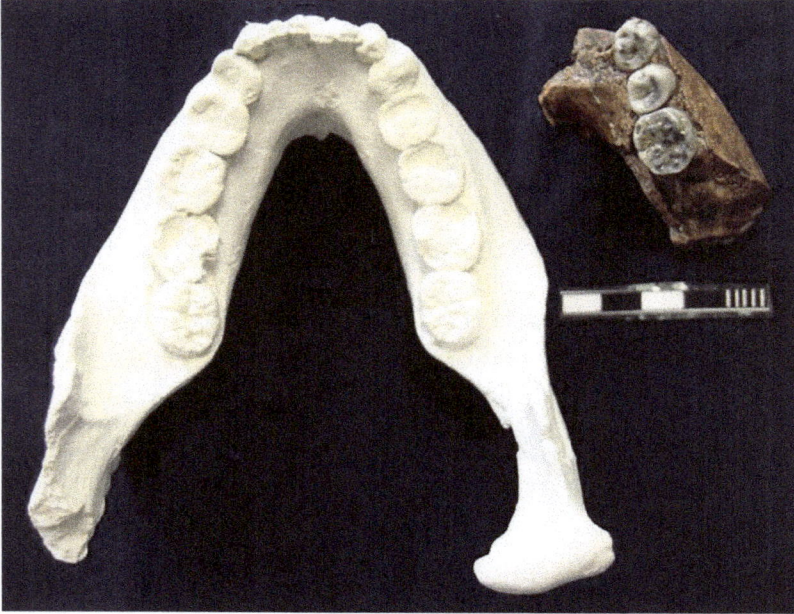

Figure 12.1b: The very thick corpus of Sangiran 6 is similar to that seen in the Peninj _Paranthropus_ mandible.

Source: Photographs taken by Michael Westaway.

Figure 12.1c: The posterior view of Sangiran 31 (cast) showing the extreme robusticity of the occipital torus.

Source: Photographs taken by Michael Westaway.

Figure 12.1d: Lateral view of Sangiran 31 (cast) showing the extreme robusticity of the occipital torus.

Source: Photographs taken by Michael Westaway.

Crossing the Wallace Line (1): The puzzle of *Homo floresiensis*

The human evolutionary context at the time of Groves' paper 'hovering on the brink' was relatively straightforward. There had been migration into the region by an archaic hominin, *Homo erectus*, which was later replaced with modern humans. The initial interpretations of the archaeological record from Flores indicated that *Homo erectus* had made it across the Wallace Line, being present from some 700 ka (Groves, 1996). Groves in this paper discussed evidence from the general palaeontological and zoogeographical record to help build a meaningful context around the stone artefacts from Flores. He suggested that *Homo erectus* in Flores was part of a general oriental dispersal along the lesser Sunda Chain.

With the discovery of *Homo floresiensis* the story became far more complicated. Initial interpretations have focused on *H. floresiensis* being derived from *H. erectus*. The principle of island dwarfing has been cited as a possible mechanism of how *H. erectus* may have initially evolved from the larger species (Brown et al., 2004). Revision of the endocranial volume of *H. floresiensis* to 426 cc by

Kubo et al. (2013) has perhaps made the proposition of dwarfism somewhat more tenable. In their comparison they consider dwarfism from the earlier known hominins from Java, provisionally called here *Homo modjokertensis* (but lumped by Kaifu and colleagues within *Homo erectus*). What the most recent study by Kaifu et al. helps establish is that *Homo floresiensis* may be derived from the earliest hominins known from Sangiran.

Cladistic analyses have suggested that *Homo floresiensis* is derived from either a *Homo rudolfensis* or *Homo habilis* like ancestor, indicating that it is part of a lineage dating back to either the Late Pliocene or Early Pleistocene (Argue et al., 2009). It is important to note at this point that *Homo erectus* is derived from the much later African species *Homo ergaster*. The evidence from the post crania also suggests that a pre-*erectus* hominin is more likely the ancestor for *Homo floresiensis*. The primitive anatomy of the wrist (Tocheri et al., 2007), the general limb proportions (Holliday and Franciscus, 2009), the pelvis (Jungers et al., 2009a), and the unusual anatomy and proportions of the feet (Jungers et al., 2009b) all seem to support this proposition. Jungers et al. (2009b) make the point that while it is possible that insular dwarfism may have resulted in some reversals to a few plesiomorphic states over a period of 800 ka, it is improbable that island dwarfism directed such dramatic change throughout so much of the cranial and postcranial anatomy of *Homo floresiensis*. One limitation that we face is the absence of fossils representative of the wrist and feet from *Homo ergaster* and *Homo erectus*.

It has recently been suggested (Collard and Wood, 2007) that the inclusion of the Late Pleistocene specimens from the site of Liang Bua, Flores, in the genus *Homo* as a new species, *H. floresiensis*, is not compatible with the commonly accepted definition of the genus *Homo* (Wood and Collard, 1999).

Regional continuity in Java: The idea of *Homo soloensis*

There has been considerable debate in Europe and Africa regarding the idea of multiple species within the *erectus*-grade, including such species as *Homo antecessor*, *Homo cepranensis*, *Homo rhodesiensis* and *Homo helmei*. The idea of intermediate species has been subject to minimal discussion within Sunda. It is possible that the long-term isolation of Java may have led to the formation of new species derived from *Homo erectus*.

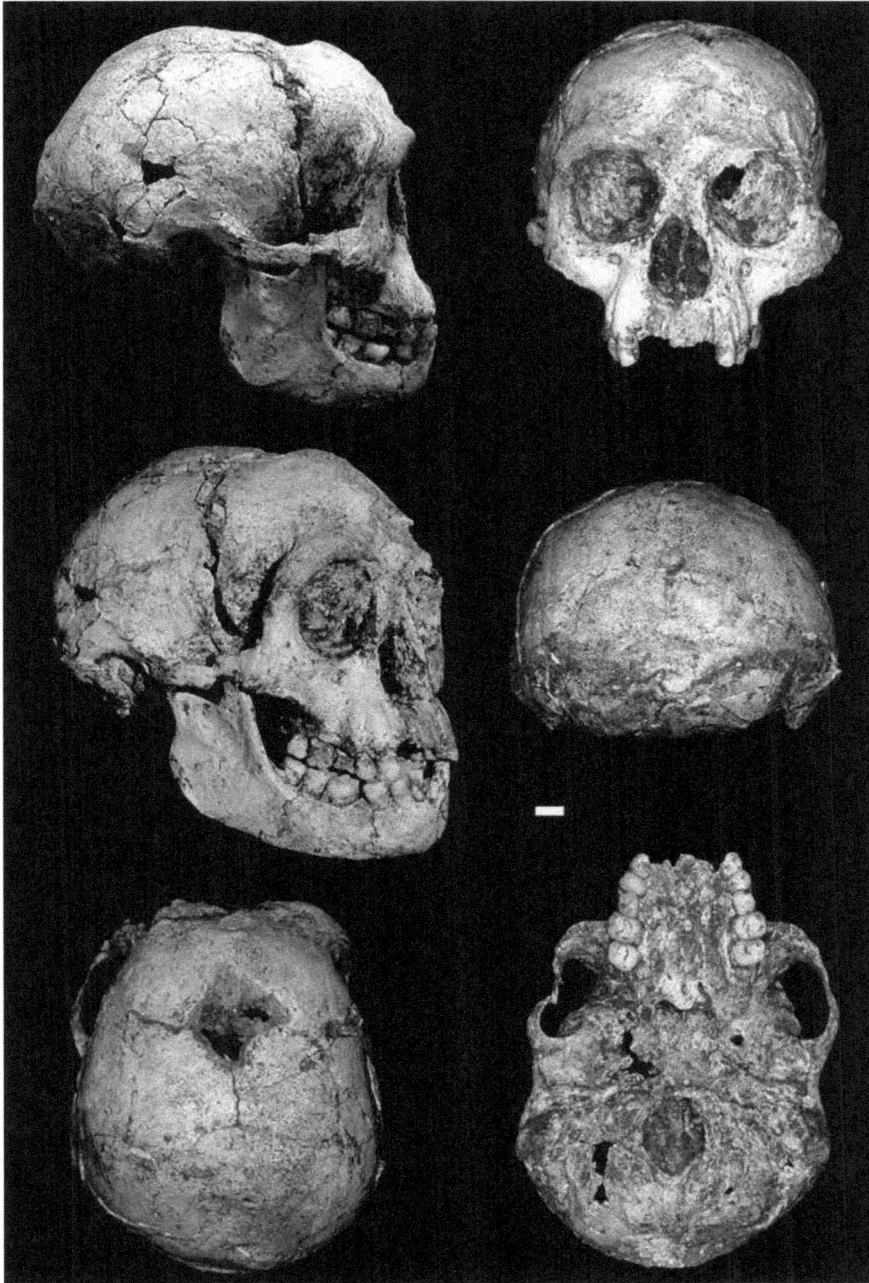

Figure 12.2: Liang Bua 1, the type specimen of *Homo floresiensis*.

Source: Morwood et al. (2004).

Alongside discussion around the proposal that archaic Javan *Homo erectus* (which we called above *Homo modjokertensis*) had evolved into the diminutive *Homo floresiensis*, there has also been a less publicised but no less important debate concerning the continuous evolution of *Homo erectus* in Java. Some workers have suggested that the later surviving individuals from sites like Ngandong and Sambungmacan, often referred to as more 'advanced' *Homo erectus* (e.g. Santa Luca, 1980; Rightmire, 1990, 1991; Lahr 1996; Anton et al., 2007) or even sometimes as 'archaic' *Homo sapiens* (e.g. Delson et al. 1977; Bräuer 1992; Frayer et al., 1993), might instead be identified as a new species dubbed *Homo soloensis*. This species name was coined by Oppenoorth (1932) in his initial descriptions of the earliest hominin material excavated from Ngandong, and has since been reconsidered by more recent authors (e.g. Widianto and Zeitoun, 2003; Zeitoun, 2009; Zeitoun et al., 2010; Durband, 2004, 2007, 2008c, 2009; Durband and Westaway, 2013). Groves (1989a) also recognised the more derived affinities of the later Javan material, referring it to its own chronosubspecies of *H. erectus soloensis*.

Kaifu and colleagues (2008) provided a detailed analysis of the evolutionary changes that accumulated in the Javan hominin lineage. Those authors recognised 'a continuous, gradual evolution of Javanese *H. erectus* from the Bapang-AG to Ngandong periods' (Kaifu et al., 2008: 578), with the Sambungmacan specimens reflecting a more intermediate morphological condition. These conclusions have been echoed by Zeitoun (2009; Zeitoun et al., 2010), and Durband (2002, 2004, 2007, 2009), who likewise found ample evidence for the accumulation of significant change within the Javanese hominin assemblage. These changes include both morphometric relationships of various cranial elements (e.g. Kaifu et al., 2008) as well as the evolution of several autapomorphic features of the cranial vault and base (Durband, 2002, 2004, 2007, 2008c). Features such as a divided foramen ovale located in a pit, the unique configuration of the mandibular fossa, and a 'teardrop' shaped foramen magnum caused by an opisthionic recess are some of the characters that have been shown to be autapomorphic in Ngandong, with most appearing in Sambungmacan and Ngawi (Durband 2004, 2007, 2008c). The faunal record of Java during this time period is indicative of relative isolation and endemism (de Vos et al., 1994; van den Bergh et al., 1996, 1999, 2001), which would be consistent with the interpretations of the patterns seen in the hominins.

While opinions vary on how to approach the taxonomy of the later Javan material (e.g. Ngandong, Sambungmacan, and Ngawi), it is becoming clear that the evolutionary history of the later Sunda hominins is considerably more complex than it is often portrayed. This evidence, particularly when considered alongside the potential for dynamic change suggested by the earliest Javan fossil material and *H. floresiensis*, suggests that there may have been more changes taking place in Sunda than has previously been appreciated. It would seem reasonable to suggest that further speciation occurred in Sunda and there was not overall stasis within the *erectus* grade, but in fact significant episodes of divergence at certain points in the Pleistocene within two hominin lineages.

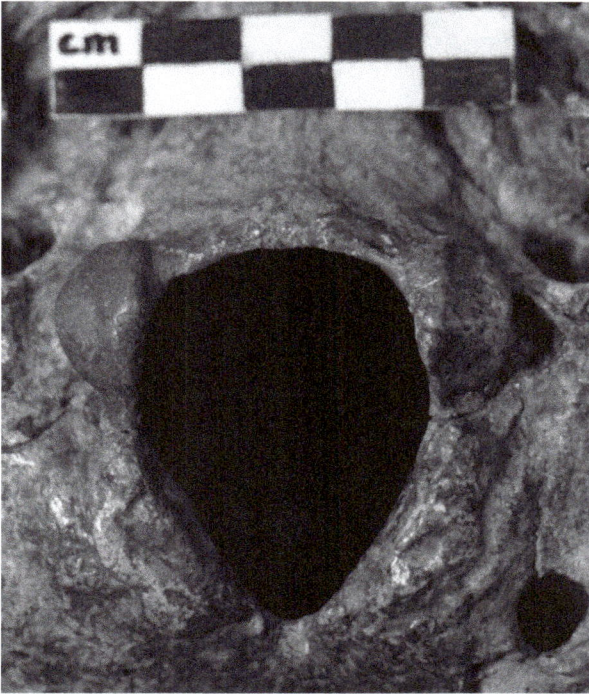

Figure 12.3a: Autapomorphic characters identified in the Ngandong series, including (a) a tear drop shaped foramen magnum (above) and (b) a divided foramen ovale (below).

Source: Photographs taken by Arthur Durband.

Figure 12.3b

Source: Photographs taken by Arthur Durband.

Crossing the Wallace Line (2): Colonisation of Sahul and morphological variation in *Homo sapiens*

The most successful migration through Sunda was that of *Homo sapiens*. The nature of this migration is now far better understood than it was in 1996, and the complexity of the migration is increasingly being revealed through studies from both ancient DNA (aDNA) and the DNA of modern populations (a topic which will be discussed in further detail in the next section of this paper). The evolution of *Homo erectus/soloensis* was probably interrupted with the arrival of *Homo sapiens* in the region, but our understanding of any possible overlap is still imprecise. It is possible that *H. sapiens* migrating into Sunda did not encounter any populations of *H. erectus/soloensis*, as some evidence suggests that they may have become extinct tens of thousands of years earlier (Storm, 2000, 2001b; see also new dates for Ngandong by Indriati et al., 2011). Certainly the replacement of *erectus/soloensis* occurred either prior to, or soon after (perhaps a matter of millennia), the arrival of *Homo sapiens*. It is unclear if *Homo sapiens* then took the southern route or the northern route to Sahul. The southern route would have brought them into contact with *Homo floresiensis*, but the earliest evidence on Flores for *H. sapiens* subfossil remains is only early Holocene in age. We still currently lack the evidence necessary in Flores to understand the interaction between these two species. In nearby Timor evidence exists for deep sea fishing activity, identified as that of modern humans, as early as 42 ka (O'Connor et al., 2011).

Homo sapiens developed as part of their cultural repertoire the capacity for long sea journeys, enabling the establishment of a viable population in Australia some 50,000 years ago. While a model identifying a significant genetic contribution from *Homo erectus* to the origin of the First Australians has been proposed in a number of formats (Thorne, 1976; Thorne and Wolpoff, 1981; Curnoe and Thorne, 2006; Webb 2006), phylogenetic analyses have demonstrated that no such signature can be demonstrated (Westaway and Groves, 2009). *Homo erectus* has a series of autapomorphic characters distinct to that species, while characters shared between *Homo heidelbergensis* and *Homo sapiens* are absent in *Homo erectus* (Groves and Lahr, 1994). Plesiomorphic retentions in *Homo sapiens* provide a clearer ancestral link to the earliest anatomically modern fossils from East Africa (Groves, 1989b).

Much of the debate in Sahul on the origins of the First Australians has focused around the meaning behind the Pleistocene robust and gracile fossil series from southeast Australia. Groves (2001) noted that it is expected that much of this variation is the result of evolutionary change over '60 ka' (we prefer a date of 50 ka from available evidence). The suggestion that cranial robusticity is representative of adaptation to the climatic stress of the Last Glacial Maximum (LGM) has been the subject of attention by a number of Australian palaeoanthropologists (Wright, 1976; Bulbeck, 2001; Stone and Cupper, 2003; Westaway and Lambert, 2013), although there remains a lack of clarity as to what the actual mechanism for such selection may be. It would seem that in the very important Willandra Lakes series cranial robusticity does not emerge until the approach to the Last Glacial Maximum (Westaway and Groves, 2009). There are five well-dated specimens that support this hypothesis. Currently these include the gracile fossils WLH1 and WLH3 dated to sometime around 40 ka (Olley et al., 2006), and the robust fossils WLH 50 (Grün et al., 2011) and the two fossils WLH 152 and 153 (Webb, 2006). The revised dates on the Kow Swamp cranial series also place this robust series of crania at the peak of the LGM (Stone and Cupper, 2003).

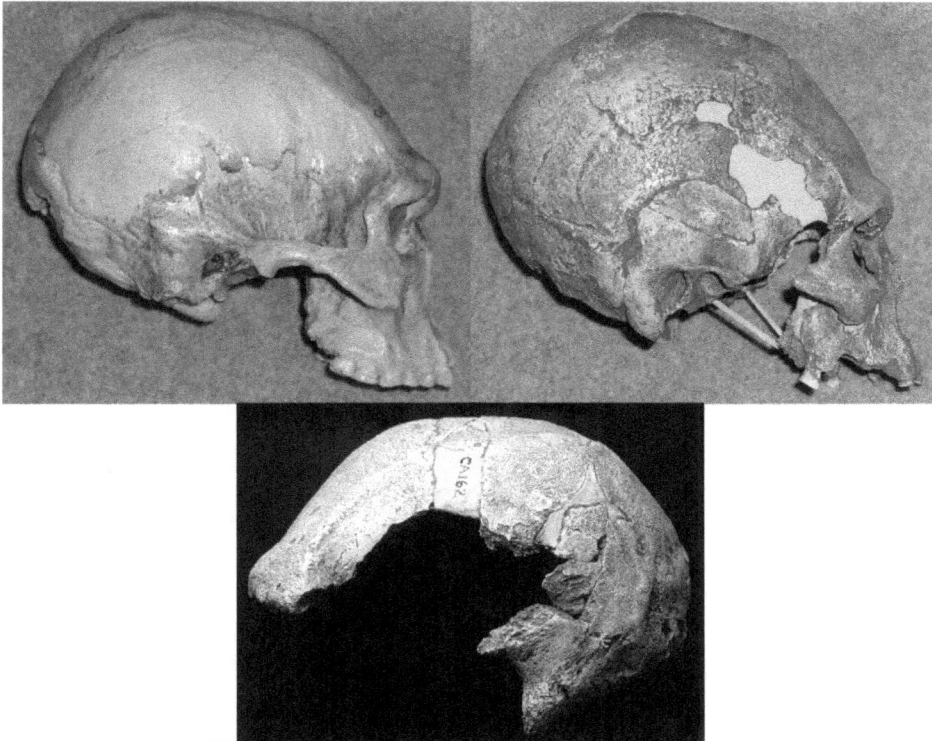

Figure 12.4a: Robust (Cohuna, Kow Swamp 5 and WLH 50) Australian fossils often discussed in the Aboriginal origins debate.

Source: Photographs taken by Arthur Durband.

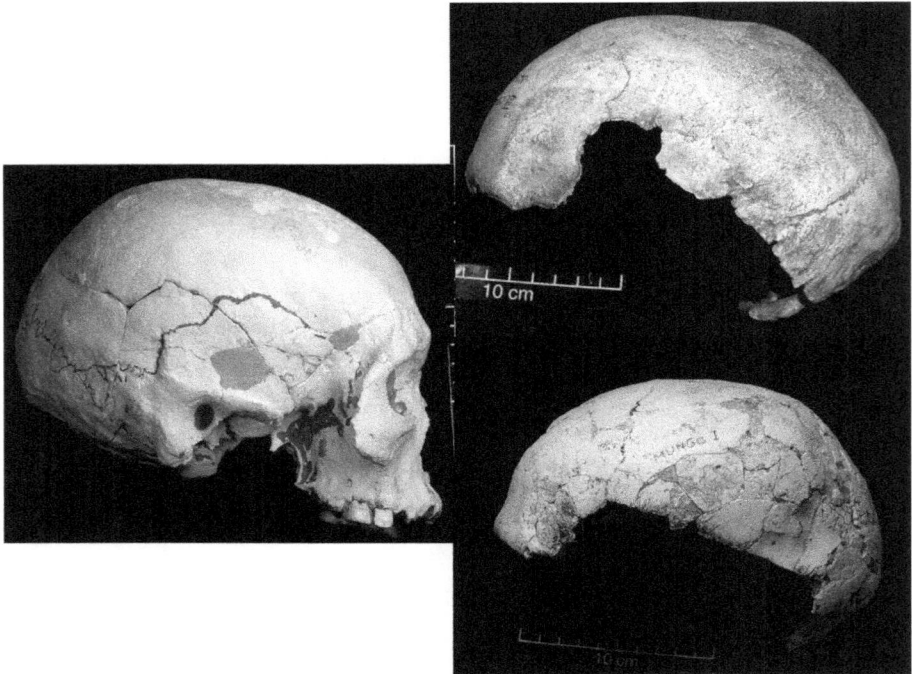

Figure 12.4b: Gracile (Keilor, WLH 3 and WLH 1) Australian fossils often discussed in the Aboriginal origins debate.

Source: Photographs taken by Arthur Durband.

Groves (2001) suggested that increased robusticity was also an artefact of artificial cranial deformation, a prospect that has been demonstrated by a number of workers (Antón and Weinstein, 1999; Durband, 2008ab). Cranial deformation contributed to the appearance of a flattened receding frontal in many of the robust fossils commonly mentioned in this debate; a feature that was initially considered to be a trait inherited from *Homo erectus*. While it is certainly true that not all of the robust fossils have been artificially deformed, many of the key fossils commonly mentioned by continuity advocates, like Kow Swamp 1, 5, and Cohuna, have been influenced by this cultural practice (Brown, 1989; Durband, 2008ab). Another key robust fossil in the debate, WLH 50, is likely pathological, which has resulted in the increased cranial vault thickness of the individual (Webb, 1990). There does remain some debate regarding the pathological diagnosis of WLH 50 (Westaway, 2006; Curnoe and Green, 2013; Durband and Westaway, 2013).

What would appear to best explain the high degree of cranial variation and the emergence of cranial robusticity is a significant amount of in situ evolution, associated with the climatic amelioration of the LGM, with a very ancient example in some regions of Sahul of the cultural practice of cranial deformation. What remains puzzling, however, is the fact that some of the robust fossils in the Australian record, such as WLH 50, WLH 19, Nacurrie 1 and 2, Kow Swamp 5 etc., do bear resemblance to some of the earliest, quite robust fossils from East Africa, such as Herto, Omo I and II and Jebel Irhoud I. This can be extended to early *Homo sapiens* fossils found outside of Africa as well, such as the fossil Skhul V. When tested phylogenetically (Westaway and Groves, 2009) these robust, circa 100 ka *Homo sapiens* fossils do sit closely to the more robust specimens from the Willandra Lakes. This suggests a situation where robusticity is present in early *Homo sapiens*, then a gracile from emerges with the Mungo and Niah individuals, and then we see a return to robust forms around the LGM in the Willandra Lakes.

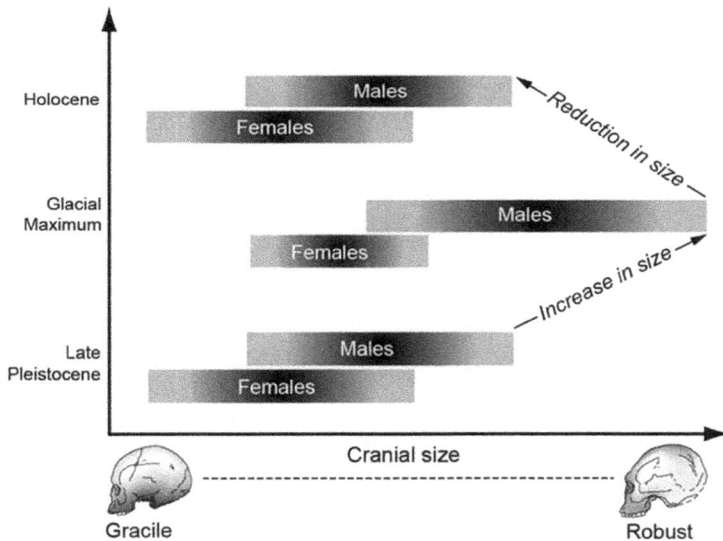

Figure 12.5: Theorised representation of the emergence of cranial robusticity associated with the onset of the Last Glacial Maximum.

Source: Westaway and Lambert (2014).

One of the big questions we now face relates to whether there was significant gene flow into Sahul later in the Pleistocene following the initial colonisation event some 50 ka. Oppenheimer (2004) suggested that a secondary migration may in fact explain the appearance of a more robust cranial form in Sahul during the LGM, which is not dissimilar to the earlier models put forward by Thorne. This is a question that is currently the subject of some scrutiny by those interested in the study of DNA. Such a migration has been suggested much later in prehistory, with the identification of gene flow from India occurring in the mid Holocene, said to have coincided with the arrival of the dingo (Pugach et al., 2013). This hypothesis will undoubtedly be the subject of reassessment through future studies of both aDNA and modern sampling, which we shall touch upon in the next section.

Revising the record through studies in DNA

DNA is rewriting our understanding of the complexity of Australia's early population history. It also serves as a catalyst to reconsider the meaning of the diversity in the Australian human fossil and osteological record. Studies on living populations have distinguished unique Y chromosome and mtDNA haplotypes in Aboriginal Australia which indicate that migrations into Australia through Asia predate 50 ka (Keinan et al., 2007; Gutenkunst et al., 2009; van Holst Pellekaan, 2013). However, like all other non-African populations, all Aboriginal Australian mtDNA studies demonstrate that their ancestry can be traced to a single L3 haplotype.

Initial forays into ancient DNA (aDNA) generated much excitement in archaeological/palaeoanthropological circles. The announcement in 2001 by Adcock et al. of the identification of an extinct mitochondrial lineage brought Lake Mungo once again into the global limelight. Their argument suggested that the 40 ka WLH 3 fossil and a later fossil from Kow Swamp (KS 8) contained a non-L3 mitochondrial lineage that was identified as no longer being present in modern populations. This suggested to the authors that two populations had co-existed in Australia, with the WLH 3 mitochondrial genome representing an extinct human lineage (Adcock et al., 2001). In a forum in *Archaeology in Oceania*, Groves noted that the triumph of the paper, that being of aDNA extraction from such ancient fossils, was overshadowed by what he regarded as the shortcomings of the authors' interpretations. His concerns were over the lack of support for the branch in their cladogram for modern humans and the other potential interpretations of the data (such as the lineage sorting within an ancient population, as opposed to the replacement of a human lineage). In a later publication he discussed with the palaeoanthropologist David Cameron the possibility of contamination in the sequence by nDNA inserts (numts), which may have resulted in the observed mtDNA sequence. They also questioned as to whether the signature represented modern contamination (Cameron and Groves, 2004).

The meaning of the aDNA record from the Willandra series is undergoing a new analysis using next generation sequencing, and a review of this more recent DNA research is providing new insights into the evolutionary past of the first Australians. The identification of archaic DNA of Neanderthals and Denisovans in the DNA of Melanesian populations has revealed that there was indeed genetic exchange between the first modern humans in Central Asia and some of the later archaic hominin species during the Late Pleistocene period (Rasmussen et al., 2011). While we have evidence that some limited DNA sequences of Neanderthals and Denisovans were assimilated into *H. sapiens*, it is feasible that interactions with *Homo erectus/soloensis* did occur, but no viable offspring resulted. The divergence from *Homo erectus* was several hundred thousand years earlier than the split between *Homo sapiens* and *Homo neanderthalensis* (the recognised common ancestor between the latter two being *Homo heidelbergensis*, which lived during the later Middle Pleistocene). Low rates of Neanderthal and Denisovan DNA can be interpreted in a number of ways. Perhaps reproductive success was not high and hence it is reasonable to assume that reproductive success would have been even less likely between *Homo sapiens* and *Homo erectus*. Current evidence suggests that there is no support for overlap between these two species (Westaway and Groves, 2009; Indriati et al., 2011).

An important component of mtDNA studies from Melanesia (Friedlaender et al., 2005), as highlighted by Davidson (2010 and 2013), is that the genetic evidence indicates two possible migration events, with a separation between the mtDNA of the colonists of north and east Papua New Guinea, with those from southern Sahul. Indeed it would appear that the haplotypes from Melanesia represent earlier haplotypes from those present in southern Sahul (Merriwether et al., 2005). The Australian mt haplogroups S, O, and M42a are not shared with New Guinea on current evidence and this, together with shared haplogroups P and Q, has been interpreted as indicating separate origins of Australian and New Guinea people, while at the same time suggesting different migration histories of these separate groups of colonisers.

More recent research has identified a genomic signature said to be representative of substantial gene flow between the Indian populations and Australia during the mid Holocene some 4,230 years ago (Pugach et al., 2013). This research group attempted to link these changes with the arrival of the dingo in Australia and the appearance of changes in stone tool technology. However, the latter of these changes have been demonstrated by archaeologists to have occurred much earlier in some regions of Australia (Hiscock, 2008), and are likely to have little relevance to a possible migration from India. The other key issue with the proposed Indian migration is how did such populations enter Australia in the mid Holocene without leaving any trace in island Southeast Asia? While an influx of additional genetic material into Australia during the Holocene possibly did occur, it is still not clear if this left any significant genetic signature. This question will require further review to establish the meaning of any such genetic signature.

Discussion

Colin Groves has made a significant contribution to our understanding of human evolution in Sunda and Sahul. One aspect of human evolution in the region that all would seem to agree upon is that resolving many of the questions discussed in the past by Groves and others will only be possible through the discovery of new specimens. Importantly, it also requires a clearer chronological sequence for the fossils, in order to place them reliably in an evolutionary framework. Refining our techniques of analysis is critical to help establish more reliable means of hypothesis testing. An approach that involves splitting fossil specimens into different taxonomically recognised species (or operationally taxonomic units, as Groves prefers) holds a great deal of potential for unravelling the nature of phylogenetic relationships in human evolution. If species are lumped together, and autapomorphies in recognisable specimens are effectively ignored, then we risk the opportunity of identifying examples of speciation within our genus.

In this chapter we have considered a number of key themes in human origins research in the regions of Sunda and Sahul. The taxonomic subdivision of various fossil discoveries of Sunda spanning some 1.5 million years into four separate species (as opposed to two) is an approach that is currently difficult to test phylogenetically, due to the fragmentary nature of the early Sangiran record. As demonstrated by Kaifu et al., it is difficult to establish with certainty if the pre-Grenzbank fossils from Sangiran truly represent a distinct species. Kubo et al. (2013) have recently demonstrated that dwarfism in *Homo floresiensis* is a less dramatic prospect when we consider this with the earlier Javan hominins, and not the later classic *Homo erectus*. Previously Kaifu et al. (2005) have raised the possibility that the earlier Javan hominins may in fact represent an earlier species. We envisage that perhaps a more parsimonious explanation for the origins of *Homo floresiensis* is that it is derived from an earlier pre-*erectus* hominin. We agree that the pre-Grenzbank fossils probably represent a more likely candidate ancestor for *Homo floresiensis*, or perhaps a sister group to a close ancestor, than do the later *Homo erectus* fossils. There are some fossils from the pre-Grenzbank, such as the very large Sangiran 6 that may even represent a different genus, as suggested by Robinson (1953 and 1955). Sangiran 6 is unlikely to represent an ancestor of either *Homo floresiensis* or *Homo erectus*. It is probable that it was the gradual divergence between species within these different populations, following the geographic isolation of one population on Flores, which led to the emergence of the new species *Homo floresiensis*. In this model then *Homo modjokertensis* was not necessarily a terminal species, and it was also not the ancestor of *Homo erectus*. The ancestral candidate for *Homo erectus* was *Homo ergaster*. We suggest that *Homo modjokertensis* on Java was eventually replaced by *Homo erectus*, the timing of which remains unknown.

Is it reasonable to continue identifying *Homo erectus* as a distinct species that remained as the same taxon for around 1 Mya-800 ka? *Homo erectus* continued along its evolutionary trajectory following the extinction of *Homo modjokertensis*. If the species evolved on Java for 800 ka or more, should we regard the late *Homo erectus* at the sites of Ngandong, Ngawi and Sambungmacan as a different species? Similar divisions have been made within the transitory grade between *Homo heidelbergensis* and *Homo neanderthalensis*. There is an increasing acceptance of evolutionary change within hominin species that did not necessarily lead to the lineage of *Homo sapiens*. This perhaps requires for consideration the designation of a later species *Homo soloensis*.

The presence of *Homo floresiensis* in the lesser Sunda Islands suggests that hominin evolution on the edge of the hominin range was no less stagnant than evolution in other regions closer to the original centre. Perhaps it is time that we begin considering that evolutionary processes in Sunda are as complex as those identified in Africa and Europe, with a model of punctuated equilibrium perhaps being relevant to Pleistocene Sunda. *Homo soloensis* was eventually replaced by *Homo sapiens* sometime after 100,000 years ago. *Homo floresiensis* was replaced soon after the arrival of *Homo sapiens* in the lesser Sunda Chain, following the arrival of modern people on the eastern side of the Wallace Line, perhaps some 60,000 years ago. It is apparent that our species was likely responsible for the extinction of two hominin lineages in Sunda, although the timing of the disappearance of *Homo erectus* remains unresolved.

Microevolutionary changes are most likely responsible for the variation that we see within *Homo sapiens* in Pleistocene Sahul. It would not appear necessary to call upon models that discuss assimilation of Sunda hominins to explain the appearance of cranial robusticity around the time of the Last Glacial Maximum. The emergence of cranial robusticity perhaps has more to do with the onset of glacial conditions in Sahul. DNA studies may be able to assist us in unravelling the gracile and robust debate, by establishing if the variation is simply just reflective of sexual dimorphism. Importantly, DNA is beginning to reveal important new information that is not available from the study of fossil morphology alone. The evidence for genetic exchange between our species and the Late Pleistocene hominins *Homo neanderthalensis* and the enigmatic Denisovans is an exciting development resulting from such studies.

Conclusion

We are very fortunate that amongst his diverse taxonomic interests Colin Groves has been able to invest some time into the question of hominin phylogeny in Sunda and Sahul. The discipline in Australia is very much richer as a result.

In this paper we have suggested a hypothesis for hominin evolution in Sunda and Sahul building on the theoretical approach of Groves. It will be tested, and potentially falsified, through the acquisition of new evidence (both fossil and DNA), the refinement of the chronology of known fossil hominins, and the development of new techniques of phylogenetic analysis on both new and old data.

Acknowledgements

Thank you very much to the editors for their kind invitation to be part of this important volume. We are grateful to Dr Christine Hertler and Dr Friedemann Schrenk from the Senckenberg, Frankfurt, for providing access to the Sangiran 6 specimen in their collection and allowing me to compare these to casts within the collection. We are also grateful to Dr Dominique Grimaud-Hervé from The Institut de paléontologie humaine, Paris, for allowing MCW access to the cast of Sangiran 31 and for discussions on the possible meaning of its morphology. Professor Mark Collard from Simon Fraser University provided many valuable comments on the possible meaning of the Sangiran 6 mandible, a jaw fragment that we hope to review with Prof Collard in further detail at some stage in the future. The Elders from the Willandra Lakes World Heritage Area have provided ongoing support for our research into the origin of their ancestors, which we are deeply grateful for. Finally, the late Professor Mike Morwood provided the images of *Homo floresiensis*, he was a very generous and supportive scholar and the discipline of palaeoanthropology in this region has suffered a great loss with his passing.

References

Adcock GJ, Dennis ES, Easteal S, Huttley GA, Jermiin LS, Peacock J, Thorne A. 2001. Mitochondrial DNA sequences in ancient Australians: Implications for modern human origins. *Proc Natl Acad Sci* 98(2):537–542.

Anton SC, Spoor F, Fellman CD, Swisher III CC. 2007. Defining *Homo erectus*: Size considered. In: Henke W, Tattersall I, editors. *Handbook of Paleoanthropology*. Berlin: Springer-Verlag. pp. 1655–1693.

Antón SC, Weinstein KJ. 1999. Artificial cranial deformation and fossil Australians revisited. *J Hum Evol* 36:195–209.

Argue D, Morwood MJ, Sutikna T, Jatmiko, Saptomo EW. 2009. Homo floresiensis: A cladistic analysis. *J Hum Evol* 57:623–629.

Bird MI, Taylor D, Hunt C. 2005. Palaeoenvironments of insular Southeast Asia during the Last Glacial Period: A savanna corridor in Sundaland? *Quat Sci Rev* 24(20–21):2228–2242.

Bräuer G. 1992. Africa's place in the evolution of *Homo sapiens*. In: Bräuer G, Smith, editors. *Continuity or replacement: Controversies in* Homo sapiens *evolution*. Rotterdam: AA Balkema. pp. 83–98.

Brown P. 1989. Coobool Creek. A morphological and metrical analysis of the crania, mandibles and dentitions of a prehistoric Australian human population. Terra Australis, 13. Canberra: Department of Prehistory, Australian National University.

Brown P, Maeda T. 2009. Liang Bua Homo floresiensis mandibles and mandibular teeth: A contribution to the comparative morphology of a new hominin species. *J Hum Evol* 57:571–596.

Brown P, Sutikna T, Morwood MJ, Soejono RP, Jatmiko Wayhu Saptomo E, Awe Due R. 2004. A new small-bodied hominin from the Late Pleistocene of Flores, Indonesia. *Nature* 431:1055–1061.

Bulbeck FD. 2001. Robust and gracile Australian Pleistocene crania: The tale of the Willandra Lakes. In: Simanjuntak T, Prasetyo B, Handini R, editors. *Sangiran: Man, culture, and environment in Pleistocene times: Proceedings of the International Colloquium on Sangiran Solo-Indonesia*. Jakarta: Yayasan Obor. pp. 60–106.

Cameron DW, Groves CP. 2004. *Bones, stones and molecules*. London: Elsevier.

Collard M, Wood BA. 2007. Defining the genus *Homo*. In: Henke W, Rothe H, Tattersall I, editors. *Handbook of Paleoanthropology*. Berlin Heidelberg: Springer. pp. 1575–1610.

Curnoe D, Thorne A. 2006. Human origins in Australia: The skeletal evidence. *Before Farming* [online version] 2006/1 Article 5.

Curnoe D, Green H. 2013. Vault thickness in two Pleistocene crania. *J Archaeol Sci* 40(2):1310–1318.

Davidson I. 2010. A lecture by the returning Chair of Australian Studies, Harvard University, 2008–09 Australian archaeology as a historical science. *J Australian Studies* 34(3):377–398.

Davidson I. 2013. Peopling the last new worlds: The first colonisation of Sahul and the Americas. *Quatern Int* 285:1–29.

Delson E, Eldridge N, Tattersall I. 1977. Reconstruction of hominid phylogeny: A testable framework based on cladisitic analysis. *J Hum Evol* 6:263–278.

de Vos J, Sondaar PY, Van Den Bergh GD, Aziz F 1994. The *Homo* bearing deposits of Java and its ecological context. *Cour Forsch Inst Senckenberg* 171:129–140.

Dubois E. 1894. Pithecanthropus erectus, *eine menschenaehnliche Ubergangsform aus Java*. Batavia: Landesdruckerei.

Durband AC. 2002. The squamotympanic fissure in the Ngandong and Sambungmacan Hominids: A reply to Delson et al. *Anat Rec* 266:138–141.

Durband AC. 2004. A test of the multiregional hypothesis of modern human origins using basicranial evidence from Indonesia and Australia. PhD dissertation, University of Tennessee.

Durband A. 2006. Craniometric variation within the Pleistocene of Java: The Ngawi 1 cranium. *Hum Evol* 21:193–201.

Durband A. 2007. The view from down under: A test of the multiregional hypothesis of modern human origins using the basicranial evidence from Australasia. *Collections in Anthropology* 3:651–659.

Durband AC. 2008a. Artificial cranial deformation in Pleistocene Australians: The Coobool Creek sample. *J Hum Evol* 54(6):795–813.

Durband AC. 2008b. Artificial cranial deformation in Kow Swamp 1 and 5: A response to Curnoe (2007). *Homo* 59:261–269.

Durband AC. 2008c. Mandibular fossa morphology in the Ngandong and Sambungmacan fossil hominids. *Anat Rec* 291:1212–1220.

Durband AC. 2009. Southeast Asian and Australian paleoanthropology: A review of the last century. *J Anthropol Sci* 87:7–31.

Durband AC, Westaway MC. 2013. Perspectives on the origins of modern Australians. In: Smith FH, Ahern J, editors. *The origins of modern humans: Biology reconsidered*. Wiley-Blackwell.

Frayer DW, Wolpoff MH, Thorne AG, Smith FH, Pope GG. 1993. Theories of modern human origins: The paleontological test. *Am Anthropol* 95:14–50.

Friedlaender J, Schurr T, Gentz F, Koki G, Friedlaender F, Horvat G, Babb P, Cerchio S, Kaestle F, Schanfield M, Deka R, Yanagihara R, Merriwether DA. 2005. Expanding southwest Pacific mitochondrial haplogroups P and Q. *Mol Biol Evol* 22:1506–1517.

Grimaud-Hervé D. 2001. Taxonomic position of the Sangiran 31 Hominid. In: Simanjuntak T, Prasetyo B, Handini R, editors. *Sangiran: Man, culture, and environment in Pleistocene times: Proceedings of the International Colloquium on Sangiran Solo-Indonesia*. Jakarta: Yayasan Obor. pp. 46–53.

Groves CP. 1989a. *A theory of human and primate evolution*. Cambridge: Cambridge University Press.

Groves CP. 1989b. A regional approach to the problem of the origin of modern humans in Australia. In: Mellars PA, Stringer CB, editors. *The human revolution: Behavioural and biological perspectives on the origins of modern humans*. Edinburgh: Edinburgh University Press. pp. 274–285.

Groves CP. 1996. Hovering on the brink but not quite getting there. *Perspect Hum Biol* 2:83–87.

Groves CP. 2001. Lake Mungo 3 and his DNA. *Archaeol Ocean* 36(3):166–167.

Groves, CP 2008. Walking with Hobbits. *Australasian Science* (March):16–19.

Groves CP, Lahr MM. 1994. A bush not a ladder: Speciation and replacement in human evolution. *Perspect Hum Biol* 4:1–11.

Groves CP, Mazák V. 1975. An approach to the taxon of Hominidae: Gracile Villafranchian hominids of Africa. *Casopsis Pro Mineralogii* 20:225–246.

Grün R, Spooner N, Magee J, Thorne A, Simpson J, Yan G, Mortimer G. 2011. Stratigraphy and chronology of the WLH 50 human remains, Willandra Lakes World Heritage Area, Australia. *J Hum Evol* 60(5):597–604.

Gutenkunst RN, Hernandez RD, Williamson SH, Bustamante CD. 2009. Inferring the joint demographic history of multiple populations from multidimensional SNP frequency data. *PLoS Genetics* 5(10):e1000695.

Hawks J, Oh S, Hunley K, Dobson S, Cabana G, Dayalu P, Wolpoff MH. 2000. An Australasian test of the recent African origin theory using the WLH-50 calvarium. *J Hum Evol* 39:1–22.

Hiscock P. 2008. *The archaeology of ancient Australia*. Routledge, London.

Holliday TW, Franciscus RG. 2009. Body size and its consequences: Allometry and the lower limb length of Liang Bua 1 (*Homo floresiensis*). *J Hum Evol* 57(3):223–228.

Hyodo M. 2011. High resolution record of the Matuyama-Brunhes transition constrains the age of Javanese *Homo erectus* in the Sangiran dome, Indonesia. *Proc Natl Acad Sci* 108.

Indriati E, Swisher III CC, Lepre C, Quinn RL, Suriyanto RA, Hascaryo AT, Grün R, Feibel CS, Pobiner BL, Aubert M, Lees W, Antón SC. 2011. The age of the 20 meter Solo River Terrace, Java, Indonesia and the survival of *Homo erectus* in Asia. *PLoS One* 6(6):e21562.

Jungers WL, Harcourt-Smith WEH, Wunderlich RE, Tocheri MW, Larson SG, Sutikna T, Awe Due R, Morwood MJ. 2009b. The foot of *Homo floresiensis*. *Nature* 459:81–84.

Jungers WL, Larson SG, Harcourt-Smith W, Morwood MJ, Sutikna T, Awe Due R, Djubiantono T. 2009a. Descriptions of the lower limb skeleton of *Homo floresiensis*. *J Hum Evol* 57:538–554.

Kaifu Y, Aziz F, Indriati E, Jacob T, Kurniawan I, Baba H. 2008. Cranial morphology of Javanese *Homo erectus*: new evidence for continuous evolution, specialization, and terminal extinction. *J Hum Evol* 55(4):551–580.

Kaifu, Y, Baba H, Aziz F, Indriati E, Schrenk F, Jacob T. 2005. Taxonomic affinities and evolutionary history of the early Pleistocene hominins of Java: dentognathic evidence. *Am J Phys Anthrop* 128:709–726.

Keinan A, Mullikin JC, Patterson N, Reich D. 2007. Measurements of the human allele frequency spectrum demonstrates greater genetic drift in East Asians than in Europeans. *Nat Genet* 29:1251–1255.

Kubo D, Kono RT, Kaifu Y. 2013. Brain size of *Homo floresiensis* and its evolutionary implications. *Proc R Soc Ser B* 280(1760):1–9.

Lahr MM. 1996. *The evolution of modern human diversity: A study of cranial variation*. Cambridge: Cambridge University Press.

Larick R, Ciochon RL, Zaim Y, Sudijono S, Rizal Y, Aziz F, Reagan M, Heizler M. 2001. Early Pleistocene 40AR/39Ar ages for Bapang Formation hominins, Central Java, Indonesia. *Proc Natl Acad Sci* 98:4866–4871.

Leigh SR. 1992. Cranial capacity evolution in *Homo erectus* and early *Homo sapiens*. *Am J Phys Anthrop* 87:1–13.

Macintosh NWG, Larnach SL. 1976. Aboriginal affinities looked at in world context. In: Kirk RL, Thorne AG, editors. *The origins of the Australians*. Canberra: Australian Institute of Aboriginal Studies Press. pp. 113–126.

Merriwether DA, Hodgson JA, Friedlaender FR, Allaby R, Cerchio S, Koki G, Friedlaender JS. 2005. Ancient mitochondrial M haplogroups identified in the Southwest Pacific. *Proc Natl Acad Sci* 102:13034–13039.

Morwood MJ, Brown P, Jatmiko, Sutikna T, Wahyu Saptomo E, Westaway KE, Rokus Awe Due, Roberts RG, Maeda T, Wasisto S, Djubiantono T. 2005. Further evidence for small-bodied hominins from the late Pleistocene of Flores, Indonesia. *Nature* 437:1012–1017.

Morwood MJ, O'Sullivan PO, Susanto EE, Aziz F. 2003. Revised age of Mojokerto 1, an early *Homo erectus* cranium from East Java, Indonesia. *Austral Archaeol* 57:1–4.

Morwood MJ, Soejono RP, Roberts RG, Sutikna T, Turney CSM, Westaway KE, Rink WJ, Zhao J-X, Van den Bergh GD, Awe Due R, Hobbs DR, Moore WM, Bird MI, Fifield LK, 2004. Archaeology and age of a new hominin from Flores in eastern Indonesia. *Nature* 431:1087–1091.

O'Connor S, Ono R, Clarkson C. 2011. Pelagic fishing at 42,000 years before the present and the maritime skills of modern humans. *Science* 334(6059):1117–1121.

Olley, J. M., Roberts, R. G., Yoshida, H., & Bowler, J. M. 2006. Single-grain optical dating of grave-infill associated with human burials at Lake Mungo, Australia. *Quaternary Science Reviews*, 25(19), 2469-2474.

Oppenheimer S. 2004. *Out of Eden: The peopling of the world*. London: Robinson Publishing.

Oppenoorth WFF. 1932. *Homo (Javanthropus) Soloensis*. Een Plistoceene mensch van Java. *Wetensch. Meded. Van den Dienst van den Mijnbouw Nederlandsch-Indië*, No. 20:49–63.

Pugach I, Delfina F, Gunnarsdóttira E, Kayserd M, Stoneking M. 2013. Genome-wide data substantiate Holocene gene flow from India to Australia. *Proc Natl Acad Sci* 110(5):1803–1808.

Rasmussen M, Guo X, Wang Y, Lohmueller KE, Rasmussen S, Albrechtsen A … & Krogh A. 2011. An Aboriginal Australian genome reveals separate human dispersals in Asia. *Science* 334(6052):94–98.

Rightmire GP. 1990. *The evolution of Homo erectus: Comparative anatomical studies of an extinct human species*. Cambridge: Cambridge University Press.

Rightmire GP. 1991. The dispersal of *Homo erectus* from Africa and the emergence of more modern humans. *J Anthropol Res* 47:177–191.

Robinson JT. 1953. *Meganthropus*, australopithecines, and hominids. *Am J Phys Anthrop* 11:1–38.

Robinson JT. 1955. Further remarks on the relationship between *"Meganthropus"* and australopithecines. *Am J Phys Anthrop* 13:429–446.

Rightmire GP. 1994. The relationship of *Homo erectus* to later middle Pleistocene hominids. *Cour Forsch-Inst Senckenberg* 171:319–326.

Santa Luca AP. 1980. The Ngandong fossil hominids: A comparative study of a Far Eastern *Homo erectus* group. *Yale Univ Publ Anthropol* 78.

Stone T, Cupper ML. 2003. Last Glacial Maximum ages for robust humans at Kow Swamp, southern Australia. *J Hum Evol* 45(2):99–111.

Storm P. 2000. The evolutionary history of humans in Australasia from an environmental perspective. *Anthropol Sci* 108:225–244.

Storm P. 2001a. Life and death of *Homo erectus* in Australasia: An environmental approach to the fate of a paleospecies. In: Sémah F, Falguères C, Grimaud-Hervé D, and Sémah A-M, editors. *Origine des Peuplements et Chronologie des cultures Paléolithiques dans Le Sud-Est Asiatique*. Paris: Semenanjung. pp. 279–298.

Storm P. 2001b. The evolution of humans in Australasia from an environmental perspective. *Palaeogeogr Palaeoclimatol Palaeoecol* 171:363–383.

Thorne A. 1976. Morphological contrasts in Pleistocene Australians. In: Kirk RL, Thorne AG, editors. *The origin of the Australians*. Canberra: AIAS Press. pp. 95–112.

Thorne A, Wolpoff MH. 1981. Regional continuity in Australasian Pleistocene hominid evolution. *Am J Phys Anthrop* 55:337–349.

Tobias PV, von Koenigswald GHR. 1964. A comparison between the Olduvai hominines and those of Java and some implications for hominid phylogeny. *Nature* 204:515–518.

Tocheri M, Orr CM, Larson SG, Sutikna T, Jatmiko, Saptomo EW, Rokus Awe Due, Djubiantano T, Morwood MJ, Jungers WI. 2007. The primitive wrist bone of Homo. floresiensis and its implications for hominin evolution. *Science* 317:1743–1745.

Tyler DE, Krantz GS, Sartono S. 1995. The taxonomic status of the '*Meganthropus*' cranial (Sangiran 31) and the '*Meganthropus*' occipital fragment III. In: Bower JRF, Sartono S, editors. *Evolution and ecology of Homo erectus*. Leiden: Pithecanthropus Centennial Foundation, Leiden University. pp. 189–201.

van den Bergh GD. 1999. The late Neogene elephantoid-bearing faunas of Indonesia and their palaeozoogeographic. A study of the terrestrial faunal succession of Sulawesi, Flores, and Java, including evidence for early hominid dispersal east of Wallace's Line implications. *Scripta Geol* 117:1–419.

van den Bergh GD, de Vos J, Sondaar PY. 2001. The late Quaternary palaeogeography of mammal evolution in the Indonesian archipelago. *Palaeogeogr Palaeoclimatol Palaeoecol* 171:385–408.

Van den Bergh GD, de Vos J, Sondaar PY, Aziz F. 1996. Pleistocene zoogeographic evolution of Java (Indonesia) and glacio-eustatic sea level fluctuations: A background for the presence of Homo. *BullIndo Pac Pre hi* 14:7–21.

Van Holst Pellekaan. 2013. Genetic evidence for the colonization of Australia. *Quatern Int* 285:44–56.

Von Koenigswald GHR. 1956. *Meeting prehistoric man*. London: Thames and Hudson.

Webb SG. 1990. Cranial thickening in an Australian hominid as a possible palaeoepidemological indicator. *Am J Phys Anthrop* 82(4):403–411.

Webb SG. 2006. *The first boat people*. Cambridge: Cambridge University Press.

Weidenreich F. 1943. The skull of *Sinanthropus pekinensis*: A comparative study on a primitive hominid skull. *Palaeontol Sin* NS D 10.

Westaway, M.C. 2006. *The Pleistocene human remains collection from the Willandra Lakes World Heritage Area, Australia, and its role in understanding modern human origins*. National Science Museum Monographs No. 34. Tokyo: National Science Museum.

Westaway MC, Groves CP. 2009 The Mark of Ancient Java is on none of them. *Archaeol Oceania* 44:84–95.

Westaway MC, Lambert D. 2014. Origins of the First Australians. In: Smith C, editor. *The global atlas of archaeology*. New York: Springer.

Widianto H, Zeitoun V. 2003. Morphological description, biometry and phylogenetic position of the skull of Ngawi 1 (East Java, Indonesia). *Int J Osteoarchaeol* 13:339–351.

Wood BA and Collard M. 1999. The human genus. *Science* 284:65–71.

Wright RVS. 1976. Evolutionary process and semantics: Australian prehistoric tooth size as a local adjustment. In: Kirk RL, Thorne AG, editors. *The origin of the Australians*. Canberra: AIAS Press. pp. 265–274.

Zeitoun V. 2009. *The human canopy:* Homo erectus, Homo soloensis, Homo pekinensis *and* Homo floresiensis. BAR International Series 1937. Oxford: John and Erica Hedges Ltd.

Zeitoun V, Detroit F, Grimaud-Herve D, Widianto H. 2010. Solo man in question: Convergent views to split Indonesian *Homo erectus* in two categories. *Quaternary International* 223-224:281–292.

PART IV

13. The domestic and the wild in the Mongolian horse and the takhi

Natasha Fijn

Introduction

> Against a strong head wind we reached this almost flat plateau, at a
> height of 6,800 ft., and immediately saw, a few kilometers off, a group of
> animals galloping away from us at full speed. My Mongolian companion,
> Namkhajdorj Balgan, recognised them at once with the naked eye as
> Przewalski wild horses, and subsequent observation with a telescope
> confirmed beyond any possible doubt that this was what they were …
> We followed them for about six or eight miles over quite open ground and
> observed them by telescope until they disappeared (Kaszab, 1966: 346).

The account above is by a scientist on a joint Hungarian–Mongolian expedition
in the summer of 1966. This was the last published sighting of horses in the wild
in the scientific literature. The final sighting of this unique horse was actually
recorded in 1969 by the Mongolian scientist N. Dovchin (Bouman and Bouman,
1994). There were, no doubt, further sightings by local Mongolian herders
travelling with their herds to waterholes in the area. These herding families
would have had an intimate knowledge about the habits of the takhi, passed
down from one generation to another over the millennia. The location, near
the border between China and Mongolia, was sparsely populated and the final
retreat for the last Asiatic wild horse population. Mongolian herders called this
mountainous location *Takhiin Shar Nuruu*, which translates as The Yellow Wild
Horse Mountain Range. With their knowledge of horses, Mongolian herders
would have recognised the significance of this area as a refuge for the takhi.

One of the earliest written references to the takhi was in *The Secret History of the
Mongols* where it describes how Chinggis Khan fell from his horse when it was
startled by takhi (Bouman and Bouman, 1994). To Mongolians the takhi is the
'father' of the Mongolian horse. The horse is an inherent part of their identity
as herders and as Mongolians. Stamina and strength are prized characteristics
in a Mongolian horse and the takhi represents the epitome of these features.
The intention of this paper is to re-examine the notions of domestic and wild in
relation to the 'domestic' Mongolian horse and the 'wild' takhi from Mongolia,

not just from a western perspective but also through a cross-cultural lens from living with herders and their herd animals in the Khangai Mountains of Mongolia.

Figure 13.1: Takhi herd with stallion to the right, Hustai Nuruu National Park, 2007.

Source: Photograph by Natasha Fijn.

Taxonomy and morphology of wild horse characteristics

During the Paleolithic, in both present-day France and Spain, humans were painting and etching horses on the walls of caves and making small clay horse figurines. These depictions of the wild horses they hunted bear a remarkable resemblance to the Asiatic wild horse, or *Equus przewalskii*. The horses had upright manes and 'beards' along the jaw line with light muzzles and underparts (see Figures 4–6 in Mohr, 1971; Figure 2 in Pruvost et al., 2011).

When Linnaeus (1707–1778) formed the classification system for scientific nomenclature, he made no mention of the Asiatic wild horse because such a horse was not yet known to exist according to the current knowledge in Europe (Bouman and Bouman, 1994). A wild horse population, *Equus ferus ferus*, known as the tarpan, was still clinging on to existence on the steppes of the Ukraine when another wild horse population was 'discovered' in 1881 (or made more widely known) through Colonel Przewalski's reports on his explorations into Central Asia. The tarpan was mouse-grey and would turn almost white in the winter to camouflage with its snowy environment. In contrast the Asiatic wild horse varied from rust or dun-coloured to bay (Groves, 1994; see Figures 13.1 and 13.2).

In 1986, Colin Groves recognised two wild horse taxa that survived into historic times, *Equus ferus przewalskii* Poliakov, 1881, and *Equus ferus ferus* Boddaert, 1785. Groves outlined key differences in external and cranial morphology between the 'wild type' Przewalski horse and that of the domestic horse from examining museum specimens, photographs, the current literature and descriptive accounts of the horse. These differences were based on a close analysis of the stature, weight, body conformation, head form, body colour, mane and tail, body hair, cranial characteristics and postcranial skeleton (described by Groves and Willoughby, 1981; Groves, 1994; Groves and Grubb, 2011).

More recently, in a revision of the taxonomy of ungulates, Groves concluded that *Equus przewalskii* should be recognised as distinct; that the domestic horse should retain its historic scientific name *Equus caballus*; while the extinct tarpan should be known in the scientific nomenclature simply as *Equus ferus* (Groves and Grubb, 2011). As Groves states, 'it is often convenient to refer to domestic animals by their own scientific name, regardless of the fact that they might, in fact, be of mixed (or hybrid) origin' (2011: 8).

The research by Colin Groves on the taxonomy and the morphology of *Equus przewalskii* and his interest in domesticated animals was one of the reasons that I began research based in Mongolia. I was initially interested in conducting cross-cultural research in Mongolia to gain a window into the domestication processes in relation to the horse. The research for my doctoral thesis broadened to encompass ethnography about multispecies hybrid communities, both human and nonhuman, including Mongolian herders, horses, cattle (including yak), sheep and goats. Throughout 2005 and during a shorter trip in the spring of 2007 I lived with Mongolian herders in two extended family encampments in the Khangai Mountains. Through daily interaction, while milking and herding, individual herders gain a remarkable knowledge about the reproduction, breeding, communication, social behaviour and ecology of these herd animals. Horses are a particularly significant part of Mongolian herding culture (see Fijn, 2011).[1]

---

1 To see footage, by the author, of the importance of the horse in relation to horse racing within a winter festival, see: http://khangaiherds.wordpress.com/about/khangai-herds-extra/.

Figure 13.2: Individual takhi at Hustai Nuruu National Park, 2007.

Source: Photograph by Natasha Fijn.

Research in the field was conducted through observational filmmaking, semi-structured interviews and the anthropological technique of participant observation, living in a herding encampment with an extended herding family (which included living amongst herd animals). Within this chapter I return to what initially sparked my interest in Mongolia in the first place, the horse, and through a multidisciplinary, cross-cultural perspective I examine the relationship between the 'domestic' Mongolian horse and the 'wild' takhi (*Equus przewalskii*) (see Figures 13.1 and 13.2).

The common name for *Equus przewalskii*

Human, Mongolian horse and takhi all co-existed in the same environment. Mongolians refer to the predominant breed of horse that is native to Mongolia as *Mongol* and to the one that is not tamed by humans as *takhi*.[2] Historically, Mongolians also distinguished between a darker mountain variety and a lighter steppe variety of takhi.[3] Colonel Przewalsky, by contrast, mistook sighting

---

2   Henceforth in this paper I refer to the tame horse (*Equus caballus*) that co-exists with Mongolian herders as the 'Mongolian horse' and to *Equus przewalskii* as 'takhi'.
3   This aligns with the views of early twentieth century authors, who named and described two different geographic populations of takhi, recognising them as distinct.

Mongolian wild asses, or *khulan*, for wild horses when they galloped off as a herd into the distance. It was only due to good fortune and luck that he happened to be given what became the type specimen, by Kirghiz hunters on the border between Russia and Mongolia.

Many academics working on research relating to the rehabilitation of *Equus przewalskii* into reserves refer to the horse by the Mongolian name of *takhi*. Van Dierendonck and Wallis de Vries (1996) state that they prefer the name takhi because Przewalski horse is misleading in that it should not be confused with the domestic Mongolian horse. The Mongolian term takhi recognises the status of this horse as a significant part of Mongolia's cultural heritage. Colonel Przewalski, sometimes spelt Przhevalsky or Prjevalsky, also had his name attributed to a species of gazelle and 80 different plant species. According to Meyer and Brysac, Przewalsky was a ruthless exploiter of the Central Asian peoples he encountered, travelling 'with a carbine [shotgun] in one hand, a whip in the other' (cited in Nalle, 2000: 199–200). The fact that the horse is still commonly referred to as 'Przewalski's horse' denotes a retention of a colonialist and imperialist form of ownership of both the name and the horse itself.

The domestic and the wild

There is a tendency in the western world to form Cartesian dualistic divisions between nature and culture, male and female, domestic and wild, native and feral. This stems from a historical basis in Judeo-Christian traditions and beliefs where the world is categorised and divided into good and evil, right and wrong, and continues into present-day Western knowledge systems, where this framework can inadvertently persist. This is not to say, however, that a categorisation according to domestic and wild animals should not be made. The intention within this chapter is to avoid viewing the domestic and the wild as 'hyper-separated' and instead think of them along a continuum, where there can be considerable crossover between the two forms. The environmental philosopher Val Plumwood provides a clear perspective on human/nature dualism in saying that this mindset 'conceives the human as not only superior to but as different in kind from the non-human, which as a lower sphere exists as a mere resource for the higher human one. This ideology has been functional for Western culture in enabling it to exploit nature with less constraint, but it also creates dangerous illusions by denying embeddedness in and dependency on nature' (Plumwood, 2012: 15).

Sandor Bökönyi's definition of a domestic animal is representative of earlier views of animal domestication in the following: 'The essence of domestication is the capture and taming by man of animals of a species with particular behavioural characteristics, their removal from their natural living area and breeding community, and their maintenance under controlled breeding conditions' (Bökönyi, 1989: 22). Juliet Clutton-Brock recently defined 'true' domestication as the 'keeping of animals in captivity by a human community that maintains *total* control over their breeding, organization of territory, and food supply' (Clutton-Brock, 2012: 3 emphasis added). These definitions may apply to a categorisation according to a western perspective but this does not apply to a Mongolian herding perspective, nor to the way herd animals live in Mongolia (see chapter by Clutton-Brock, this volume).

The definition is of course correct in relation to how horses are kept in a western scenario, where the territory is decided upon by humans with often only one or two horses within a grassed area bounded by fences, or indoors on their own within a stable. Many are strictly bred according to studbooks through artificial insemination. Thoroughbreds are perhaps a good example of this definition of the domestic, in terms of *total* human control over their breeding, territory and food supply.[4]

The Mongolian horse is not held in captivity because the land in Mongolia is not fenced and they are free to roam without human-made barriers. They have not been removed from their natural environment, as the grassland steppe habitat was still occupied by wild horses until historic times and it is likely that their ancestors in the Pleistocene roamed over a similar grassland landscape, perhaps even in the same geographical location. Mongolian horses choose where to forage, they protect themselves from predators and even make their own way to new seasonal pastures ahead of the herding encampment. Mongolian horses exist within a social structure that they would naturally adopt, for example, the main herd generally consists of a stallion and up to 25 mares with their young. Herders do not have *total* control over their breeding, as a stallion can potentially impregnate mares from another herd if he can outcompete another stallion.

In contrast, for 13 generations the entire takhi population were socially disrupted, captured, removed from their natural living area by being transported from Mongolia to Europe and other parts of the world, separated into human controlled breeding communities, or zoos, and maintained under controlled living conditions. Humans had *total* control of their breeding and organisation of territory because not one individual remained on the grassland steppe of Mongolia. Because the population spent over 90 years in captivity, the takhi required protective enclosures before being released back into reserves

---

4   For an ethnography related to the breeding of thoroughbreds and horse racing, see Cassidy (2002).

in Mongolia, as individuals had not learnt how to survive the extremely harsh Mongolian winters without supplementary feed, nor that they could be prey to wolves.

The 'domestic' Mongolian horse is not closely monitored by herders against wolf attack (see Hovens and Tungalaktuja, 2005). During the harsh winter they do not require additional feed but survive on the standing fodder that grows only during the short summer season (see Figure 13.3). When snow covers the ground, Mongolian horses know how to paw through the snow to get at the vegetation beneath. Van Dierendonck and others (1996) and Linklater (2000) found that the behaviour and social structure of the takhi, once they had been rehabilitated back onto the grassland steppe, was remarkably similar to feral horse populations. This is also the case in Mongolian 'domestic' horse herds. I observed that their social organisation was remarkably similar to findings in relation to the reintroduced takhi and to 'feral' horse populations (see Fijn, 2011: 65–69).

Figure 13.3: A Mongolian horse herd in a snowstorm in spring, Arkhangai Province, 2005.

Source: Photograph by Natasha Fijn.

It seems ironic that, according to the dualistic categorisation of domestic and wild, the way the 'domestic' Mongolian horse exists does not fit the criteria of a domesticated animal, whereas the 'wild' Mongolian horse living within a captive zoo fits the accepted definition of a domestic animal well. It reveals a difference in attitude whereby '[t]he behaviour of a species is often one of the last areas of their biology to be studied and one of the last aspects to be considered in making management decisions' (Boyd and Houpt, 1994: 267); whereas for Mongolian herders the behaviour of the horse is a priority. I agree with Kaczensky and others' (2007) recommendation that local Mongolian herders need to be more actively involved in the management and conservation of the takhi, including recognising their depth of knowledge in relation to the behaviour and social structure of the horse. Through the takhi's reintroduction back into three separate reserves in Mongolia since 1992, the herds have gradually been given some of their own agency back and, hopefully, will ultimately be released from *total* human control and management.[5]

In comparison to many rangelands throughout the world, the pastoral ecosystem in Mongolia is relatively intact (see Mallon and Jiang, 2009). The herders, herd animals and their wild counterparts have all co-existed and adapted within a functioning ecosystem for thousands of years. A sign of a healthy ecosystem is if top predators still persist in the landscape and in Mongolia there are still wolves (*Canis lupus*) and, although endangered, other large predators, such as the Gobi bear (*Ursus arctos*) and snow leopard (*Uncia uncia*). Large ungulates still inhabit Mongolia including wild representatives that are related to the domestic animals, such as argali (*Ovis ammon*), ibex (*Capra sibirica*) and wild Bactrian camels (*Camelus ferus*). In many of these cases Mongolia functions as one of the last vestiges for these large mammals that during the Pleistocene would have lived in far greater numbers across large parts of Eurasia (Bedunah and Schmidt, 2000).

A cross-cultural re-examination of the accounts of capturing takhi

Cultural differences in approach to the human–horse relationship have been evident from ancient and classical history. These differences persist to the present day. There are two main approaches, a co-operative approach based upon understanding the behaviour of the horse, and an alternative approach based on human dominance and equine submission. Social interactions and contact between humans and horses have reflected these differences in approach (Van Dierendonck and Goodwin, 2005: 65).

---

5 There has been a large body of scientific research focusing on the behaviour and ecology of the takhi after reintroduction into reserves in Mongolia (see King, 2002; King and Gurnell, 2005, 2007; Van Dierendonck and Wallis de Vries, 1996; Van Dierendonck et al., 1996).

This nicely summarises the difference between a historically Euro-American approach toward horses and a Mongolian herding one. In this section I draw upon my observations of how herders captured and tamed Mongolian horses and integrate these ethnographic details with earlier historic accounts to consider how the recruitment of wild individuals into a tame Mongolian horse herd could potentially occur.

Figure 13.4: A prized Mongolian horse with elaborate saddle and bridle and wild-type characteristics, note the coat colour, stripes on the legs and eel-stripe down the back, 2005.

Source: Photograph by Natasha Fijn.

While in the field, I asked herders about their favourite horses. They were often selected on the basis of their fine appearance, distinctive markings, or quiet behaviour. One particular aspect that was particularly favoured were ancestral, or wild-type, characteristics, reminiscent of ancient cave paintings: a dun coat, dark mane and tail, dark eel-stripe down the back and faint stripes on the upper legs and neck. As early as 1868 Charles Darwin perceptively recognised these features as possible 'wild-type' characteristics in certain breeds of horse. Lusis (1943) described how the 'wild-dun colour' is common in Mongolian domestic horses and how it is always combined to some degree with striping on the legs. Bökönyi noted the resemblance of some domestic horses in Central Asia to the takhi, not only in the colouration of the horse but also the shape of the skull. Some Kazakh horses 'faithfully present the colour of the takhi, they have a well-developed dorsal stripe ... and they often have stripes on their legs' (1975: 85). From noting these characteristics in the field, I began to consider that perhaps these markings were an indication that the Mongolian horse represents an ancient breed of domestic horse and is evidence of their long association and bond with Mongolian herders; or perhaps even an indication that on rare occasions takhi may have interbred with Mongolian domestic horses and contributed to their genetic heritage (see Figure 13.4).

Bökönyi pointed out that a good motivation for herders to occasionally introduce individuals from a wild herd was to increase the number of herd animals, as the size of the herd is representative of a herding family's wealth. He states, 'even at the end of the last century and the beginning of ours it happened that Mongolian animal breeders would capture Przevalsky foals, admit them to their herds and rear them there: that is to say, they domesticated them' (1974: 85). From observing how herders in Mongolia catch foals with an *urgaa*, or lasso pole, I can readily envisage how individuals from a wild population could be captured. One way is through herders capturing foals by chasing the takhi herd on horseback until a foal tires, separating the foal from the herd, and then the herder could merely slip the *urgaa* over the exhausted foal's head.[6]

An account from one of the brothers Gram-Grshimailo, credited as being the first Europeans to see takhi in the wild, related such a scenario:

> [T]he Mongols have often attempted to tame [adult] wild horses but always in vain. The wild horse will not accept human contact, is terrified of them and will not allow himself to be used. Wild horses are caught in a fairly simple manner. During the foaling season the Kalmucks [western Mongolians in Russia] take two horses into the desert. As soon as they have found a herd, they chase them until the exhausted foals fall over. These foals are picked up and put into the domesticated herd (quoted in Mohr, 1971: 68).

This technique does correspond with techniques still used by Mongolian herders today, as I have observed herders use a similar technique on a number of occasions by relying on exhaustion to break in their two-year-old horses. One adult herder may ride the mare (or the horse's mother), while a lighter teenager leaps onto the two-year-old bareback, inducing the young horse to buck and gallop alongside the mare until it is too exhausted to attempt to get rid of the rider and becomes calm (also see Fijn, 2011: 144–147).

---

6   It is general knowledge within the oral history of Mongolian herding families that the two kinds of horse readily interbred in the past. I was informed that they produced 'good stock' but that herders kept their techniques a secret (Ranger at Hustai Nuruu, pers. comm., 13 April 2005).

Unfortunately, this technique of capturing the foals was misappropriated by European hunters to capture takhi in order to take them to zoos in Europe. The hunters were additionally instructed by Falz-Fein (a wealthy zoo owner who engaged in the business transactions with the hunters) to shoot the mares in the herd before capturing the foals. Falz-Fein was proud of this technique and wrote 'I laid much stress on the animals not being chased before capture, but rather by shooting their mother. As we could not get milking mares from the Mongolians living in the area, we had to buy them in Bijsk and have them covered so they foaled at the same time as the wild mares. In order to feed the wild foals we had to kill the tame ones' (Mohr, 1971: 95–96). Bökönyi regretted such a practice, as 'to kill the mother mares with a view to catch the foals more easily was an appalling procedure indeed … The killing of the mares put an end to the possibility of a number of further foals being born' (1974: 98).

A Mongolian scientist, S. Dulamtseren, commented that one of the reasons for the extinction of the last population of takhi in the wild was due to the breakdown of the social structure of the herd by these hunters activities (van Dierendonck and Wallis de Vries, 1996). On just one expedition in 1901 the hunters captured 51 foals. Of these, only 28 survived the stressful journey to a European zoo (Bökönyi, 1974). This practice would have been counter to the ethics of Mongolian herders, as they do not kill female animals that are still able to reproduce.

Even though horses are generally able to roam free within their own herd in Mongolia, during the summer season foals are tethered along pegged lines, so that the herders can obtain milk from the mares.[7] In the instance of the takhi foals, the herder would be able to encourage the foal to feed from a surrogate mare, just as herders do today. Often at least one mare loses her own foal to wolf predation and is encouraged to adopt another, or perhaps even two. The adopted foal is handled many times a day when it is taken to feed from the mare prior to milking and in this way the foal becomes quite tame and easily handled later in life. If a takhi foal grew up within a horse herd, nurtured by a surrogate mare, then the takhi is more likely to integrate with the herd as an adult and subsequently reproduce further (hybrid) offspring.

Bökönyi (1974) provided an archaeological example from kurgans (burial mounds) from the Turkic period in Mongolia. One of the horses sacrificed within the burial mound corresponded with the skull of an interbred horse from a takhi mare and a Mongolian stallion. It would not be a great biological leap for a takhi to breed with a Mongolian horse, as they do produce viable, fertile offspring. A greater biological distance exists between the Mongolian yak

---

7   To see footage by the author of a young herder catching adult horses with an urgaa, followed by foals and two-year-olds being handled during milking, see: www.khangaiherds.wordpress.com/about/khangai-herds-part-3/.

and the Mongolian cow, yet I observed how readily they breed within a mixed herd of a yak bull (or bulls), Mongolian cows and yak–cow hybrids (see Fijn, 2011: 87–89). In summary, it is highly probable that throughout history wild horses, in this instance takhi, were on occasion integrated and bred with tame horse herds, specifically Mongolian horse herds.

Bökönyi acknowledged that the takhi may have interbred with tame Mongolian horses in the past, but this 'does not at all reduce their quality as genuine wild horses' (1974: 45) and still means that they are a unique population of horse in their own right. With the inclusion of my ethnographic observations, it is feasible that individuals could have been caught from this distinct population and integrated within the domestic horse population as foals, ultimately contributing to the genetic heritage of the domestic Mongolian horse (see Figure 13.5).

Figure 13.5: Mongolian herder with Mongolian horse, note the coat colour, eel-stripe down the back and stripes on the legs, Bugat Province, 2005.

Source: Photograph by Natasha Fijn.

Recent evidence from DNA analysis on the domestic and the wild horse

From the genetic evidence, it is an unlikely scenario that horse domestication sprang up in just one location, but very likely that it occurred in a number of different geographical locations and that wild horses were repeatedly integrated within the domestic sphere even up until recent history when the last wild herds disappeared. To take a parallel hominid example, research on the Neanderthal genome found that Neanderthals have contributed toward the present day genetic ancestry of most humans beyond Africa (Green et al., 2010). Through studies into DNA, scientific ideas of what makes up a species, including humans and domestic animals, are rapidly changing.

Prior to modern day genetics it was thought that *Equus przewalskii* might represent the original wild ancestral stock from which all domestic horses stem. Through more recent DNA analysis it is now clear that they diverged as a distinct population before horses began to co-exist with humans (Ishida et al., 1995; Wallner et al., 2003). Early studies of their genetics indicate that takhi have a different number of chromosomes (2N = 66) in comparison to domestic horses (2N = 64) (Ryder, 1994), the difference being due to a simple Robertsonian translocation. They can readily interbreed and produce fertile offspring, whereas in other members of the genus *Equus* there is far greater chromosomal variation and the hybrid offspring are often infertile (for example, donkeys and horses produce infertile mules).

A large number of mares have contributed to a great variation of mitochondrial DNA in horses, indicating an 'extensive utilization and taming of wild horses' (Vila et al., 2001: 474). Jansen and others concluded perhaps more than 77 successfully breeding mares contributed to the domestic horse, leading them to suggest that 'the horse has been domesticated on numerous independent occasions' (2002: 531). Lindgren and others found that 'the maternal gene pool may have been diversified by the capture of only wild females from local populations (while backcrosses with wild stallions were prevented)' (2004: 335). Cieslak and others (2010) also found multiple introgressions of females adding to the genetic diversity, particularly from the Iron Age onwards. In the instance of Mongolia, this would have meant that once the captured takhi foals became adults, herders bred from takhi females with their Mongolian horse stallion.

Many genetic studies on the horse have focused specifically on the difference between the takhi and the domestic horse but the findings are controversial. Part of the lack of clarity surrounding the results of these analyses is that more than one domestic mare was interbred with takhi while in captivity (see Bökönyi, 1974; Mohr, 1971). This fact has been taken into consideration within genetic analyses (see for example, Oakenfull et al., 2000) but little consideration has been given to the likelihood of hybridisation with takhi occurring within the domestic horse population *before* the takhi disappeared from the grassland-steppe.

It is interesting that as early as 1995, Ishida and others had tentatively suggested that some gene flow occurred between the ancestral populations of the Przewalski wild horse and the Mongolian native horse. Since then a number of authors have found that there is substantial overlap in terms of their mitochondrial DNA, or the maternal line (see for example, Myka et al., 2003; Vila et al., 2001). Analysis of the nucleotide sequence diversity along the Y chromosome, or male line, has revealed that the takhi is distinct from the domestic horse, suggesting that incorporation of takhi genes into the Mongolian horse population was by mares, not stallions (Wallner et al., 2003; Lindgren et al., 2004). This could be explained by the Mongolian herding practice of castrating most male horses in their second year, so that they remain tame and easy to handle. The males could have been kept for riding and racing purposes. Only the female takhi would have become reproductive adults and mated with the domestic Mongolian stallion within the herd, producing fertile hybrid offspring. By 2009, Lau and others felt able to state with more certainty that there was a constant flow of female mediated genetic information between takhi and domestic horses, and used the term 'unidirectional hybridisation' to describe interbreeding between takhi and domestic horses.

Mongolian horses show the largest amount of genetic diversity in comparison with other Asian and European horse breeds (Ishida et al., 1995; Tozaki et al., 2003). This diversity means that closely related Japanese and Korean horse breeds are derived from Mongolian horses. The Norwegian Fjord horse resembles the takhi in many physical features (Groves, pers. comm.). Bjørnstad and others suggested a close genetic relationship between the Mongolian and Norwegian horse. They concluded that the 'Mongolian horse has had a major impact on a wide range of breeds, and that Central Asia could have been a centre of dispersal of horses through trading and human migration' (2003: 57). Dispersal of Mongolian horses as far as Western Europe would have occurred particularly during the time of the Mongol Empire when the Khans ruled the largest empire in history (Morgan, 1986). It is, therefore, important to retain this genetic diversity and a robust population of Mongolian horses. The Mongolian horse is unique, as representatives of ancient ancestral stock, and should be valued for its role, not only as a partner in the unprecedented migration during the Khan Empire, but in providing a valuable contribution to the genetics of horse breeds across Eurasia today (see Figure 13.6).

Figure 13.6: A saddled piebald (*kharlag*) Mongolian horse, Arkhangai Province, 2005.

Source: Photograph by Natasha Fijn.

Conclusion

The relationship between Mongolian herders and the Mongolian horse is unique. Through Mongolian herders allowing the horse greater agency, in allowing them to behave and live within a social structure that they would adopt of their own volition, they are allowing horses greater autonomy. Natural selection can still occur to a notable degree in the Mongolian horse and the population is likely to be more resilient as a result. Mongolian horses have retained their anti-predator behaviour against wolves and an ability to forage on the grassland throughout the seasons, without the dependence found in Euro-American breeds.

To Mongolian herders, the difference between a 'domestic' and a 'wild' animal is not considered so much according to a physiological or morphological condition but relates more to behavioural disposition: domestic animals are tame and associate with humans and the herding encampment, while those that are wild are afraid and choose not to live with humans. King and Gurnell describe how before the takhi were reintroduced into reserves, 'there was no [*Western*] knowledge about the ecology of takhi in the wild before they became extinct, and so it was not clear how the released animals would cope in their new surroundings' (2005: 278). It would have come as no surprise to Mongolian herders, however, that the reintroduced takhi would be capable of surviving on the grassland steppe, as their horse herds have been successfully doing so for tens of hundreds, if not thousands of years. It took me some months to adjust to life in Mongolia, the takhi also took time to adjust after their forebears had spent 13 generations in captivity in Europe. The takhi population is now thriving and have adapted smoothly to living on the grassland steppe,[8] just as many horses throughout the world have adjusted to become 'feral' and no longer require an association with humans to survive. Instead of viewing domestic and wild as a hyper-separated dichotomy, these different states should be thought of as being more fluid, where there can be considerable cross-over, or interbreeding, between the two.

Colin Groves is the world expert on the morphology and taxonomy of *Equus przewalskii*, or the takhi. It should be noted that he encourages those he has mentored to explore different avenues and he readily embraces interdisciplinary research. This has meant that I have not been constrained in my exploration of the relationship of the takhi and the Mongolian horse, through not only the taxonomic and morphological analysis, but also my ethnographic findings from Mongolia, previous historical accounts and the current wealth of scientific literature on the genetics of the horse. This has led me to conclude, in agreement with Colin Groves' previous findings, that the takhi are morphologically and genetically distinct from other horses and merit our conservation. Any conservation project with the goal of reintroducing 'wild' takhi herds back into Mongolia should seriously take into account Mongolian herding knowledge and expertise.

---

8 Reintroductions to Hustai Nuruu National Park began in 1992. By 2000 the population at this reserve alone consisted of 120 horses (King, 2002).

Prior to the capture and confinement of the entire takhi population in zoos across the globe, it is highly likely that there was integration of individual takhi within Mongolian horse herds and this would have inevitably influenced the diverse genetic makeup of the native Mongolian horse. This evidence should help to dispel the popular portrayal of the 'origin myth' in relation to animal domestication: the view that the domestication of the horse was a singular instance, occurring at some stage during the Neolithic, and that the scientific goal should be to pinpoint this particular point in time with a specific date. The process of the domestication of the horse from the 'wild' is not a dichotomous one, where a binary line is crossed, but is a fluid, adaptable and ongoing process that still occurs into the present day.

Figure 13.7: Mongolians greeting by passing snuff. Note the colour diversity within one large herd of Mongolian horses.

Source: Photograph by Natasha Fijn.

References

Bedunah DJ, Schmidt SM. 2000. Rangelands of Gobi Gurvan Saikhan National Conservation Park, Mongolia. *Rangelands* 22(4):18–24.

Bjørnstad G, Nilsen NØ, Røed K. 2003. Genetic relationship between Mongolian and Norwegian horses? *Anim Genet* 34:55–58.

Bökönyi S. 1974. *The Przevalsky horse*. London: Souvenir Press.

Bouman I, Bouman J. 1994. The history of Przewalski's horse. In: Boyd L, Houpt KA, editors. *Przewalski's horse: The history and biology of an endangered species*. New York: SUNY. pp. 5–38.

Boyd L, Houpt KA. 1994. *Przewalski's horse: The history and biology of an endangered species*. New York: SUNY.

Cassidy R. 2002. *The sport of kings: Kinship, class and thoroughbred breeding in Newmarket*. Cambridge: Cambridge University Press.

Cieslak M, Pruvost M, Benecke N, Hofreiter M et al. 2010. Origin and history of mitochondrial DNA lineages in domestic horses. *PLoS ONE* 5(12): e15311. doi:10.1371/journal.pone.0015311.

Clutton-Brock J. 2012. *Animals as domesticates: A world view through history*. Michigan: Michigan State University Press.

Fijn N. 2011. *Living with herds: Human–animal coexistence in Mongolia*. Cambridge, New York and Melbourne: Cambridge University Press.

Green RE, Krause J, Briggs AW, Maricic T et al. 2010. A draft sequence of the Neandertal genome. *Nature* 328:710–722.

Groves CP. 1986. The taxonomy, distribution and adaptations of recent equids. In: Meadow RH, Uerpmann HP, editors. *Equids in the Ancient World*. Wiesbaden: Ludwig Reichert Verlag. pp. 11–65.

Groves CP. 1994. Morphology, habitat and taxonomy. In: Boyd L, Houpt KA, editors. *Przewalski's horse: The history and biology of an endangered species*. New York: SUNY. pp. 39–60.

Groves C, Grubb P. 2011. *Ungulate taxonomy*. Baltimore: Johns Hopkins University Press.

Groves C, Willoughby DP. 1981. Studies on the taxonomy and phylogeny of the genus *Equus* 1. Subgeneric classification of a recent species. *Mammalia* 45(3):321–354.

Hovens JPM, Tungalaktuja K. 2005. Seasonal fluctuations in the wolf diet in the Hustai National Park (Mongolia). *Mammal Biol* 70(4):210–217.

Ishida N, Oyunsuren T, Mashima S, Mukoyama H, Saitou N. 1995. Mitochondrial DNA sequences of various species of the genus *Equus* with special reference to the phyogenetic relationship between Przewalskii's wild horse and domestic horse. *J Mol Evol* 41:180–188.

Jansen T, Forster P, Levine MA, Oelke H et al. 2002. Mitochondrial DNA and the origins of the domestic horse. *PNAS* 99(16):10905–10910.

Kaczensky N, Enkhsaihan O, Ganbaatar O, Walzer C. 2007. Identification of herder-wild equid conflicts in the Great Gobi B strictly protected area in NSW. *Erforsch biol Ress Mongolei* 10:1–18.

Kaszab Z. 1966. New sightings of Przewalski Horses. *Oryx* 8(6):345–347.

King SRB. 2002. Home range and habitat use of free-ranging Przewalski horses at Hustai National Park, Mongolia. *Appl Anim Behav Sci* 78:103–113.

King SRB, Gurnell J. 2005. Habitat use and spatial dynamics of takhi introduced to Hustai National Park, Mongolia. *Biol Conserv* 124:277–290.

King SRB, Gurnell J. 2007. Scent-marking behaviour by stallions: An assessment of function in a reintroduced population of Przewalski horses (*Equus ferus przewalskii*). *J Zool* 272:30–36.

Lau AN, Peng L, Goto H, Chemnick L et al. 2009. Horse domestication and conservation genetics of Przewalski's horse inferred from sex chromosomal and autosomal sequences. *Mol Biol Evol* 26(1):199–208.

Lindgren G, Backström N, Swinburne J, Hellborg L et al. 2004. Limited number of patrilines in horse domestication. *Nat Genet* 36(4):335–336.

Linklater WL. 2000. Adaptive explanation in socio-ecology: Lessons from the Equidae. *Biol Rev* 75:1–20.

Lusis JA. 1943. Striping patterns in domestic horses. *Genetica* 23(1):31–62.

Mallon DP, Jiang Z. 2009. Grazers on the plains: Challenges and prospects for large herbivores in Central Asia. *J Appl Ecol* 46:516–519.

Mohr E. 1971. *The Asiatic wild horse*. London: J. A. Allen & Co.

Morgan D. 1986. *The Mongols*. Oxford: Basil Blackwell.

Myka JL, Lear TL, Houck ML, Ryder OA, Bailey E. 2003. Fish analysis comparing genome organization in the domestic horse (*Equus caballus*) and the Mongolian wild horse (*E. przewalskii*). *Cytogenet Genome Res* 102:222–225.

Nalle D. 2000. Tournament of shadows: The great game and the race for empire in Central Asia. *Middle East Policy* 7(3):199–200.

Oakenfull EA, Ryder OA. 1998. Mitochondrial control region and 12S rRNA variation in Przewalski's horse (*Equus przewalskii*). *Anim Genet* 29(6):456–459.

Plumwood V. 2012. *The eye of the crocodile*. Canberra: ANU E Press.

Ryder OA. 1994. Genetic studies of Przewalski's horses and their impact on conservation. In: Boyd L, Houpt KA, editors. *Przewalski's horse: The history and biology of an endangered species*. New York: SUNY. pp. 75–92.

Pruvost M, Bellone R, Benecke N, Sandoval-Castellanos E et al. 2011. Genotypes of predomestic horses match phenotypes painted in Paleolithic works of cave art. *PNAS* 108(46):18626–18630.

Tozaki T, Takezaki N, Hasegawa N, Ishida M et al. 2003. Microsatellite variation in Japanese and Asian horses and their phylogenetic relationship using a European horse outgroup. *J Hered* 94(5):374–380.

van Dierendonck MC, Bandi N, Batdorj D, Dügerlham S, Munkhtsog B. 1996. Behavioural observations of reintroduced Takhi or Przewalski horses (*Equus ferus przewalskii*) in Mongolia. *Appl Anim Behav Sci* 50:95–114.

van Dierendonck MC, Goodwin D. 2005. Social contact in horses: Implications for human–horse interactions. In: de Jong FH, van den Bos R, editors. *The human–animal relationship*. Assen, The Netherlands: Van Gorcum. pp. 65–81.

van Dierendonck MC, Wallis de Vries MF. 1996. Ungulate reintroductions: Experiences with the takhi or Przewalski Horse (*Equus ferus przewalskii*) in Mongolia. *Conserv Biol* 10(3):728–740.

Vila C, Leonard JA, Götherström A, Marklund S et al. 2001. Widespread origins of domestic horse lineages. *Science* 291(5503):474–477.

Wallner B, Brem G, Müller M, Achmann R. 2003. Fixed nucleotide differences on the Y chromosome indicate clear divergence between *Equus przewalskii* and *Equus caballus*. *Anim Genet* 34:453–456.

14. Rhino systematics in the times of Linnaeus, Cuvier, Gray and Groves

Kees Rookmaaker

History inspires innovation

Most scientists agree that it is imperative to read the literature in their immediate field of interest hot off the press to keep abreast of the latest advances. They build on a more extensive body of knowledge which is the result of past explorations and discoveries by persons working in the same discipline. Generally it is almost irrelevant how the current consensus was reached or which people were responsible for it, unless a discovery or theory represented a major breakthrough. The history of a given subject is of course highly interesting, and any scientist would be advised to pay attention to the lives of their predecessors in their speciality. While for the majority of scientists such historical insight is largely optional, it often appears that students of animal taxonomy have much less choice. Animal and plant biodiversity is so immense that a revision must necessarily take into account all previous studies of the specimens and populations now combined in a certain species or higher ranking group. We are bound to respect the decisions and models of previous generations of taxonomists, although we have every right to fine-tune them or even dismiss them completely.

The historical component of taxonomy is nowadays often as quickly recognised as it is dismissed. Searching old and dusty books, examining skulls and hides in museum storerooms, locating old geographical placenames, understanding obscure concepts of classification, is it really needed to understand the evolutionary species groups? When a bibliographer or historian finds a scientific name which had long remained hidden in some obscure or rarely consulted journal, there is often an immediate call to relegate it to the growing list of 'forgotten names' – names often forgotten merely because the contemporary colleagues of the naturalist who proposed the name had only a partial knowledge of the new literature. Nomenclature may be a part of taxonomy which in its rules has an historical dimension, but taxonomy is one science where history cannot safely be ignored.

Colin Groves is one of those taxonomists who is sensitive to past research. He has shown this in his 2008 book *Extended family: Long-lost cousins (a personal look at the history of primatology)*. In the same vein, the present chapter is a personal look, this time mine, at four different phases in the history of the systematic research of the small group of currently existing species of rhinoceros. Given that there are only a handful, or maybe just over a handful of qualifying rhino taxa, the history of the understanding of their relationships is unexpectedly treacherous and complicated, as well as ever changing. As this is a personal look, I have only quoted the most relevant literature in the bibliography. Much has been written about the rhinoceros. In 1983 I bravely published a *Bibliography of the Rhinoceros*, and was proud to state that I had been able to extract 3,106 references (pleased enough even to count them manually). Continuing ever since to archive the world's output on these animals, the bibliography has now extended to over 18,000 references (fortunately counted by a computer). All these are globally accessible on the Rhino Resource Centre, which provides the references as well as the text on the website: www.rhinoresourcecentre.com, a stable and hopefully not too ephemeral source of information, current as well as historic, on everything related to rhino studies.

Although history does not often allow a saltationary approach, I will focus on just four periods in the story of rhino systematics. This will show how the classification of the living world changed over time, from the mid-eighteenth century to the present.

Rhino systematics in the time of Carl Linnaeus

Uppsala, Sweden: 1758

Carl Linnaeus (1707–1778), the famous professor of botany in the small university town of Uppsala, is unlikely ever to have seen a rhinoceros alive. For most of his work on the animal kingdom, he had to depend on occasional specimens in a few collections, and of course on a thorough knowledge of the renaissance literature. Given the enormous scope of his pioneering endeavours to classify all known animals and plants, the results were remarkably fair to the prevailing sentiments of his time. His work was of course highly innovative, which makes it easy to understand that it took him several adjustments along the way to find the right format to fit all available information.

His main work on animals was the *Systema Naturae*, first published in 1735 and updated through a series of revisions to the most authoritative tenth edition of 1758. Linnaeus masterfully condensed the entire animal kingdom: at first he needed only a pamphlet of just 12 pages, in 1758 expanded to about 800 pages, to list all known vertebrates and invertebrates. His list of mammals occupied just 63 pages, nevertheless was remarkably comprehensive. His species were inclusive of all variants, defined strictly morphologically rather than zoogeographically.

This emphasis on morphological and anatomical characters led to some rather surprising results. Because the rhinoceros was known to have incisor teeth, the animal came to be classed together with rodents of the genera *Hystrix*, *Lepus*, *Castor*, *Mus* and *Sciurus*, away from other pachyderms or ungulates. But Linnaeus had relied on insufficient evidence, inasmuch as only one of the two species of rhinoceros recognised in his book actually has these incisors. The only type of rhinoceros well-known in his time was the species with a single horn, now known as *Rhinoceros unicornis*, the Indian or greater one-horned rhinoceros.

There were of course plenty of rumours of rhinos having two horns, both in Asia and in Africa. There were even a number of specimens in European collections where the two horns were still attached to each other and obviously belonged to the same animal. The academic world, for some reason still poorly understood, was reluctant to admit the existence of a second species of rhinoceros, which would have two horns on the nose instead of one. The reasoning at the time was as ingenuous as it was convoluted, with the number of horns being ascribed to climate, age, sex, size, to just about anything except to the possibility that there was more than one species of rhinoceros. In an age when unicorns and dragons were still very much in discussion, this reluctance seems excessive.

Linnaeus was a brave taxonomist. He was certain that there was a rhinoceros species with two horns on the nose, separated from the one with a single horn, but remarkably, he did appear to see the need to explain himself. 'I have seen the complete head of one of these animals', he stated, therefore he could not doubt its existence. However, when we carefully read his paragraphs in the *Systema Naturae* of 1758, we end up with a species of rhinoceros with two incisors on either side of the jaws, a double horn on the nose, living in India. If Linnaeus actually examined the head which he claimed to have seen, with his extensive knowledge of animal morphology, it must be concluded that he saw a skull of an Indian rhinoceros where a second horn was artificially added to the specimen. If this interpretation is correct, it exonerates Linnaeus, but did nothing to help the study of rhino systematics in the remainder of the eighteenth century.

Rhino systematics in the time of Georges Cuvier

Paris, France: 1816–1836

There were a number of new discoveries in the half century separating the works of Linnaeus and Georges Cuvier (1769–1832), professor of natural history in Paris. The vicinity of the Cape of Good Hope was carefully explored and yielded exhaustive morphological and anatomical descriptions of the resident black rhinoceros by people like Robert Jacob Gordon and Anders Sparrman (Figure 14.1). Travellers in other parts of Asia had long known about the existence of a rhinoceros outside the Indian subcontinent, said to bear either one or two horns on the nose. The evidence soon became overwhelming enough to recognise that changes were needed.

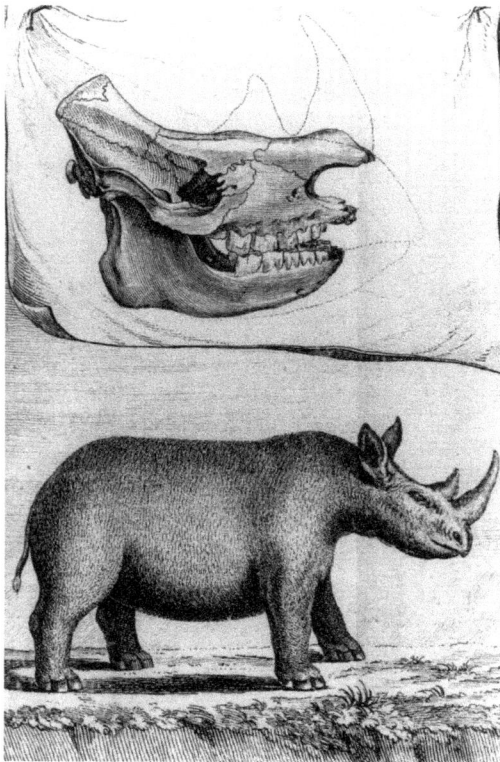

Figure 14.1: Black rhinoceros and skull, sketched by Anders Sparrman after a specimen which he shot in South Africa's Cape Province, December 1775. Published in his *Resa till Goda-Hopps-Udden*, Stockholm, vol. 1 (1783), plate 5.

Source: From Resa till Goda-Hopps-Udden, Stockholm, vol. 1 (1783), plate 5.

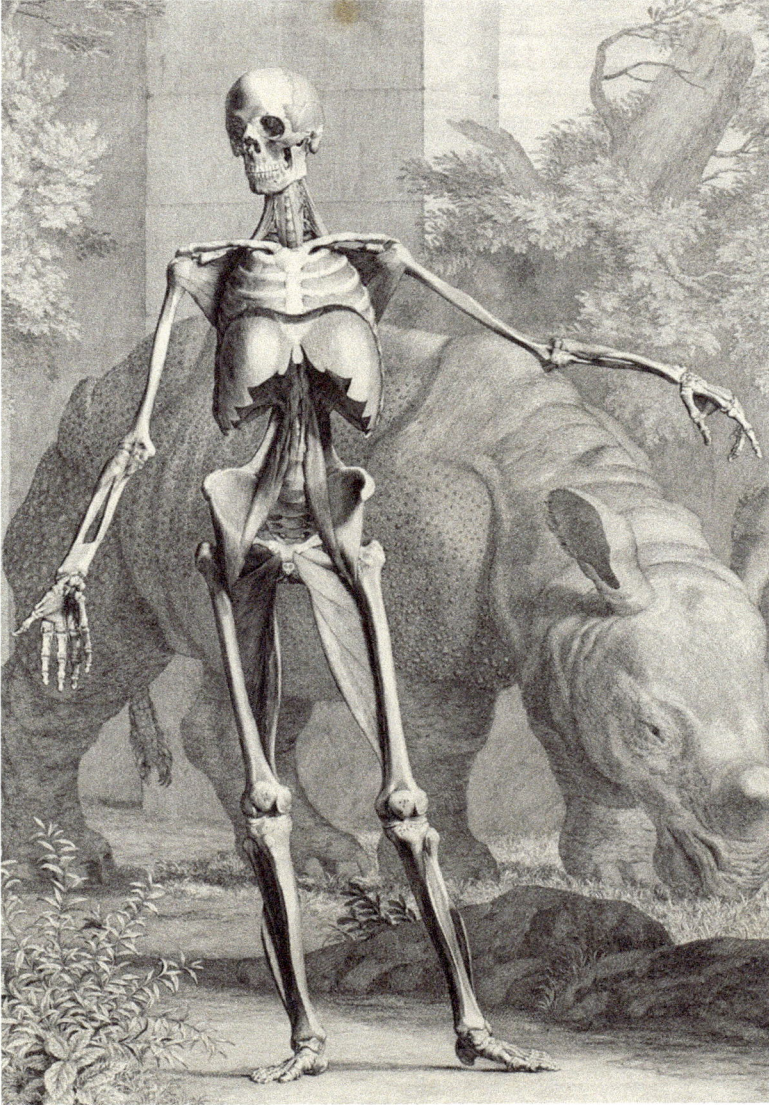

Figure 14.2: The famous Indian rhinoceros 'Clara', shown all around Europe between 1741 and 1758, depicted behind a human skeleton in the anatomical atlas by Bernard Siegfried Albinus, *Tabulae sceleti et musculorum humanis corporis* (1747), pl. 4.

Source: From Bernard Siegfried Albinus, Tabulae sceleti et musculorum humanis corporis (1747), pl. 4.

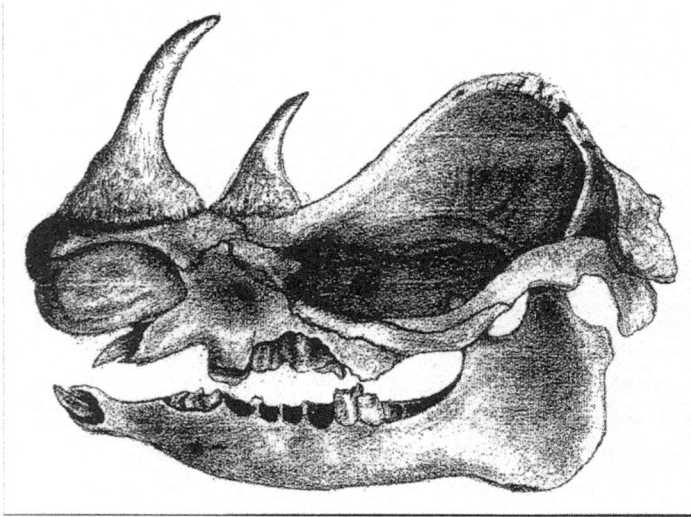

Figure 14.3: A rhinoceros skull drawn by the English physician James Parsons in or near London in the early eighteenth century. It is an Indian rhinoceros (*Rhinoceros unicornis*) with incisors, indicating that the posterior horn must have been an artificial addition.

Source: Hunterian Library, Glasgow, Av.1.17 folio 13.

William Bell (d.1792), trained as a zoological draughtsman by the great collector John Hunter in London, spent a few months in the British Fort Marlborough on the west coast of Sumatra. Some 10 miles away from the town, he found a rhinoceros, made drawings, collected the skull, and sent all off with a useful description to London, where his work was duly published by the Royal Society. Hence the existence of *Rhinoceros sumatrensis* could no longer be doubted.

Rhino specimens sent from Java had already convinced Petrus Camper (1722–1789) in the 1780s that this animal really differed from the better known single-horned rhinoceros. Camper was not only a famous professor of human anatomy, but he was also interested in the morphological structures of the larger mammals like reindeer, orangutan, elephant and rhinoceros. At the end of his life, his studies convinced him that the rhinoceros of Java differed materially from the other species. Although possibly his death kept him from writing a treatise on the subject, he did engage a local artist to show the different types of rhinoceros in an engraving, which he could send to his colleagues at home and abroad.

In southern Africa, the British explorer William John Burchell (1781–1863) had penetrated far enough into the unknown interior to find a new species of rhinoceros, larger than the known kind and showing a broad rather than pointed upper lip. He carried a skull back to the coast together with drawings made on the spot where he first saw them. Remarkably, and for reasons still poorly understood, he sent word of his greatest discovery not to one of his academic friends in Britain, but to a professor of natural history in Paris, Henri de Blainville (1777–1850), who duly published his letter and drawing of the new *Rhinoceros simus*.

Figure 14.4: The first description of the double horned rhinoceros of Sumatra by William Bell, Surgeon in the Service of the East India Company at Bencoolen. It was published in the *Philosophical Transactions of the Royal Society of London* in 1793.

Source: From Bell, the *Philosophical Transactions of the Royal Society of London* in 1793.

Georges Cuvier (1769–1832) was one of the most influential zoologists in Europe in the first quarter of the nineteenth century, where he taught at the university and studied animals in the Jardin des Plantes. When he came to write a general overview of the animal kingdom, it was up to him to evaluate all the different strands of new information which had come to light since the time of Linnaeus. In 1816 he published the first part of his *Règne Animal* (*Animal Kingdom*), where, among the pachyderms, he distinguished three living species of rhinoceros: the African two-horned and Indian one-horned types of Linnaeus, to which he added the animal discovered by Bell in Sumatra. We must note the absence of Camper's rhinoceros from Java and Burchell's new African sort, the last of which is explained by the actual date of publication of Cuvier's book, late 1816 rather than 1817 as stated on the title-page. There is still an important role to play by bibliographers to unravel the more intricate puzzles in the development of science.

It must be said that Cuvier appears to have been particularly careful in his definition of the species, not wanting to recognise what in the end might be difficult to separate as a different species on morphological grounds alone. His colleague de Blainville was a little more liberal in a paper published in 1817 where he enumerated three Asian and four African species of rhinoceros.

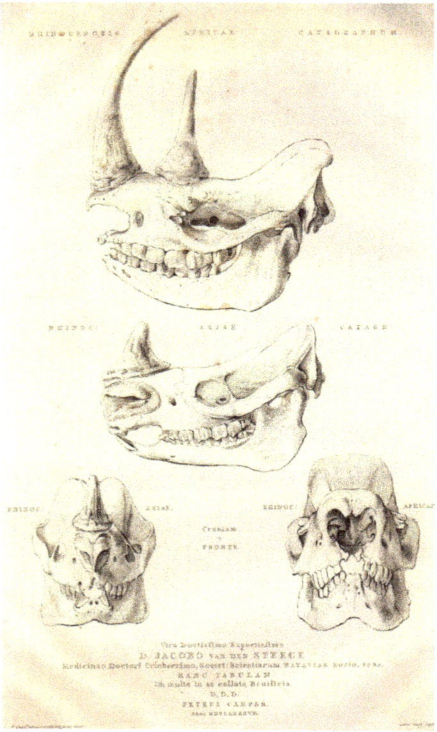

Figure 14.5: Broadside published by the Dutch professor Petrus Camper to show the differences between skulls of a black rhinoceros obtained from the Cape of Good Hope and of a rhinoceros from Java sent to him by Jacob van der Steege. Engraved by Reinier Vinkeles: 'Rhinocerotis Africae Catagraphum', 1787.

Source: British Museum, London.

Figure 14.6: Depiction of a rhino hunt at Chué Spring (Heuningvlei) in South Africa by William Burchell in 1812.

Source: Aquatint in Library of Parliament, Cape Town.

Twelve years later, in 1829, when Cuvier published the second edition of his definitive *Règne Animal*, he repeated most of the text, but he did add the Javan rhinoceros as a fourth species. He could not do differently, as his brother had described the animal after specimens and a written treatise had been received from Alfred Duvaucel (1793–1824), who had been sent to collect materials on behalf of the museum in Paris. His classification was of course very close to what we would recommend today. I should note, however, the absence of *Rhinoceros simus*, the white rhinoceros, from his writings, and one wonders why he did not feel inclined to add this as a fifth kind. Maybe the lack of original material at his disposal led to this course of action.

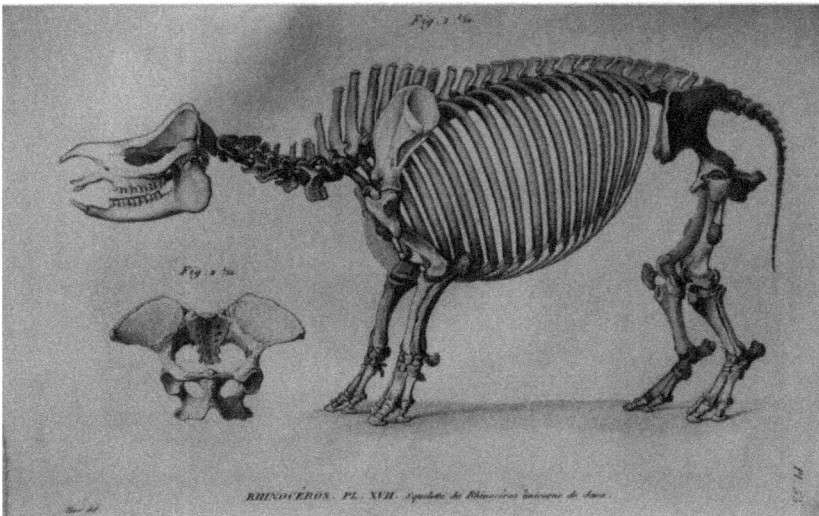

Figure 14.7: Skeleton of the 'Rhinoceros unicorne de Java' in the Paris Museum of Natural History. From Georges Cuvier, *Recherches sur les ossemens fossiles* (1836), Atlas, pl. 17.

Source: From Georges Cuvier, Recherches sur les ossemens fossiles (1836), Atlas, pl. 17.

Rhino systematics in the time of John Edward Gray

London, United Kingdom: 1862–1875

After Cuvier's final edition of the *Règne Animal* in 1836, the focus of rhino taxonomy definitely shifted to the British sphere of influence. There was a great influx of specimens from all of the range states which had to be diagnosed and named. It was a time when new species would be recognised after reading a report in a travel journal or examining just one or even part of a specimen which had just arrived from abroad. Though the aim was obviously to understand the great biodiversity in nature, the result was an array of species and varieties which, if viewed in their totality, was often bewildering.

On 1 June 1835, the British surgeon and naturalist Andrew Smith (1797–1872) was travelling in the African interior near present-day Mafikeng (North-West Province, South Africa) when his hunters alerted him that a different kind of rhinoceros had been shot. Hurrying to the spot to examine this exciting trophy, Smith carefully looked over the animal and that evening, sitting at the campfire, decided to follow the assessment of the assembled crowd. The animal differed from the common black rhino by its greater ferocity and the shape of the horns which were of equal length, and he called it the *keitloa*. When Smith published his *Illustrations of the Zoology of South Africa* in 1838, the first instalment started with the new *Rhinoceros keitloa*, obviously seen as the greatest prize of his expedition into the unknown parts of South Africa.

Figure 14.8: Lateral view of black rhinoceros called 'Rhinoceros keitloa' drawn by Gerald Ford for Andrew Smith, *Illustrations of Zoology* (June 1838), vol. 1, plate 1.

Source: Drawn by Gerald Ford for Andrew Smith, Illustrations of Zoology (June 1838), vol. 1, plate 1.

The description of this new species of African rhinoceros may not appear particularly momentous, but in a way it opened the flood gates, when an experienced and respected zoologist like Andrew Smith without hesitation was willing to denote horn shape and temperament as characteristics useful enough to warrant specific distinction. In the African bush, this led to widespread speculation and endless discussions. Soon it was not unusual for big game hunters to allude to the existence of six or seven rhino species, all in the southern parts of Africa. Four were almost universally recognised, two types of black (*borele* and *keitloa*) and two types of white (*mohoohoo* and *kobaaba*). This practice did not remain restricted to the realm of campfire tales, these species were duly named, immortalised and generally accepted from the time of the revisions by John Edward Gray (1800–1875). As the main zoologist in the British Museum, Gray published a series of catalogues of the collections, which were influential and highly regarded. In 1862, Gray had no hesitation to list four African species which he called *Rhinoceros bicornis, keitloa, simus* and *oswellii* – and in some ways, he was relatively conservative in his assessment.

In this catalogue of 1862, Gray remained close to the classification of Cuvier by listing the same three species of rhinoceros inhabiting Asia, but he added a fourth one (*Rhinoceros crossii*) on the basis of an unlocalised strangely shaped horn. However, the time of change had arrived. In a new revision of 1868, Gray had five species of Asian one-horned rhinos as well as one species of Asian two-horned rhino. Edward Blyth (1810–1873) of the Asiatic Museum in Calcutta in 1862 found that skulls differed in broadness and suggested that these characters had specific status. The Secretary of the Zoological Society of London, Philip L. Sclater (1829–1913), compared a two-horned rhino from Chittagong in the London Zoo from 6 February 1872 with another from Malacca which arrived on 2 August 1872. The animals differed, especially in the length of the hairs fringing the ears, and were declared separate species (*R. sumatrensis* and *R. lasiotis*).

Considering that the above account is a rather watered-down version of the changing taxonomies of rhino species, leaving aside several spurious and even more ill-defined additions to the list, somehow time had come to put some of these distinctions to rest. Maybe it was hardly a coincidence that stabilisation was brought to the field soon after the death of John E. Gray in March 1875. William Henry Flower (1831–1899) had used the extensive collections of the Royal College of Surgeons in London to study the variations in cranial and dental characters. His revision almost miraculously brought new sense to the chaos of conflicting ideas and interpretations. In essence, he reverted back to Cuvier's last views, and recognised just five extant species, in Asia *unicornis, sondaicus, sumatrensis,* and in Africa *bicornis* and *simus,* but was unsure about the status of *lasiotis*. It was not quite the end to the era of superfluous descriptions and convoluted classifications, but there certainly was a path towards a workable taxonomy.

Figure 14.9: Images of the female Sumatran rhino 'Begum' shown in London Zoo from 15 February 1872 to 31 August 1900. She was the type of *Rhinoceros lasiotis*. From Sclater, On the rhinoceroses now or lately living in the Society's Menagerie (1877), pl. 98.

Source: From Sclater, On the rhinoceroses now or lately living in the Society's Menagerie (1877), pl. 98.

Rhino systematics in the time of Colin Groves

UK and Australia: From 1965

The intensity of the debates around rhino systematics in the mid-nineteenth century appears to have scared away any newcomers to the scene. It can truthfully be said that very little change was advocated for just about a century. The only notable exception was the rhinoceros shot by Major Percy Horace Gordon Powell-Cotton (1866–1940) in the Lado Enclave in the central parts of Africa. In this remote and unknown district, he found an animal very much like the white rhinoceros inhabiting regions further south, which differs particularly in the width of the nasal bones. Richard Lydekker (1849–1915) announced the discovery in a short notice tucked away in one of the issues of *The Field*, a magazine intended for the gentleman interested in field sports. On 22 February 1908 he named the animal *Rhinoceros simus cottoni*, after the discoverer, using a subspecific epithet, becoming more popular at the time, due to the perceived similarity of the two types of wide-mouthed rhinos.

Figure 14.10: Nile rhino from the Lado Enclave in Sudan. Plate drawn by J Terrier and printed by J Pitcher. It was used to illustrate the account of this new species by Edouard Louis Trouessart, Le rhinoceros blanc du Soudan (*rhinoceros simus cottoni*) (1909), plate 1.

Source: From Edouard Louis Trouessart, Le rhinoceros blanc du Soudan (rhinoceros simus cottoni) (1909), plate 1.

A reasonably stable situation had emerged and remained unchallenged, maybe partly because there was no excitingly new material that warranted a new revision, probably also due to a general lack of interest. In order to compare fossil bones with those of recent African rhinos, Arthur Tindell Hopwood (1897–1969) at the Natural History Museum in London had a fresh look at black rhinoceros systematics. His paper of 1939 had little news to offer, differentiating almost every population examined as a separate subspecies (*bicornis, holmwoodi* and *somaliensis*). Inevitably, many new rhino specimens had come to museums and zoos during the century. The African black rhinoceros is widespread and animals from different regions differ in size or colour or temperament or dentition or in a variety of other ways.

The first to undertake a much-needed revision of the species was Ludwig Zukowsky (1888–1965) in Germany, published in *Der Zoologische Garten* in 1965, which is of course a journal of international rank, but not particularly one where a major taxonomic study would be expected. Zukowsky had been able to compare many living animals during his work in German zoos like Hamburg and Leipzig, to which he added pictures taken in the field and a comprehensive survey of literature. Zukowsky's work, although largely ignored, remains in fact one of the major monographs written about a single rhino species. However, he worked in the tradition of taxonomic splitters, who generally used very minor differences to denote new species or subspecies, thereby increasing the number of forms to untenable levels. It is therefore not surprising that Zukowsky recognised 16 subspecies of *Diceros bicornis*, eight new and others resurrected from older works.

This is where Colin Groves first entered the scene. While pursuing his PhD studies, he found that the rhinoceros of Borneo differed enough from those of other parts of the range to warrant its description as a new subspecies, which he named after Tom Harrisson (1911–1976), one of those intrepid scholars who combine field work and museum studies. Next, on reading Zukowsky's study of 1965, he was aware that either the book would be totally forgotten and misunderstood, or needed to be put in a context of modern taxonomic theories. With an interest in rhinos which continues to this day, he set out to redress the excessive splitting by Zukowsky and divided the black rhino species in just seven subspecies. It is unfortunate that the international community of conservationists found it hard to cope with this sudden increase in subspecies, from three to seven, as theoretically there had been really no more than three types across the African continent for most of the century. An increase was inevitable though, in view of the fact that the black rhino in eastern and central Africa shows much variation and gradation and sudden morphological changes according to habitats and other patterns.

Groves has made several adjustments to rhino taxonomy during his long career. Together with Claude Guérin, the renowned expert in rhino palaeontology, he looked at data from Indochina and described the material as a subspecies of *Rhinoceros sondaicus*, correctly resurrecting a forgotten name *annamiticus* used only once earlier. More recently, while pursuing his belief in the Phylogenetic Species Concept, he has shown that in many ways the two subspecies of white rhino in southern and central Africa differ enough to be separated as species: *Ceratotherium simum* and *Ceratotherium cottoni*. This new understanding of their phylogenetic, genetic and biological relationship comes sadly at a time when the last examples in central Africa are being slaughtered by poachers and other opportunists. *Ceratotherium cottoni*, previously also distinguished in the vernacular as the northern white rhino, really deserves its own name, for which Nile rhinoceros may be one of the better historical choices.

Although the rhino taxonomy proposed by Groves certainly has its critics, there is no published alternative arrangement, which equally takes into account the available museum specimens, genetic material, histories of zoo animals, nomenclatorial rules and a thorough knowledge of the historical as well as current distribution of the rhino species. New insights will certainly come, and taxonomy is an evolving field which hopefully one day will become popular again with the increased need to understand global biodiversity. There will always be room for adjustments, but until that time, here is the latest classification of the recent rhinoceroses in *Ungulate Taxonomy* by Groves and Grubb (2011). Six species are recognised: *Rhinoceros unicornis*, *Rhinoceros sondaicus* (three subspecies, one extant), *Dicerorhinus sumatrensis* (three subspecies, two extant), *Diceros bicornis* (eight subspecies, four extant), *Ceratotherium simum* and *Ceratotherium cottoni*.

Taxonomy in the service of conservation

One of the reasons to look at the interpretation of rhino diversity over a period of several centuries is to weigh the impact of differing taxonomic interpretations on conservation initiatives. Taxonomy, it is of course recognised, is a field of academic pursuit no different from other scientific disciplines. New facts are constantly added, theories are adjusted, discarded or discovered, and inevitably systematic arrangements will constantly remain in a state of flux. If in 1758 a need had been felt to manage the remaining rhino populations in Africa, there would have been no scientific impediment to translocate anywhere in the continent and breed animals from different regions at will, because only one species was known. If in 1860 similar problems would have been addressed, the managers would have needed to understand a plethora of rather poorly defined species (not even subspecies), which probably would have thwarted the ingenuity of even the best minds given a practical need in the field. And it remains true that changes in systematics can cause any number of awkward or unwanted situations to occur, where animals are translocated to areas outside a range as understood at a given time.

Taxonomy should not obstruct conservation. Conservation should not ignore taxonomy. The goal is to understand biodiversity in all its wonderful facets and to preserve all its elements for future generations. Anybody working for rhino conservation has difficult choices to make on a daily basis. Rhino populations are not just dwindling, they are actively threatened to be wiped out completely. It is a war – to fight people greedily exploiting wildlife, it is a war – to stop encroachments on forest and bush. The possibility that this war will be lost is not unrealistic at all, with ever decreasing resources and ever increasing threats. Maybe the study of taxonomy will soon turn into a study of past biodiversity, gone before it is properly understood. Most people would say that this should not happen, yet only too few are willing to make the sacrifices needed to keep the world stocked with rhinos, and all the other beautiful creatures still precariously kept away from the brink of extinction.

Acknowledgements

A long line of people have devoted some of their considerable intellectual or practical expertise to the study and conservation of the rhinoceros. I have had the privilege to meet some and correspond with others, always learning and often enjoying small discoveries in unexpected corners. I acknowledge their expertise and willingness to share their findings. Colin Groves is one of the busy professionals who finds time to guide a younger generation and his advice to me over the years has been invaluable. My wife Sandy and I look back with pleasure on the few times we could meet Colin and Phyll. There is still much to learn about rhinoceros taxonomy, ecology, history, management and conservation, and there is no time to lose as rhinos are under enormous pressure in their wild habitat. In my view, no new study or new conservation project should be undertaken without first understanding what has been done in the distant and recent past in order that we build on the expertise of past generations of dedicated researchers. In the case of the rhinoceros, the Rhino Resource Centre provides an almost unparalled platform which allows access to literature unimpeded by funding or locality, globally available without restriction. I thank all authors who have shared their publications by this means.

References

Note that all titles are available to view on the website of the Rhino Resource Centre (sponsored by the International Rhino Foundation, SOS Rhino, Rhino Carhire, Save the Rhino International and WWF-Areas).

Bell W. 1793. Description of the double horned rhinoceros of Sumatra. *Phil Trans R Soc Lond* 1793:3–6, pls. 2–4.

Blyth E. 1862. A memoir on the living Asiatic species of rhinoceros. *J Asiatic Soc Bengal* 31(2):151–175, pls. 1–4.

Braun A, Groves CP, Grubb P, Yang Qi-Sen, Xia Lin. 2001. Catalogue of the Musée Heude collection of mammal skulls. *Acta Zootaxonomica Sinica* 26(4):608–660.

Burchell WJ. 1817. Lettre a M. H. de Blainville. In: Blainville HMD de, Lettre de M. W.J. Burchell sur une nouvelle espèce de rhinoceros, et observations sur les différentes espèces de ce genre. *Journal de Physique, de Chimie et d'Histoire Naturelle* 85:163–168, pl. 1.

Cave AJE, Rookmaaker LC. 1977. Robert Jacob Gordon's original account of the African black rhinoceros. *J Zool, Lond* 182:137–156, pls. 1–7, figs. 1–2.

Cuvier G. 1817. *Le règne animal distribué d'après son organisation pour servir de base a l'histoire naturelle des animaux et d'introduction à l'anatomie comparée.* Paris: Deterville.

Cuvier G. 1829. *Le règne animal*, 2nd edition. Paris: Deterville et Crochard.

Cuvier G. 1836. *Le règne animal*, 3rd edition. Bruxelles: Louis Haumann.

Flower WH. 1876. On some cranial and dental characters of the existing species of rhinoceroses. *Proc Zool Soc Lond* 1876 May 16:443–457, figs. 1–4.

Gentry A, Clutton-Brock J, Groves CP. 2004. The naming of wild animal species and their domestic derivatives. *J Archaeol Sci* 31:645–651.

Geoffroy Saint-Hilaire E, Cuvier G. 1824. Rhinoceros de Java. In: Geoffroy St. Hilaire E et al. *Histoire naturelle des mammifères*, vol. 6(45). Paris: Blaise. pp. 1–2, pl. 309.

Gray JE. 1868. Observations on the preserved specimens and skeletons of the Rhinocerotidae in the collection of the British Museum and Royal College of Surgeons, including the description of three new species. *Proc Zool Soc Lond* 1867:1003–1032, figs. 1–6.

Gray JE, Gerrard E. 1862. *Catalogue of the bones of mammalia in the collection of the British Museum.* London: Trustees of the British Museum.

Groves CP. 1965. Description of a new subspecies of rhinoceros, from Borneo, *Didermocerus sumatrensis harrissoni*. *Säugetierkundliche Mitteilungen* 13(3):128–131.

Groves CP. 1967a. Comment on the proposed decision on the validity of *Didermocerus* Brookes, 1828. *Bull Zool Nomencl* 24(5):279.

Groves CP. 1967b. Geographic variation in the black rhinoceros, *Diceros bicornis* (L, 1758). *Z Säugetierk* 32(5):267–276, figs. 1–2, tables 1–2.

Groves CP. 1967c. On the rhinoceroses of South-East Asia. *Säugetierkundliche Mitteilungen* 15(3):221–237, figs. 1–4, tables 1–5.

Groves CP. 1971. Species characters in rhinoceros horns. *Z Säugetierk* 36(4):238–252, figs. 1–22.

Groves CP. 1972. *Ceratotherium simum*. *Mammalian Species* no. 8:1–6, figs. 1–5.

Groves CP. 1975. Taxonomic notes on the white rhinoceros *Ceratotherium simum* (Burchell, 1817). *Säugetierkundliche Mitteilungen* 23(3):200–212, fig. 1, tables 1–3.

Groves CP. 1982a. Asian rhinoceroses: Down but not out. *Malayan Naturalist* 36(1):11–17, 20–22, figs. 1–4, table 1.

Groves CP. 1982b. The skulls of Asian rhinoceroses, wild and captive. *Zoo Biology* 1:251–261, tables 1–4.

Groves CP. 1983. Phylogeny of the living species of rhinoceros. *Z Zool Syst Evol* 21(4):293–313, figs. 1–11.

Groves CP. 1993a. Bad medicine for wildlife. *The Skeptic* 13(1):12–14.

Groves CP. 1993b. Testing rhinoceros subspecies by multivariate analysis. In: Ryder OA, editor. *Rhinoceros biology and conservation: Proceedings of an international conference*. San Diego, USA: San Diego, Zoological Society. pp. 92–100, figs. 1–5.

Groves CP. 1995a. A comment on Haryono et al.'s Report. *Asian Rhinos* 2:9.

Groves CP. 1995b. What is wrong with the captive population of Sumatran rhinos? *Asian Rhinos* 2:12.

Groves CP. 1995c. Why the Cat Loc (Vietnam) rhinos are Javan. *Asian Rhinos* 2:8–9, fig. 1.

Groves CP. 1997. Die Nashörner – Stammesgeschichte und Verwandtschaft. In: Anonymous. *Die Nashörner: Begegnung mit urzeitliche Kolossen*. Fuert: Filander Verlag. pp. 14–32, figs. 1–5.

Groves CP. 2003. Taxonomy of ungulates of the Indian subcontinent. *J Bombay Nat Hist Soc* 100(2/3):341–362.

Groves CP. 2008a. *Extended family: Long-lost cousins. A personal look at the history of primatology*. Arlington, VA: Conservation International.

Groves CP. 2008b. Review of L.C. Rookmaaker, Encounters with the African rhinoceros. *J Mammal* 89(6):1570.

Groves CP, Chakraborty S. 1983. The Calcutta collection of Asian rhinoceroses. *Rec Zool Surv India* 80:251–263, tables 1–2.

Groves CP, Fernando P, Robovsky J. 2010. The sixth rhino: A taxonomic re-assessment of the critically endangered northern white rhinoceros. *PLoS One* 5(4) e9703:1–15

Groves CP, Grubb P. 2011. *Ungulate taxonomy*. Baltimore: Johns Hopkins University Press.

Groves CP, Guérin C. 1980. Le Rhinoceros sondaicus annamiticus d'Indochine: Distinction taxinomique et anatomique; rélations phylétiques. *Géobios* 13(2):199–208, figs. 1–4, tables 1–2.

Groves CP, Kurt F. 1972. *Dicerorhinus sumatrensis*. *Mammalian Species* 21:1–6, figs. 1–4.

Groves CP, Leslie Jr. DM. 2011. *Rhinoceros sondaicus* (Perissodactyla: Rhinocerotidae). *Mammalian Species* 43 (887):190–208.

Groves CP, Robovsky, J. 2011. African rhinos and elephants: Biodiversity and its preservation. *Pachyderm* 50:69–71.

Hillman Smith K, Groves CP. 1994. *Diceros bicornis*. *Mammalian Species* 455:1–8, figs. 1–3.

Laurie WA, Lang EM, Groves CP 1983. *Rhinoceros unicornis*. *Mammalian Species* 211:1–6, figs. 1–3.

Linnaeus C. 1758. *Systema naturae per regna tria naturae, secundum classes, ordines, genera, species, cum characteribus, differentiis, synonymis, locis.* Editio decima, reformata. Holmiae: Laurentii Salvii.

Lydekker R. 1908. The white rhinoceros. *Field* 22 February 1980 (2878:319).

Prins HHT. 1990. Geographic variation in skulls of the nearly extinct small black rhinoceros *Diceros bicornis michaeli* in northern Tanzania. *Z Säugetierk* 55:260–269, figs. 1–3, tables 1–5.

Robovsky J, Fernando P, Groves CP. 2010. Misto peti nosorozcu sest. Novy druk, ktery uz viasne neesistuje. *Vesmir* 89:368–371, figs. 1–4.

Rookmaaker LC. 1978. Two collections of rhinoceros plates compiled by James Douglas and James Parsons in the eighteenth century. *Journal of the Society for the Bibliography of Natural History* 9(1):17–38, figs. 1–7

Rookmaaker LC. 1982. The type locality of the Javan Rhinoceros (*Rhinoceros sondaicus* Desmarest, 1822). *Z Säugetierk* 47(6):381–382

Rookmaaker LC. 1983a. *Bibliography of the rhinoceros: An analysis of the literature on the recent rhinoceroses in culture, history and biology.* Rotterdam and Brookfield: A.A. Balkema.

Rookmaaker LC. 1983b. Historical notes on the taxonomy and nomenclature of the recent Rhinocerotidae (Mammalia, Perissodactyla). *Beaufortia* 33(4):37–51, figs. 1–4.

Rookmaaker LC. 1983c. Jamrachs Rhinozeros. *Bongo, Berlin* 7:43–50, fig. 1

Rookmaaker LC. 1984. The taxonomic history of the recent forms of Sumatran Rhinoceros (*Dicerorhinus sumatrensis*). *J Malays Branch R Asiatic Soc* 57(1):12–25, figs. 1–6.

Rookmaaker LC. 1989. *The zoological exploration of Southern Africa 1650–1790*. Rotterdam and Brookfield: A.A. Balkema.

Rookmaaker LC. 1998. The sources of Linnaeus on the rhinoceros. *Sven Linnesallskap Arsskr* 1996/97:61–80, figs. 1–15.

Rookmaaker LC. 1999. Specimens of rhinoceros in European collections before 1778. *Sven Linnesallskap Arsskr* 1998/99:59–80, figs. 1–7.

Rookmaaker LC. 2003a. The last white rhinoceros in Zimbabwe. *Pachyderm* 35:100–114, figs. 1–8.

Rookmaaker LC. 2003b. Why the name of the white rhinoceros is not appropriate. *Pachyderm* 34:88–93, pl. 1.

Rookmaaker LC. 2004. *Rhinoceros rugosus* – a name for the Indian rhinoceros. *J Bombay Nat Hist Soc* 101(2):308–310.

Rookmaaker LC. 2005. Review of the European perception of the African rhinoceros. *J Zool, Lond* 265:365–376, figs. 1–9.

Rookmaaker LC. 2005. The black rhino needs a taxonomic revision for sound conservation. *International Zoo News* 52(5):280–282.

Rookmaaker LC. 2008. *Encounters with the African rhinoceros: A chronological survey of bibliographical and iconographical sources on rhinoceroses in southern Africa from 1795 to 1875: Reconstructing views on classification and changes in distribution*. Münster: Schuling Verlag.

Rookmaaker LC. 2010. The sixth living rhino species? *International Zoo News* 57(3):171.

Rookmaaker LC. 2011a. A review of black rhino systematics proposed in *Ungulate Taxonomy* by Groves and Grubb (2011) and its implications for rhino conservation. *Pachyderm* 50:72–76.

Rookmaaker LC. 2011b. The early endeavours by Hugh Edwin Strickland to establish a code for zoological nomenclature in 1842–1843. *Bull Zool Nomencl* 68(1):29–40.

Rookmaaker LC, editor. 2013. www.rhinoresourcecentre.com. (accessed 2013).

Rookmaaker LC, Kraft R. 2011. The history of the unique type of *Rhinoceros cucullatus*, with remarks on observations in Ethiopia by James Bruce and William Cornwallis Harris (Mammalia, Rhinocerotidae). *Spixiana* 34(1):133–134, figs. 1–8.

Rookmaaker LC, Groves CP. 1978. The extinct Cape rhinoceros, *Diceros bicornis bicornis* (Linnaeus, 1758). *Säugetierkundliche Mitteilungen* 26(2):117–126, fig. 1, tables 1–3.

Rookmaaker LC, Visser RPW. 1982. Petrus Camper's study of the Javan rhinoceros (*Rhinoceros sondaicus*) and its influence on Georges Cuvier. *Bijdragen tot de Dierkunde, Amsterdam* 52(2):121–136, figs. 1–8.

Sclater PL. 1872. Notes on *Propithecus bicornis* and *Rhinoceros lasiotis*. *Ann Mag Nat Hist* (4)10:298–299.

Sclater PL. 1877. On the rhinoceroses now or lately living in the Society's Menagerie. *Trans Zool Soc Lond* 9:645–660, pls. 95–99, figs. 1–9.

Smith A. 1838. *Illustrations of the zoology of South Africa; consisting chiefly of figures and descriptions of the objects of natural history collected during an expedition into the interior of South Africa*, vol. 1: *Mammalia*, part 1. London: Smith, Elder and Co.

Toit R du. 1987. The existing basis for subspecies classification of black and white rhinos. *Pachyderm* 9:3–5.

Trouessart E-L. 1908. Le rhinoceros blanc du Bahr-el-Gazal. *La Nature, revue des sciences et de leurs applications aux arts et à l'industrie* 37:50–53, figs. 1–2.

Trouessart E-L. 1909. Le rhinoceros blanc du Soudan (*rhinoceros simus cottoni*). *Proc Zool Soc Lond* 1909 February 16:198–200.

Zukowsky L. 1965. Die Systematik der Gattung Diceros Gray, 1821. *Zoologische Garten* 30:1–178, figs. 1–8.

15. Conservation consequences of unstable taxonomies: The case of the red colobus monkeys

John Oates and Nelson Ting

Introduction

Species are the common primary 'currency' used in biodiversity conservation planning. Regions and ecosystems are often prioritised for conservation action based on measures of species richness and endemism (e.g. Myers et al., 2000; Olson and Dinerstein, 2002), and species judged to be in danger of extinction are usually given special attention (e.g. with focused conservation action plans produced by the International Union for Conservation of Nature's Species Survival Commission (IUCN SSC); see www.iucn.org/about/work/programmes/species/publications/species_actions_plans/). Such species-based thinking is quite understandable. From long before there was any science of biology or taxonomy, people around the world have recognised sets of similar organisms as distinct entities (and given names to these sets); this 'natural' species concept provided the basis for the work of Linnaeus and those who have followed him. With or without scientific classification, people would recognise horses as different from asses and lions as different from tigers, and factor this recognition into their world view and decision making. When science is brought fully into play in conservation planning, however, a species-based approach can lead to serious difficulties in determining conservation priorities in those cases where a group of organisms has a poorly resolved or unstable species-level classification. Difficulties arise both in establishing relative conservation priorities within that group, and in relation to other groups. Unstable classification also creates problems for communicating information to policy-makers and managers who may have little knowledge of taxonomy.

Among primates, the taxonomy of Africa's red colobus monkeys has been particularly unstable and contentious. A great number of different classifications have been published in the last 45 years, recognising between one and 16 species. Several of these different classifications have been produced by Colin Groves, who has long been interested in this group of monkeys (e.g. Groves, 1989, 2001, 2007). Groves' classificatory changes have been influenced not only by new research findings, but also by his move from using the Biological Species Concept (BSC) as the basis of taxonomic analysis to the Phylogenetic

Species Concept (PSC). The frequent, and often quite radical, changes in red colobus classifications have led to confusion both among field workers studying behaviour and undertaking surveys, and in conservation assessments published by national and international organisations. This has been particularly problematic because red colobus monkeys are among the most endangered primates in Africa with numerous populations in danger of extinction due to hunting or habitat modification by humans (Oates, 1996). With no consensus on their classification and on which forms are particularly distinct, it has been difficult to designate conservation priorities for this group of primates.

In this chapter we consider some of the causes and consequences of this example of taxonomic instability. For instance, could particular colobus populations, such as the Critically Endangered Tana River red colobus of Kenya and the probably recently extinct Miss Waldron's red colobus of West Africa, have suffered from a lack of sufficient conservation attention in part through their ambiguous distinctiveness? And could the use of different classifications have influenced the relative priority given to different regions of Africa for primate conservation? Finally, using the red colobus example, we consider what taxonomic practices might most beneficially be applied to conservation without a loss of scientific integrity.

Systematics theory and background

While no evolutionary biologist would debate the importance of the species concept in the development of evolutionary theory, there has been a lack of consensus on how a species should be defined (Frankham et al., 2012). This is one of the reasons why red colobus monkeys have been so difficult to classify. In fact, the taxonomic issues within this group are related to a larger theoretical debate that dominated the field of systematics in the latter half of the twentieth century. A comprehensive review of the history of this debate is beyond the scope of this chapter, but it is worthwhile highlighting some aspects of the debate particularly relevant to the problem we are discussing here.

A major theoretical divide has arisen between the two approaches that have come to be known as Evolutionary Systematics and Phylogenetic Systematics. The former is rooted in the union of evolutionary theory and population genetics that occurred in the 1940s, now known as 'The Modern Synthesis' (Huxley, 1942), while the latter finds its origins in the cladistic approach advocated by Willi Hennig in his book *Phylogenetic Systematics*, which was translated into English in 1966. One of the major differences between these two taxonomic schools has been in how they have viewed species. Those following Evolutionary Systematics have typically used process-based species concepts (i.e. considering the process leading to population divergence), while those who have supported Phylogenetic Systematics have generally employed pattern-based species concepts (i.e. the patterns resulting from divergence). The best-known process-based species concept is the biological species concept (BSC), and the most commonly cited pattern-based concept is the phylogenetic species concept (PSC).

Usually credited to Mayr (1942), the BSC only applies to sexually reproducing organisms and defines species as *'groups of interbreeding natural populations that are reproductively isolated from other such groups'* (Mayr, 1996: 264). Some authors (e.g. Bock, 2004; Coyne and Orr, 2004) have elaborated on this definition to allow for limited gene flow between two species as long as their respective gene pools are protected from one another. Several other species concepts have been formulated as modifications to the BSC, but it has been argued that many of these are redundant (e.g. Evolutionary Species, Mate Recognition Species; Szalay, 1993; Mayr, 1996). In regard to the PSC, Cracraft's (1983) definition is the most commonly accepted: *'the smallest diagnosable cluster of individual organisms within which there is a parental pattern of ancestry and descent'*. This has been further refined to a group of populations with *shared and fixed* character combinations that represent minimal units appropriate for cladistic analysis (Davis and Nixon, 1992; Groves, 2004). In this sense, it is not so much a 'species concept' but a criterion for the diagnosis of species (Mayr, 1996; Goldstein and DeSalle, 2000). In fact, Groves (2012) states that a more appropriate name for the PSC might have been the 'Diagnosability Species Concept'.

Evolutionary Systematics (and thus the BSC and its derivatives) dominated the field of systematics for decades following the Modern Synthesis. However, the BSC has been criticised on several grounds (see Sokal and Crovello, 1970). One of its biggest shortcomings is the difficulty it creates for species diagnosis when populations do not overlap in distribution, precluding complete confidence in whether or not they would interbreed if brought into contact. In such circumstances, species status is typically given to a population when its differences (usually morphological) from other populations exceed the amount of variation seen within a typical species of the larger taxonomic group to which it belongs. Species diagnosis can thus change depending on what traits are compared and what is regarded as 'typical variation' within a species. This subjectivity has led to a great deal of confusion in the classification of many taxa, including the red colobus monkeys, as we outline below. Frustration over this subjectivity, in combination with the rise of molecular phylogenetics, has led to an increasing acceptance of Phylogenetic Systematics and the PSC over the past couple of decades. No one exemplifies this paradigm shift better than Colin Groves himself; his early classifications of red colobus monkeys (and other taxa) were consistent with the BSC, but more recently he has fully adopted the PSC in his classifications and has advocated for its use (Groves, 2001, 2004, 2012). Table 15.1 displays how red colobus classifications have changed over the years, including Groves' classifications.

Table 15.1: Taxonomic arrangements of red colobus forms by different authors.

Taxon	Thorington and Groves 1970	Dandelot 1971	Napier 1985	Oates 1986	Groves 1989	Oates et al. 1994	Groves 2001	Grubb et al. 2003	Groves 2007
temminckii	badius	badius	badius	badius	badius	badius	badius	badius	badius
badius	badius	badius	badius	badius	badius	badius	badius	badius	badius
waldroni	badius	waldroni	badius	badius	badius	badius	badius	badius	waldroni
epieni	UN	UN	UN	UN	UN	UN	pennantii	pennantii	epieni
preussi	badius	preussi	badius	pennantii	preussi	badius	preussi	pennantii	preussi
pennantii	badius	pennantii	badius	pennantii	pennantii	badius	pennantii	pennantii	pennantii
bouvieri	badius	pennantii	badius	pennantii	pennantii	badius	pennantii	pennantii	bouvieri
tholloni	badius	tholloni	badius	badius	pennantii	badius	tholloni	?	tholloni
rufomitratus	badius	rufomitratus	badius	rufomitratus	rufomitratus	badius	rufomitratus	?	rufomitratus
tephrosceles	badius	rufomitratus	badius	rufomitratus	pennantii	badius	tephrosceles	?	tephrosceles
oustaleti	badius	rufomitratus	badius	rufomitratus	pennantii	badius	foai	?	oustaleti
foai	badius	rufomitratus	badius	rufomitratus	pennantii	badius	foai	?	foai
lulindicus	syn foai	-	-	-	-	-		?	lulindicus
ellioti	badius	rufomitratus	badius	rufomitratus	pennantii	badius	foai	?	ellioti?*
semlikiensis	syn. ellioti	-	-	-	-	-	foai	syn. ellioti	semlikiensis
langi	syn. ellioti	-	-	-	-	-	-	?	langi
parmentieri	UN	-	UN	UN	UN	badius	-	?	parmentieri
gordonorum	badius	rufomitratus	badius	gordonorum	pennantii	badius	gordonorum	gordonorum	gordonorum
kirkii	kirkii	kirkii	kirkii	kirkii	pennantii	badius	kirkii	kirkii	kirkii
No. species recognised	2	7	2	5	4	1	9	at least 5	17–18

Notes: Species/subspecies names of geographically distinct forms are listed in the first column, followed by each author's species-level diagnosis in subsequent columns. Last row shows total number of species recognised in each classification.

'UN' indicates taxon undescribed at time of publication; 'syn.' means 'synonymised with'; '?' means population recognised as distinct, but not allocated to a species.

*Groves, 2007 lists *oustaleti*, *langi* and *semlikiensis* as separate species, and considers the population labeled *ellioti* to constitute a 'three-way hybrid swarm' of those three species.

Sources: Listed in table.

Red colobus monkey distribution and variation

Red colobus monkeys are commonly regarded as belonging to the subfamily Colobinae of the family Cercopithecidae (Old World monkeys) and they are closely related to the other two living African colobine groups – the olive colobus and the black-and-white colobus. Based largely on pelage differences, 16–18 different forms of red colobus are recognised in many recent classifications (see, e.g. Grubb et al., 2013), distributed across equatorial Africa in a primarily allopatric manner, with the exception being a putative hybrid zone in the eastern Democratic Republic of Congo (Figure 15.1). All populations of red colobus have varying amounts of red, black, white, brown and grey in their pelage, with certain forms showing considerable intra-populational variation, while others are relatively uniform (Kingdon, 1997; Struhsaker, 2010). Red colobus also have a complex and graded vocal system that makes it difficult to classify their vocalisations into discrete categories, unlike the calling array of the black-and-white colobus group (Marler, 1970). Furthermore, their crania display a clinal pattern of size and shape variation across Africa (Cardini and Elton, 2009). These features of red colobus biology have made their classification one of the thorniest issues in African primate taxonomy (Grubb et al., 2003).

Nearly all recently published classifications recognise the same 16–18 different forms of red colobus, with each form regarded as either a subspecies or species; there is thus broad agreement and stability in terms of the recognition of different geographic populations as being taxonomically distinct. The only major exceptions involve populations that occupy the putative hybrid zone in Central Africa. There has been little agreement, however, regarding how many species are present among these 16–18 taxa, and into which species each different form should be classified (see Table 15.1). This is because most classifications of these monkeys have attempted to diagnose species under the BSC, and because the distinct populations are distributed allopatrically objective diagnosis of biological species is extremely difficult, if not impossible. Furthermore, most of the systematic work done has involved comparisons of pelage patterns, which is problematic, given that red colobus coat colour varies at populational and even social group levels. Other research, involving vocalisations or craniometrics, has suffered from incomplete sampling and been confounded by the complex patterns of variation in these monkeys. This has often led to 'giving up' (in the words of Groves, 2001) and a decision to combine all of the red colobus into one species, except sometimes for a few particularly distinct forms. Despite this, most authors have recognised that the level of variation among the different forms does exceed that which is typical for a single primate species under the BSC.

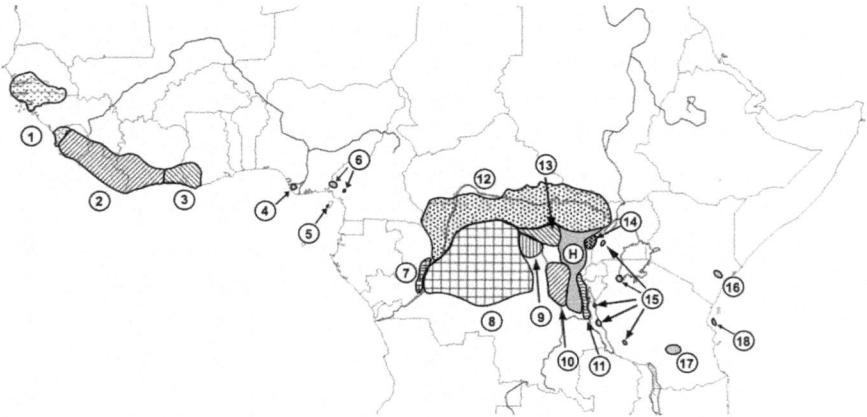

Figure 15.1: Distribution of 18 allopatric populations of red colobus monkeys that have been given taxonomic names of subspecies or species rank. 1, *temminckii*; 2, *badius*; 3, *waldroni*; 4, *epieni*; 5, *pennantii*; 6, *preussi*; 7, *bouvieri*; 8, *tholloni*; 9, *parmientieri*; 10, *lulindicus*; 11, *foai*; 12, *oustaleti*; 13, *langi*; 14, *semlikiensis*; 15, *tephrosceles*; 16, *rufomitratus*; 17, *gordonorum*; 18, *kirkii*. 'H' is a putative hybrid population in the eastern Democratic Republic of Congo.

Source: Distribution map created by authors using published accounts of red colobus population locations.

A history of red colobus monkey classification

Before describing the contentious species-level history of red colobus monkey classification, it is worth noting that the genus-level classification of this group too has changed substantially over the years. Such changes have also in part been caused by paradigm shifts in systematics and likewise may have influenced conservation policy by introducing further confusion. Briefly, most classifications prior to 1980 placed the red colobus monkeys with other African colobines in the genus *Colobus*. Since then, they have been recognised as (1) the subgenus *Piliocolobus* within the genus *Procolobus*, reflecting a close relationship to the olive colobus, or (2) members of a distinct genus, *Piliocolobus*. These different arrangements result from differences of opinion about what criteria (e.g. morphological variation, genetic variation, time) should be used to diagnose taxa above the species level; there is no current consensus as to which is most appropriate (Goodman, 1996; Groves, 2001).

An important benchmark in the classification of primates in modern times was the publication in 1967 of *A Handbook of Living Primates* by John and Prudence

Napier. Napier and Napier followed Verheyen (1962) in recognising just two species of red colobus monkeys, *Colobus badius* and *C. kirkii*, which they grouped together in the subgenus *Piliocolobus*. In the same year, Kuhn (1967) also followed Verheyen in separating *C. kirkii* of Zanzibar from *C. badius* as a monotypic species, and also placed the red colobus in the subgenus *Piliocolobus*.

Colin Groves, in addition to his doctoral research on gorilla systematics and ecology (Groves, 1966, 1967, 1970a), took an early interest in the systematics of gibbons and leaf-eating monkeys, and co-authored an influential classification of Old World monkeys with Richard Thorington (see also Groves, 1970b). For red colobus, Thorington and Groves (1970) used the classification of Kuhn (1967), but noted that the recognition of several species might be 'more in line with taxonomic practice'. In the same volume as Thorington and Groves, Rahm (1970) recognised only one species, *Colobus badius*, with 14 subspecies, including *C. b. kirkii*, and said that 'no definite answer can be given from the point of view of species and subspecies'. Not long after this, Dandelot (1971) produced a five-species classification of red colobus (*Colobus badius*, *C. pennantii*, *C. rufomitratus*, *C. tholloni* and *C. kirkii*), noting that more extensive research would undoubtedly lead to an increase in the number of species recognised, and suggesting *C. ellioti*, *C. preussi* and *C. waldroni* as 'potential' species. Struhsaker (1975) analysed the call repertoires of five different populations generally regarded as subspecies and found *C. b. preussi* to have the most divergent repertoire.

Confusion continued into the 1980s. Extending his earlier analysis of vocalisations to include additional red colobus populations, Struhsaker (1981) identified four clusters of subspecies based on degree of vocal similarity: (1) *badius* and *temminckii*; (2) *preussi*; (3) *tholloni*, *tephrosceles* and *rufomitratus*; and (4) *gordonorum* and *kirkii*. Wolfheim (1983) recognised only a single species of red colobus, while P. Napier (1985) retained the two-species arrangement (*C. badius* and *C. kirkii*) of Kuhn and of Thorington and Groves, citing Verheyen's observation (1962) that *kirkii* had a relatively small cranial capacity. In an IUCN SSC conservation action plan Oates (1986) – taking account of Struhsaker (1981) – regarded the red colobus monkeys as members of a single superspecies, *Procolobus badius*, provisionally containing five species: *P. badius*, *P. pennantii*, *P. rufomitratus*, *P. kirkii* and *P. gordonorum*; all the forms found in the central and eastern Congo Basin, together with *tephrosceles* and *rufomitratus* of eastern Africa, were grouped together in *P. rufomitratus*. Groves (1989), however, citing unpublished research by himself and Pierre Dandelot, moved to a four-species arrangement, recognising a central species (*Colobus pennantii*, but which also now included *C. kirkii*), a species restricted to Kenya's Tana River (*C. rufomitratus*), an Upper Guinea species (*C. badius*), and *C. preussi* of Cameroon and Nigeria (said to be 'very distinct').

In a review of colobine monkey diversity in the mid-1990s, Oates and others (1994) decided to treat all the red colobus as a single species, *Procolobus* (*Piliocolobus*) *badius*, based on the lack of consensus in other classifications. This publication did not contain any new analysis, and did not influence later taxonomic studies, but it did influence conservation listings, as we describe below. Kingdon (1997), by contrast, in a widely used field guide, placed the red colobus in their own genus, *Piliocolobus*, and used an eight-species classification: *P. kirkii*, *P. gordonorum*, *P. rufomitratus*, *P. tholloni*, *P. oustaleti*, *P. pennanti* [sic], *P. preussi* and *P. badius*.

Since the year 2000, several substantially different taxonomic arrangements of red colobus monkeys have been published, adding to the confusion. The IUCN SSC Primate Specialist Group convened a meeting of primate biologists in Orlando, Florida, in 2000 in an attempt to produce a taxonomic consensus that could be used in conservation planning. This meeting, in which Colin Groves participated, failed to reach a clear consensus on the species-level classification of red colobus. A classification of African primates resulting from the meeting placed the red colobus in the genus *Procolobus* (subgenus *Piliocolobus*); it recognised five distinct species (*P. badius*, *P. kirkii*, *P. gordonorum*, *P. rufomitratus* and *P. pennantii*), and left an additional 5–8 subspecies in a poorly defined 'central assemblage' on which the working group recommended further research to establish relationships (Grubb et al., 2003). This central assemblage of populations is the same group of taxa referred to as *P. rufomitratus* by Oates (1986), except that Grubb and others excluded *rufomitratus* itself, treating this Tana River red colobus as a separate species. Meanwhile, Groves (2001) had published his influential book *Primate Taxonomy*, which listed nine species of red colobus, allocated to the genus *Piliocolobus*: *P. badius*, *P. pennantii*, *P. preussi*, *P. tholloni*, *P. foai*, *P. tephrosceles*, *P. gordonorum*, *P. kirkii* and *P. rufomitratus*.

Later, Groves revised his nine-species classification to a 16-species arrangement by additionally recognising *Piliocolobus waldronae*, *P. epieni*, *P. bouvieri*, *P. parmientieri*, *P. oustaleti*, *P. langi* and *P. semlikiensis* as full species (Groves, 2007). Shortly after this, Ting (2008) presented the results of the first thorough comparison of mitochondrial DNA in red colobus monkeys and proposed instead a five-species arrangement within the genus *Procolobus* (subgenus *Piliocolobus*): *P. badius*, *P. pennantii*, *P. kirkii*, *P. rufomitratus* and *P. epieni*.

Struhsaker (2010), paying special attention to patterns of vocal similarity and difference, recognised seven groups of 'taxa' (*badius* and relatives, *preussi*, *pennantii*, *bouvieri*, *rufomitratus* and relatives, *gordonorum* and *kirkii*), but was not prepared to allocate these groups, or the populations within them, to named species. In an appendix to Struhsaker's book, Grubb and others (2010) listed 18 subspecies as belonging to a single species, *Procolobus badius*.

In the recently published *Mammals of Africa*, Grubb and others (2013) 'very provisionally' recognise six species: *Procolobus badius*, *P. preussi*, *P. pennantii*, *P. rufomitratus*, *P. gordonorum* and *P. kirkii*. This is the same arrangement used by Struhsaker (2010), except that these taxa are given species rank, and *bouvieri* of the Congo Republic is included within *P. pennantii*.

Effects on conservation planning

One of the earliest attempts to provide an inventory of threatened species to guide conservation planning was the publication of the Red Data Books by IUCN's Survival Service Commission (known since 1980 as the Species Survival Commission). These publications began to appear in 1966 as loose-leaf datasheet volumes giving information on rare and endangered animals which had come to the attention of IUCN. The first Red Data Book on mammals (Simon, 1966) included 25 species and 22 subspecies of primates judged to be rare or endangered; among these were three red colobus monkeys, listed as *Colobus badius kirkii*, *C. b. rufomitratus* and *C. b. gordonorum*. In 1978 *Colobus badius preussi* was added to the list (Goodwin et al., 1978). In 1980, datasheet publications were superseded by bound volumes, with different volumes covering different groups of taxa, and in 1986 the Red Data Books became the Red List. Table 15.2 compares a selection of IUCN's threat ratings of red colobus taxa from 1978 to 2012.

Table 15.2: Selected threat status listings of red colobus taxa by IUCN.

Taxon	Red Data Book 1978	Lee et al.1988	Red List1996	Red List2012
temminckii	–	R	EN	EN
badius	–	VU	–	EN
waldroni	–	EN	CR	CR
epieni	UN	UN	EN	CR
preussi	EN	EN	EN	CR
pennanttii	–	EN	EN	EN
bouvieri	–	EN	EN	CR
tholloni	–	K	–	NT
rufomitratus	EN	EN	EN	EN
tephrosceles	–	VU	–	EN
oustaleti	–	K	–	LC
foai	–	K	DD	–
lulindicus	–	–	DD	–
ellioti	–	K	–	–
semlikiensis	–	–	DD	–
langi	–	–	DD	–
parmentieri	UN	–	DD	–
gordonorum	R	EN	EN	EN
kirkii	R	EN	EN	EN

Notes: CR = Critically Endangered; DD = Data Deficient; EN = Endangered; K = Insufficiently Known; LC = Least Concern; NT = Not Threatened; R = Rare; UN = Undescribed; VU = Vulnerable; a dash indicates that the taxon was not given an individual listing.

Source: Data from IUCN Red Data Books and Red Lists.

During the 1980s two conservation assessments appeared that focused on African primates. The IUCN SSC Primate Specialist Group's *Action Plan for African Primate Conservation: 1986–90* (Oates, 1986) made an assessment of the status of every African primate species. The action plan used the five-species classification of red colobus referred to above (see Table 15.1): *P. badius* and *P. rufomitratus* were rated as Vulnerable, *P. pennantii* as Endangered, and *P. kirkii* and *P. gordonorum* as Highly Endangered. The same species-level classification was followed by Lee and others (1988) in the *Threatened Primates of Africa: The IUCN Red Data Book*. Lee and others included both species and subspecies; each of the five species was listed as either Vulnerable (*P. badius* and *P. rufomitratus*) or Endangered (the remaining three species), and nine subspecies were also given attention through being regarded as of conservation concern. IUCN's 1988 Red List also employed this five species arrangement (IUCN, 1988).

A significant change in the classification of red colobus monkeys for conservation purposes occurred with IUCN's 1996 Red List (IUCN, 1996). Here, all red colobus were lumped into one species, *Procolobus badius*, with 14 subspecies. The 1996 primate assessments were made by the Primate Specialist Group; Oates and others (1994) is almost certainly the source of the classification employed. Although several subspecies were listed as Endangered or Critically Endangered in the 1996 Red List, the species as a whole was rated as only Near Threatened, based on the new system of threat categories and criteria adopted by IUCN in 1994. The same one-species classification was employed in the revised edition of the Primate Specialist Group's African primate action plan, which appeared in the same year (Oates, 1996).

The IUCN Red List is now published in digital form (www.iucnredlist.org). At the time of writing, the Red List assesses the status of 13 different geographically and taxonomically distinct forms of red colobus, including six species and 10 subspecies (IUCN, 2012). The six species are: *Procolobus badius*, *P. gordonorum*, *P. kirkii*, *P. pennantii*, *P. preussi* and *P. rufomitratus*. This arrangement is based in part on IUCN's Global Mammal Assessment of 2008, for which primates were initially assessed at a workshop in 2005; that workshop used Grubb and others (2003) as a primary reference for classification. Grubb and others treat *preussi* as a subspecies of *P. pennantii*, but the latest Red List elevates this taxon to species level, following Butynski and Kingdon (2013). Of the 13 taxa on the Red List, four are rated as Critically Endangered and seven as Endangered. The taxa *foai*, *lulindicus*, *ellioti*, *langi* and *parmentieri* are not individually assessed in the 2012 Red List; they are listed as subspecies of *P. rufomitratus* which, as a species, is given a rating of Least Concern. The 2012 Red List does not list *semlikiensis*, following Grubb and others (2003) in treating this taxon as synonymous with *ellioti*.

How might changes in classification have affected the attention given by conservationists to the rarest and most threatened of red colobus monkey populations? We will highlight the cases of three forms of red colobus, one probably extinct, one possibly extinct, and one verging on extinction.

Miss Waldron's red colobus of eastern Côte d'Ivoire and western Ghana was referred to as the subspecies *Procolobus badius waldroni* by Oates (1986), Lee and others (1988) and IUCN (1988). Groves (1989) did not specifically mention *waldroni*, but also implied that it should be considered as a subspecies of *Procolobus badius*. The 1996 IUCN Red List gave separate treatments to species and subspecies; all red colobus were treated as one species (*Procolobus badius*, listed as Near Threatened), while the subspecies *P. badius waldroni* was listed as Critically Endangered (IUCN, 1996). A few years later, Oates and others (2000) reported that *P. b. waldroni* was probably extinct, and suggested that even if a few individuals survived, no viable population remained. Subsequently, no reliable record has emerged of any Miss Waldron's red colobus having been seen in the wild, although the remains of a few individuals were found with hunters in Côte d'Ivoire, most recently in 2006 (Oates, 2011). Since 2006, Groves (2007) has elevated this monkey to species status (as *Piliocolobus waldronae*), and the genetic study by Ting (2008) has also indicated that this might be a reasonable course, confirming the suggestion of Dandelot (1971). We are left to wonder whether recognition of *waldroni* as a species during the 1980s and 1990s might have directed more conservation attention to this monkey, and averted its extinction. In other words, did taxonomy 'kill' this monkey in the sense used by Morrison and others (2009)?

Bouvier's red colobus is known only from a handful of specimens collected in the nineteenth and early twentieth centuries in the former French Congo (today's Republic of the Congo, or Congo-Brazzaville). Dandelot (1971) classified this monkey as *Colobus pennantii bouvieri*, based on the similarity of its colour pattern and arrangement of hair on the front of the head to Pennant's red colobus of Bioko. Most subsequent classifications have kept Bouvier's colobus as a subspecies of Pennant's colobus (where that taxon is regarded as a species), or recognised it as one among many subspecies of red colobus in single-species classifications. Exceptions are Ting (2008), who could not confidently place this form into a species because a lack of biomaterials precluded its inclusion in DNA analysis, and Groves (2007), who elevates this form to species rank as *Piliocolobus bouvieri*, and says that the 'status of this extremely poorly known monkey needs urgent investigation'. The lack of specimens for both morphological and genetic comparative study (and the fact that there have been no substantial scientific observations in the wild) has led this monkey to be seriously neglected. Even the exact locations from where the museum specimens originated are in some doubt, but they seem to lie mostly in the swamp forests on the right bank of the lower Sangha River and near the mouth of the Likouala-

Mossaka River. A report of red monkeys with light faces and white underparts from the Lefini Reserve in the 1970s (quoted in Groves, 2007) doubtfully refers to this colobus. In the 1996 Red List, *bouvieri* was rated as an Endangered subspecies of *Procolobus badius* (IUCN, 1996) and the current Red List includes it as a Critically Endangered subspecies of *Procolobus pennantii*. Given the lack of any convincing observational reports of this colobus for many decades, there must be a strong possibility that it is extinct; R. Dowsett (pers. comm. to JFO, 1974) noted that monkeys in general are very heavily hunted in this part of Congo, although he added that the area from which *bouvieri* is known is difficult to access. As with Miss Waldron's colobus we speculate that a clearer taxonomic definition of Bouvier's colobus might have led to it receiving more attention.[1]

The Tana River red colobus of Kenya was listed in the original IUCN Red Data Book as a subspecies, *rufomitratus*, of the species *Colobus badius* and rated as Endangered (Simon, 1966). In 1972 the surviving population was estimated at around 1900 individuals (Goodwin et al., 1978). Oates (1986), Lee and others (1988) and IUCN (1988) continued to list the Tana colobus as Endangered, but classified it as a subspecies of *Procolobus rufomitratus*, a species considered to occupy the Congo Basin and the Western Rift Valley as well as the Tana River, on the basis of the vocal patterns reported by Struhsaker (1981). Kingdon (1997), however, regarded the Tana population as a distinct species, *Piliocolobus rufomitratus*, a course later followed by Groves (2001, 2007) and (using the name *Procolobus rufomitratus*) by Grubb and others (2003). In 1999–2001, Meikle and Mbora (2004) recorded a total of 613 individuals in the forests along the Tana River, 50% of the number estimated present in 1994, and referred to the Tana red colobus as 'the most endangered primate species in Africa.' Mbora and Butynski (2009) describe the long-time survival prospects of the Tana colobus as very bleak, especially since the High Court of Kenya ruled in 2007 that the Tana River National Primate Reserve was not properly established by law. However, the current Red List (IUCN, 2012) treats the Tana colobus as the subspecies *Procolobus rufomitratus rufomitratus*, with the status of Endangered, apparently using an older population estimate of 1,100–1,300 individuals and an assessment that there has not been a significant population decline since 1975.

In addition to the three forms of red colobus we have highlighted, a majority of the remaining 13–15 forms must be regarded as threatened, based on having small, fragmented, and/or rapidly declining populations. Of particular concern are Pennant's red colobus of Bioko Island, the Niger Delta red colobus, and Preuss' red colobus of western Cameroon and eastern Nigeria. Each of these monkeys is rated as Critically Endangered on the current Red List (IUCN, 2012) where they are called *Procolobus pennantii pennantii*, *P. pennantii epieni* and

---

1   Note added in proof: Lieven Devreese (pers. comm.) planned to conduct a field survey to locate any surviving populations of Bouvier's red colobus in the early part of 2015. At the time of writing, no information from this survey was available.

P. preussi respectively – the same classification employed by Grubb and others (2013). Groves (2007) calls these taxa *Piliocolobus pennantii, P. epieni* and *P. preussi.*

At least six forms of red colobus monkey could readily be regarded, therefore, as among the most endangered primates in Africa, along with the roloway monkey of Ghana and Côte d'Ivoire, the kipunji of Tanzania, and the mountain and Cross River gorillas. Some combination of these primates has featured for some years, with others, on the list of the *World's 25 Most Endangered Primates* compiled by the IUCN Primate Specialist Group, Conservation International and the International Primatological Society (see, e.g. Mittermeier et al., 2009). No more than two forms of red colobus (listed as species or subspecies) have ever appeared on this list, however, because of a perceived need (in terms of raising support for primate conservation) to distribute the 25 primates selected as the 'Most Endangered' relatively evenly across Africa, Asia, Madagascar and the Neotropics, as well as across higher taxonomic groups (including strepsirrhines and great apes). This less than objective approach has served to diminish a general awareness of how many red colobus forms are in trouble, and the fact that red colobus often appear on this list as subspecies (in contrast to a large majority of full species occupying the other slots) may also diminish a sense of the crisis faced by these primates.

Differences among red colobus monkey classifications could also potentially affect the prioritisation of areas for conservation. Area-based conservation planning typically compares geographic regions based on their levels of species endemism and/or richness. In general, regions with greater numbers of species, and particularly endemic species, are judged to warrant higher conservation priority and therefore may have a greater chance of being designated for protection efforts. However, levels of species endemism and richness can change depending on what species classification is used, so that a simple change in species concept can alter priority areas for conservation (Agapow et al., 2004). Figures 15.2–15.4 display the distribution of red colobus monkey species according to four different classification schemes. Under a single-species classification (e.g. Oates et al., 1994), there is no area of red colobus species endemism. Using Groves' 2001 nine-species classification (Figure 15.2) East and Central Africa become priority areas for conservation, with six endemic species. Figure 15.3 displays Groves' full application of the PSC to these primates (Groves, 2007) and would give conservation priority to the Congo Basin, which contains nearly half of the red colobus forms. Alternatively, Ting's 2008 classification recognises five species, three of which are endemic to west Central Africa (Figure 15.4). These four different classifications emphasise how differences in taxonomy can create tangible differences in the selection of conservation priority areas. For example, although Myers and others (2000) did not consider the Congo Basin as a biodiversity hotspot, Olson and Dinerstein (2002) do consider the Northeastern Congo Basin Moist Forests as a special ecoregion with many endemic species, including '*Piliocolobus oustaleti*'.

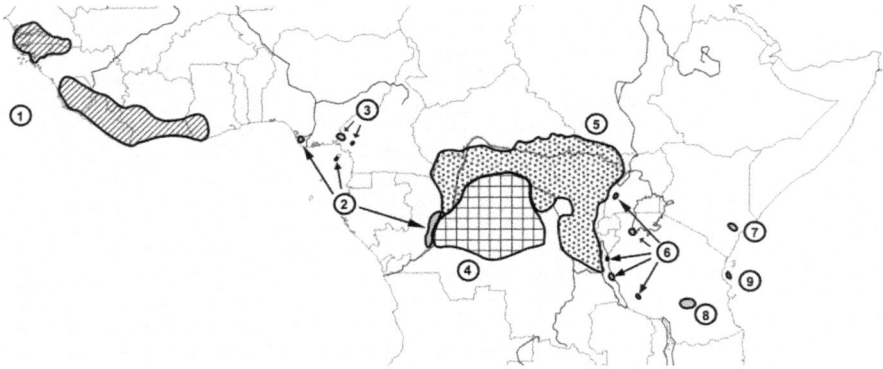

Figure 15.2: Geographical distribution of red colobus according to nine-species arrangement of Groves (2001); 1 = *P. badius*, 2 = *P. pennantii*, 3 = *P. preussi*, 4 = *P. tholloni*, 5 = *P. foai*, 6 = *P. tephrosceles*, 7 = *P. rufomitratus*, 8 = *P. kirkii*, 9 = *P. gordonorum*.

Source: After Groves (2001); Ting (2008).

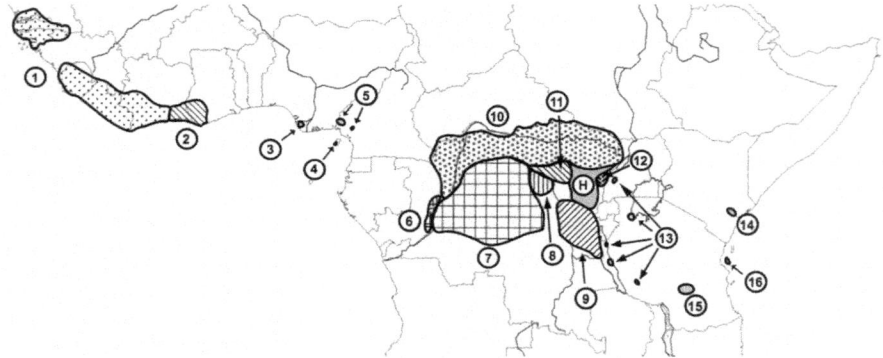

Figure 15.3: Geographical distribution of red colobus according to 16-species arrangement of Groves (2007); 1 = *P. badius*, 2 = *P. waldroni*, 3 = *P. epieni*, 4 = *P. pennantii*, 5 = *P. preussi*, 6 = *P. bouvieri*, 7 = *P. tholloni*, 8 = *P. parmentieri*, 9 = *P. foai*, 10 = *P. oustaleti*, 11 = *P. langi*, 12 = *P. semlikiensis*, 13 = *P. tephrosceles*, 14 = *P. rufomitratus*, 15 = *P. gordonorum*, 16 = *P. kirkii*.

Source: After Groves (2007); Ting (2008).

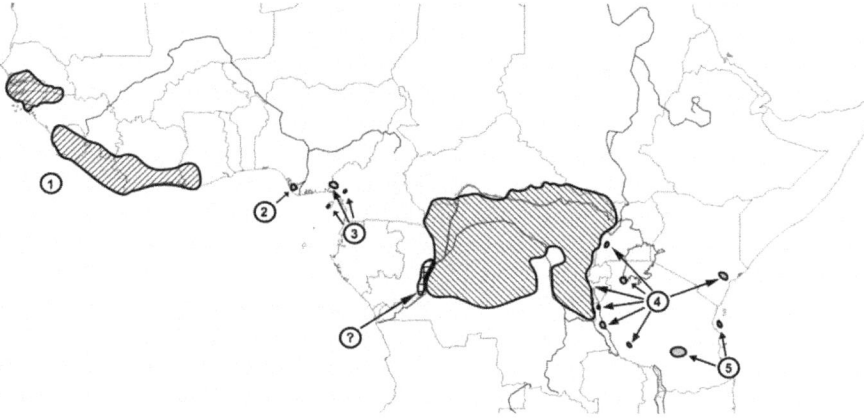

Figure 15.4: Geographical distribution of red colobus according to five-species arrangement of Ting (2008): 1 = *P. badius*, 2 = *P. epieni*, 3 = *P. pennantii*, 4 = *P. rufomitratus*, 5 = *P. kirkii*; '?' is bouvieri which could not be considered in Ting's analysis due to lack of material.

Source: After Ting (2008).

Discussion and conclusions

The points reviewed in this chapter raise broader issues regarding systematics and conservation biology. It has long been argued that the two fields require better integration if conservation priorities are to be set in the most effective way (e.g. see Rojas, 1992; Dubois, 2003; Mace, 2004; Agapow et al., 2004). Despite progress in this area over the past 20 years, however, large areas of debate remain. One of these is the extent to which conservation concerns should be taken into account when diagnosing taxa. After all, more money and higher levels of protection are commonly directed to endangered populations recognised as distinct species or subspecies. Understanding this, Groves (2001) recognised the Cross River gorilla as the subspecies *Gorilla gorilla diehli* based more on an appreciation of conservation concern, and as a stimulus to further research, 'than anything else'. However, although a concern for the survival of a population suspected to be more distinctive than is generally recognised can be a useful spur to new taxonomic investigation (which in turn may produce new conservation attention), our view is that elevating populations to higher taxonomic ranks solely due to conservation concern is not valid and can undermine the credibility of instruments such as the Red List in the larger world.

As we have discussed, different species concepts in the field of systematics lead to different ways of classifying organisms, and most systematists (or at least those who are not strongly wedded to a single concept) would agree that the choice of concept is subjective. In an ideal world, we would all agree on a single species concept that can be applied consistently across all taxa. However, such agreement is very unlikely, and we are thus stuck with a plurality of species concepts. Given this circumstance, we believe it would be sensible to choose species concepts based on their applicability to the particular study organisms, as long as researchers are transparent about which species concept they are using. This would help ensure that taxonomic revisions are due to new data and discoveries rather than to a simple change in species concept, and it would prevent populations being forced into species concepts that are poorly applicable.

Choosing species concepts based on circumstance would also have an effect on conservation planning by generating stable classifications more quickly. For example, the BSC has been argued to be most appropriate for conservation planning purposes because it is process-based, grounded in population genetics, and looks to the present and future (Frankham et al., 2012). While we believe this to be true *in theory*, conservation decisions have to be made by balancing what is best in theory with what can be implemented *in practice*. For example, application of the BSC to allopatric populations can be very time and labour intensive and produces subjective classifications. Use of the BSC can therefore be a hindrance to urgent action in these circumstances, especially if multiple types of data are required and the organisms concerned are rare and/or live in remote areas. The red colobus monkeys are an example of this problem. While scientists have spent several decades attempting to delimit BSC boundaries in these monkeys, one form has probably gone extinct (Miss Waldron's red colobus), one may be extinct (Bouvier's red colobus), and several more have declined to precarious states.

Use of the PSC for allopatric populations has advantages because it produces objective and unambiguous classifications (Vogler and DeSalle, 1994; Gippoliti and Groves, 2013). Some have suggested that phylogenetic species are not necessarily evolutionarily meaningful and that the PSC undermines the importance of species in the evolutionary process (e.g. Tattersall, 2007, 2013). While we appreciate these concerns, it is important to point out that the most significant unit of evolution is the population, and the PSC recognises species as *populations* that have diverged in some manner. Even if this divergence is not enough to produce reproductive isolation, the evolutionary significance of whether or not two populations can potentially interbreed is moot if they are allopatric and will never come into contact. There is thus more recognition of evolutionary theory in the Phylogenetic Species Concept than some acknowledge.

The use of phylogenetic species does not preclude further research into, and incorporation of, more process-based and adaptive frameworks in conservation, such as grouping certain species together into larger management units or identifying divergent taxa as high conservation priorities (Gippoliti and Groves, 2013).

We consider that the instability in the taxonomic treatment of red colobus has been one factor that has led these monkeys to be relatively neglected in conservation planning compared to some other primates, and that this neglect may have led to a lack of sufficient action to halt the decline and possible extinction of some distinctive populations. Unstable classifications may have caused confusion, and led to less focused conservation action than has been needed. The primary cause of the taxonomic instability has been the inability of scientists to diagnose species in this group according to the Biological Species Concept and the resulting gradual transition to the application of the Phylogenetic Species Concept. Red colobus monkey conservation might have benefitted if the PSC had been applied at a much earlier point. We do not believe that application of the PSC undermines the scientific credibility of either systematics or conservation as long as those using it are transparent regarding its use. While the case of the red colobus is only a single example of an advantage of applying the PSC over the BSC, it is possible that application of the PSC across other taxonomic groups could prevent similar dire situations from arising in other organisms.

We suggest that the application of taxonomy to conservation planning could also be improved if conservation authorities such as the IUCN SSC Primate Specialist Group used standardised species lists updated at regular intervals (e.g. five years); such lists should be accompanied by a clear statement on the species concept used to produce the list. The IUCN Red List would then follow the standardised lists formulated by specialist groups. Meanwhile, normal research in systematics would continue, and its findings be considered during reviews of standard lists.

However, although taxonomic instability may have contributed to the lack of conservation attention given to red colobus monkeys, despite the precarious status of many forms, it is almost certainly not the only factor leading to their neglect. Even though their classification has been highly unstable, red colobus have long featured in some way on the IUCN Red List. We conclude that taxonomy is probably only one factor in their neglect compared to, for instance, great apes and lemurs. Morrison and others (2009) have found that the 'charisma' of animals like red wolves, polar bears and green turtles has meant that there has been no reduction in the conservation efforts devoted to them despite taxonomic research findings that question their species status. Red colobus lack the charisma that great apes gain from their close similarity to humans, their intelligence and their size. They lack the 'cuteness' of furry lemurs. Both great apes and lemurs are readily seen close-up in many zoos, and have been the focus of a great deal of media attention. Red colobus, which have never survived very long in captivity, lack all these attributes.

While a case could be made for conservation action to be undertaken largely independently of current taxonomic opinion, as happens today with some particularly charismatic animals, if this course was generally followed then objective conservation planning would be almost impossible. Those organisms less charismatic to the general public would have a low priority for conservation attention. Thus good taxonomy is essential for effective conservation. By dedicating a lifetime to describing biological diversity, Colin Groves has greatly aided conservation efforts in a wide range of taxa. His relatively recent endorsement of the PSC has helped illuminate the issue of the role of taxonomy in conservation and produced a healthy debate.

Acknowledgements

We thank Tom Struhsaker for first introducing us to red colobus, and for his huge contribution to an understanding of these monkeys' biology. We are also grateful to Eric Delson and Cliff Jolly for stimulating discussions about systematics and taxonomic theory, and to Colin Groves and the late Peter Grubb for many thought-provoking conversations about primate and especially red colobus taxonomy. As we made final edits to this chapter we learned of the death (on 5 July 2014) of Dr Peter Marler. Marler briefly studied red colobus in Uganda's Kibale Forest in 1965 and encouraged Tom Struhsaker – who had been his PhD student – to undertake a more comprehensive study of these monkeys. JFO joined Tom in Kibale in 1970, leading to an affiliation with the Marler lab at Rockefeller University which lasted into the 1980s and was a life-changing experience. Our contribution is therefore dedicated to the memory of Peter Marler.

References

Agapow PM, Bininda-Emonds ORP, Crandall KA, Gittleman JL, Mace GM, Marshall JC, Purvis A. 2004. The impact of species concept on biodiversity studies. *Q Rev Biol* 79:161–179.

Bock WJ. 2004. Species: The concept, category and taxon. *J Zool Syst Evol Res* 42:178–190.

Butynski TM, Kingdon J. 2013. *Procolobus preussi* Preuss' red colobus. In: Butynski TM, Kingdon J, Kalina, J, editors. *Mammals of Africa*. Vol. II: *Primates*. London: Bloomsbury Press. pp. 134–136.

Cardini A, Elton S. 2009. Geographical and taxonomic influences on cranial variation in red colobus monkeys (Primates, Colobinae): Introducing a new approach to 'morph' monkeys. *Global Ecol Biogeogr* 18:249–263.

Coyne JA, Orr HA. 2004. *Speciation*. Sunderland, MA: Sinauer Associates.

Cracraft JL. 1983. Species concepts and speciation analysis. *Curr Ornithol* 1:159–187.

Dandelot P. 1971. *Order Primates*. Part 3, *The mammals of Africa: An identification manual*. Washington, DC: Smithsonian Institution Press.

Davis JI, Nixon KC. 1992. Populations, genetic-variation, and the delimitation of phylogenetic species. *Syst Biol* 41:421–435.

Dubois A. 2003. The relationships between taxonomy and conservation biology in the century of extinctions. *C R Biol* 326:S9–S21.

Frankham R, Ballou JD, Dudash MR, Eldridge MDB, Fenster CB, Lacy RC, Mendelson JR III, Porton IJ, Ralls K, Ryder OA. 2012. Implications of different species concepts for conserving biodiversity. *Biol Conserv* 153:25–31.

Gippoliti S, Groves CP. 2013. 'Taxonomic inflation' in the historical context of mammalogy and conservation. *Hystrix* 23:1–7.

Goldstein PZ, DeSalle R. 2000. Phylogenetic species, nested hierarchies, and character fixation. *Cladistics* 16:364–384.

Goodman M. 1996. Epilogue: A personal account of the origins of a new paradigm. *Mol Phylogenet Evol* 51:269–285.

Goodwin HA, Holloway CW, Thornback J. 1978. *Red data book*. Vol. 1: *Mammalia*. Revised edition. Morges, Switzerland: IUCN.

Groves CP. 1966. Variation in the skulls of gorillas with particular reference to ecology. PhD thesis, University College London.

Groves CP. 1967. Ecology and taxonomy of the gorilla. *Nature* 213:890–893.

Groves CP. 1970a. Population systematics of the gorilla. *J Zool, Lond* 161:287–300.

Groves CP. 1970b. The forgotten leaf-eaters, and the phylogeny of the Colobinae. In: Napier JR, Napier PH, editors. *Old World monkeys: Evolution, systematics, and behavior.* New York: Academic Press. pp. 555–587.

Groves CP. 1989. *A theory of human and primate evolution.* Oxford: Oxford University Press.

Groves CP. 2001. *Primate taxonomy.* Washington, DC: Smithsonian Institution Press.

Groves CP. 2004. The what, why, and how of primate taxonomy. *Int J Primatol* 25:1105–1126.

Groves CP. 2007. The taxonomic diversity of the Colobinae in Africa. *J Anthropol Sci* 85:7–34.

Groves CP. 2012. Species concepts in primates. *Am J Primatol* 74:687–691.

Grubb PJ, Butynski TM, Oates JF, Bearder SK, Disotell TR, Groves CP, Struhsaker TT. 2003. Assessment of the diversity of African primates. *Int J Primatol* 24:1301–1357.

Grubb PJ, Glander K, Siex K. 2010. Annotated list and measurements of red colobus taxa. Appendix 1.1. In: Struhsaker TT, editor. *The red colobus monkeys.* Oxford: Oxford University Press. pp. 277–284.

Grubb PJ, Struhsaker TT, Siex KS. 2013. Subgenus *Piliocolobus.* In: Butynski TM, Kingdon J, Kalina J, editors. *Mammals of Africa.* Volume II: *Primates.* London: Bloomsbury Press. pp. 125–128.

Hennig W. 1966. *Phylogenetic systematics.* Champaign, IL: University of Illinois Press.

Huxley J. 1942. *Evolution. The modern synthesis.* London: Allen and Unwin.

International Union for Conservation of Nature (IUCN). 1988. 1988 *IUCN Red List of threatened animals.* Gland, Switzerland: IUCN.

IUCN. 1996. 1996 *IUCN Red List of threatened animals.* Gland, Switzerland: IUCN.

IUCN. 2012. *IUCN Red List of threatened species.* Version 2012.2. www.iucnredlist.org. accessed 7 May 2013.

Kingdon J. 1997. *The Kingdon field guide to African mammals.* San Diego: Academic Press.

Kuhn H-J. 1967. Zur Systematik der Cercopithecidae. In: Starck D, Schneider R, Kuhn H-J, editors. *Neue Ergebnisse der Primatologie*. Stuttgart: Gustav Ficher. pp. 25–46.

Lee PC, Thornback J, Bennett EL. 1988. *Threatened primates of Africa: The IUCN Red Data book*. Gland, Switzerland: IUCN.

Mace GM. 2004. The role of taxonomy in species conservation. *Phil Trans R Soc Lond B:Biol Sci* 359:711–719.

Marler P. 1970. Vocalizations of East African monkeys. *Folia Primatol* 13:81–91.

Mayr E. 1942. *Systematics and the origin of species*. New York: Columbia University Press.

Mayr E. 1996. What is a species, and what is not? *Philos Sci* 63:262–277.

Mbora DNM, Butynski T. 2009. Tana River red colobus. In: Mittermeier RA et al. Primates in peril: The world's 25 most endangered primates 2008–2010. *Primate Conserv* 24:15.

Mbora DNM, Meikle DB. 2004. Forest fragmentation and the distribution, abundance and conservation of the Tana River red colobus (*Procolobus rufomitratus*). *Biol Conserv* 118:67–77.

Mittermeier RA, Wallis J, Rylands AB, Ganzhorn JU, Oates JF, Williamson EA, Palacios E, Heymann EW, Kierulff MCM, Yongcheng L, Supriatna J, Roos C, Walker S, Cortés-Ortiz L, Schwitzer C. 2009. Primates in peril: The world's 25 most endangered primates 2008–2010. *Primate Conserv* 24:1–57.

Morrison WR III, Lohr JL, Duchen P, Wilches R, Trujillo D, Mair M, Renner SS. 2009. The impact of taxonomic change on conservation: Does it kill, can it save, or is it just irrelevant? *Biol Conserv* 142:3201–3206.

Myers N, Mittermeier RA, Mittermeier CG, da Fonseca GAB, Kent J. 2000. Biodiversity hotspots for conservation priorities. *Nature* 403:853–858.

Napier JR, Napier PH. 1967. *A handbook of living primates: Morphology, ecology and behaviour of nonhuman primates*. London: Academic Press.

Napier PH. 1985. *Catalogue of primates in the British Museum (Natural History) and elsewhere in the British Isles*. Part III: *Family Cercopithecidae, Subfamily Colobinae*. London: British Museum (Natural History).

Oates JF. 1986. *Action plan for African primate conservation: 1986–90*. Stony Brook, NY: IUCN/SSC Primate Specialist Group.

Oates JF. 1996. *African primates: Status survey and conservation action plan.* Revised edition. Gland, Switzerland: IUCN.

Oates JF, Abedi-Lartey M, McGraw WS, Struhsaker TT, Whitesides GH. 2000. Extinction of a West African red colobus monkey. *Conserv Biol* 14:1526–1532.

Oates JF, Davies AG, Delson E. 1994. The diversity of living colobines. In: Davies AG, Oates JF, editors. *Colobine monkeys: Their ecology, behaviour and evolution.* Cambridge: Cambridge University Press. pp. 45–73.

Olson DM, Dinerstein E. 2002. The Global 200: Priority ecoregions for global conservation. *Ann Missouri Bot Gard* 89:199–224.

Rahm UH. 1970. Ecology, zoogeography, and systematics of some African forest monkeys. In: Napier JR, Napier PH, editors. *Old World monkeys: Evolution, systematics, and behavior.* NewYork: Academic Press. pp. 589–626.

Rojas, M. 1992. The species problem and conservation: What are we protecting? *Conserv Biol* 6:170–178.

Simon N. 1966. *Red data book. Volume 1: Mammalia, a compilation.* Morges, Switzerland: IUCN.

Sokal RR, Crovello TJ. 1970. The biological species concept: A critical evaluation. *Am Nat* 104:127–153.

Struhsaker TT. 1975. *The red colobus monkey.* Chicago: University of Chicago Press.

Struhsaker TT. 1981. Vocalizations, phylogeny, and palaeogeography of red colobus monkeys (*Colobus badius*). *Afr J Ecol* 19:265–283.

Struhsaker TT. 2010. *The red colobus monkeys.* Oxford: Oxford University Press.

Szalay FS. 1993. Species concepts: The tested, the untestable, and the redundant. In: Kimbel WH, Martin LB, editors. *Species, species concepts, and primate evolution.* New York: Plenum Press. pp. 21–41.

Tattersall I. 2007. Madagascar's lemurs: Cryptic diversity or taxonomic inflation? *Evol Anthropol* 16:12–23.

Tattersall I. 2013. Species-level diversity among Malagasy lemurs. In: Masters J, Gamba M, Génin F, editors. *Leaping ahead: Advances in prosimian biology.* New York: Springer. pp. 11–20.

Thorington RW Jr, Groves CP. 1970. An annotated classification of the Cercopithecoidea. In: Napier JR, Napier PH, editors. *Old World monkeys: Evolution, systematics, and behavior.* New York: Academic Press. pp. 629–647.

Ting N. 2008. Molecular systematics of red colobus monkeys (*Procolobus* [*Piliocolobus*]): Understanding the evolution of an endangered primate. PhD Thesis, City University of New York.

Verheyen WN. 1962. Contribution à la craniologie comparée des Primates: Les genres *Colobus* Illiger 1811 et *Cercopithecus* Linné 1758. *Ann Musée Roy Afr Centr Ser 8 Sci Zool* 105:1–255.

Vogler AP, Desalle R. 1994. Diagnosing units of conservation management. *Conserv Biol* 8:354–363.

Wolfheim JH. 1983. *Primates of the world: Distribution, abundance, and conservation.* Seattle: University of Washington Press.

16. The phylogenetic species concept and its role in Southeast Asian mammal conservation

Erik Meijaard and Benjamin Rawson

Introduction

Taxonomy may appear a somewhat old-fashioned or even outdated science, but recent heated debate has blown some significant dust off the subject. There is an ongoing dispute between propenents of the Biological Species Concept (BSC) (Zachos, 2013; Zachos et al., 2013) and the Phylogenetic Species Concept (PSC) (Gippoliti and Groves, 2013; Groves, 2013). More precisely the disagreement centres on the perceived implications of the use of either concept for species conservation. Users of the PSC have been accused of causing irresponsible taxonomic inflation and with that dilution of the conservation efforts allocated to each species (Zachos, 2013). The defence of PSC proponents has been that, without the use of the PSC, taxa of potentially significant conservation importance would be overlooked. Such taxa would be ranked as subspecies, and the assumption is that few people or organisations allocate efforts and funding to their conservation. In return, the counterargument has been that subspecies are rarely phylogenetically distinct and do therefore not warrant major conservation efforts (Zink, 2004). The real debate thus appears to be about the best method to allocate an appropriate taxonomic rank to taxa that is in line with their evolutionary distinctness and thus the need to prevent their extinction.

The BSC dates back to the 1930s and remains the most commonly used species concept today. The concept depends on the inference that species are those entities which do not interbreed. When two taxa occur sympatrically (i.e. their ranges overlap) and do not interbreed, the BSC is straightforward to apply. Things get complicated, however, when species occur allopatrically (i.e. no range overlap), because opportunities for interbreeding do not then naturally occur. In fact, many phylogenetic studies indicate a high frequency of gene flow between distinct species, even sympatric ones, with hybridisation often leading to speciation (Bell and Travis, 2005; Mallet, 2007).

As opposed to the BSC, 'the PSC depends on evidence, not on inference' (Groves, 2013: 7). Under the PSC, species are considered to be populations differing by at least one taxonomic character from all others, and within each of which there is 'a parental pattern of ancestry and descent' (Cracraft, 1989; Isaac et al., 2004). An absence of interbreeding is not required under the PSC.

Because of its use of evidence (e.g. measurable differences in a sample) rather than inference (e.g. an assumption that two taxa are unlikely to interbreed and produce fertile offspring), the PSC is more likely to lead to taxonomic instability. Two taxa can initially be separated when only a small sample is available, possibly missing some of the variation in the populations of the two taxa (e.g. by not fully sampling a cline). Once more material becomes available, broader variation within either taxon may result in overlap and the conclusion that, after all, the two taxa are not separated as species. Such taxonomic instability concerns conservation organisations that want to be sure that investments in a particular taxon will not be undermined by changes in a species taxonomy.

Taxonomic inflation, or the elevation of many existing subspecies to species level (cf. Isaac et al., 2004), is considered another unwanted side-effect of the use of the PSC (Zachos et al., 2013). With many subspecies being raised to species level, and concomitant shrinking of species' ranges, the list of species of conservation concern would rapidly grow through the use of PSC. It has been argued however that such taxonomic inflation is needed because it was preceded by irresponsible taxonomic deflation. In the words of Brandon-Jones and colleagues, 'an increase in recognized species is a desirable reversal of the regrettable trend from about 1920 to 1980, when specific recognition was excessively restrained, with correspondingly reckless subspecific recognition' (Brandon-Jones et al., 2004: 98). A transition from the dominant paradigm of the BSC to the PSC is therefore likely to lead to an increased number of recognised species, an increase of relative threat level at the species level and possibly a reduction in the available conservation resources devoted to each.

We here assess how the application of the PSC has influenced large mammal conservation in Southeast Asia. We look specifically at two groups, the Southeast Asian wild pigs (Suidae) and gibbons (Hylobatidae), because we are most familiar with their conservation through our work as, respectively Chair of the IUCN SSC Wild Pig Specialist Group and Vice-chair of the Section on Small Apes (SSA) of the IUCN SSC Primate Specialist Group. We try to answer three questions: (1) Has conservation attention increased for a taxon since it received full species status? (2) If so, would this increase in conservation attention likely have happened if the species would have remained a subspecies? (3) Has the increase in the recognised number of species in these families led to a detrimental dilution in funding available for all of them.

Southeast Asian pig taxonomy and conservation

A history of taxonomic change in Southeast Asian pigs

Southeast Asia has long been recognised as the center of diversity of wild pig species (Meijaard et al., 2011). At present, 18 extant species of pig are recognised in the family Suidae, the majority of which occur in Southeast Asia. There are two genera in the region, *Sus* and *Babyrousa*. Taxonomic revision of the Suidae is ongoing, especially in the Philippines and Indonesia where several taxa are waiting to be described. Even without additional species this is a species-rich family, reflecting the evolutionary success, ecological versatility, and, considering their endemic presence on a great number of islands, the extensive dispersal ability of pigs.

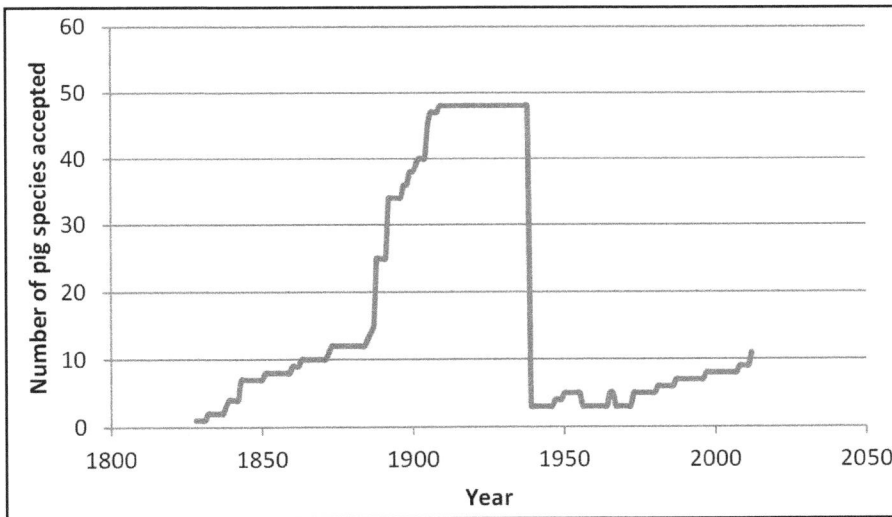

Figure 16.1: Change over time of the number of Southeast Asian pig species described and generally recognised.

Source: Figure created by authors using authors' own data and data provided in published accounts of species numbers.

The number of Southeast Asian Suidae species has fluctuated greatly over time. In 1758, Linneaus described the first species, *Sus scrofa*, which ranges over much of the Eurasian landmass, as well as parts of insular Southeast Asia (Meijaard et al., 2011). Following this, there were 80 years of taxonomic stagnation until *Sus barbatus*, the bearded pig of Sumatra, Borneo and Peninsular Malaysia was described. A general interest in natural history, and specific curiosity about the fauna of newly explored parts of Southeast Asia resulted in additional collection of specimens and description of new species (Figure 1). From 1888 onwards, a rapid increase in the number of species was primarily caused by Pierre Marie Heude, a French Jesuit missionary and zoologist based in China. Within the space of a few years, he described some 32 pig species (Heude, 1888, 1892), and by 1909, 48 species of pig had been described in Southeast Asia.

New thinking about what constitues a species, much influenced by evolutionary biologists such as Ernst Walter Mayr, led to major taxonomic lumping of species around the 1940s. The number of pig species in Southeast Asia was reduced to three, and then hovered around the four or five species until the 1980s. A subsequent revision of these taxa by Groves (1981), based on a morphological, ecological and karyological review, reinstated several of the species that had been lumped in the 1940s. This resulted in the following species: *Sus scrofa, S. barbatus, S. verrucosus,* and *S. celebensis.* In addition to these there is the babirusa (*Babyrousa babirussa*) which Groves did not consider in his 1981 review.

Further application of the phylogenetic species concept led to the recognition of additional species of pig, which in 1993 were officially accepted by the IUCN SSC Pigs, Peccaries and Hippos Specialist Group (Oliver, 1993). In addition to the species above, *Sus cebifrons* and *S. philippensis* from the Philippines were now recognised as distinct. The process did not stop there and, in 1997, the Indochinese species *S. bucculentus* was resurrected (Groves et al., 1997), followed by the separation of the babirusa into three distinct species (Meijaard and Groves, 2002), the description of *S. oliveri* from Mindoro in the Philippines (Groves, 2001b), and the separation of *S. ahoenobarbus* from Palawan in the Philippines from *S. barbatus* (Groves, 2001b; Lucchini et al., 2005). These processes have now resulted in 12 suid species in Southeast Asia being recognised by the IUCN.

We note that subsequent work indicates further taxonomic changes, with *Sus bucculentus* being subsumed into *S. scrofa,* but *S. scrofa* itself being split into several distinct species (Groves and Grubb, 2011; Meijaard and Groves, 2013). *Sus blouchi* from Bawean Island appears to be distinct enough to be allocated to full species level (Groves and Grubb, 2011). Also, an unpublished manuscript found that the pigs from the Sulu Islands in the Philippines were not *Sus barbatus,* as had been assumed, but a distinct taxon, and in addition, morphological and genetic variation within *Sus scrofa* indicates consistent differences between, what are presently different subspecies. Finally, recent morphological and genetic studies have indicated that the genus *Babyrousa* consists of six distinct species rather than the three that are presently recognised (A. MacDonald, pers. comm). These revisions will ultimately result in 15 to 20 recognised species of Suidae in Southeast Asia.

The impact of taxonomic change on pig conservation

It is obvious that suid taxonomy is a field in flux. The more we learn about the variation within and between pig taxa and the more we understand the general drivers of such variation, the more likely it has become that taxonomies are revised. Conservation practitioners and scientists generally do not like such taxonomic instability. It interferes with long term conservation planning for species survival and the development of captive breeding and zoo programs. As argued by the opponents of the PSC, the splitting of species leads to a dilution of conservation effort (Zachos et al., 2013). Is this really true however?

The Southeast Asian pig taxa of most conservation concern are the Critically Endangered Visayan warty pig, *S. cebifrons*, and the Endangered Javan warty pig, *S. verrucosus*, Mindoro warty pig, *S. oliveri*, and Togean Babirusa, *Babyrousa togeanensis*. Three of these (*S. cebifrons*, *S. oliveri* and *B. togeanensis*) were relative recently described using the phylogenetic species concept.

In 1993, the IUCN SSC Pigs, Peccaries and Hippos Specialist Group and other supporting partner agencies (including the Zoological Society of San Diego and, subsequently, the Rotterdam Zoo) devised and initiated a 'Visayan Warty Pig Conservation Programme' with the Philippine Government (Meijaard et al., 2011). Prior to its recognition as a full species, captive populations of Philippine pigs were maintained as single individuals or small groups of animals of mixed origin, and no attempts had been made at captive breeding (Oliver et al., 1993). Since then, two captive breeding and rescue centres for *S. cebifrons* have been established on Negros Island, and by 2011 at least 87 animals were kept in European zoos alone (Glatston, 2011). The successful breeding programs now require the identification of a safe release site, and this process is ongoing. The species remains highly threatened, but the immediate extinction risk seems to have been averted.

One could argue that the same kind of conservation action could have been undertaken if *S. cebifrons* would have remained a subspecies of *S. barbatus*, where it was originally placed (Groves, 1981). The IUCN presently lists 401 mammal subspecies as threatened, most of which are primates, so it is possible to focus global and local conservation authorities on taxa at the subspecies level. It is doubtful, however, that the Philippine authorities would have been as supportive for the conservation of a subspecies of *S. barbatus*, which also occurs in Indonesia and Malaysia. An interesting comparison for this exists in the Sulu Islands, where there was a locally popular appeal to the Philippine Government to take all due steps to 'eradicate' wild pigs. The government was initially supportive because it treated these populations, which they assumed to be *S. barbatus*, as non-native to the Philippines. Only when studies indicated that the Sulu pig might be an endemic new species to these islands (K Rose and P Grubb, unpublished manuscript) were the eradication plans cancelled. The above strongly suggest that recognition of *S. cebifrons* as a full species has played a vital role in preventing its extinction. Another species that appears to have benefited from their recognition as full species is *S. philippensis*, which is now treated as distinct in at least some zoos (e.g. the Amsterdam Zoo).

The two other Endangered species that were only recently elevated to species level appear to have benefited less from this taxonomic change. There are no specific *in situ* or *ex situ* programs yet for *S. oliveri* and *B. togeanensis*. In fact, the Indonesian Government is presently developing a babirusa action plan, which has retained the original taxonomy that combines the three IUCN-recognised species of babirusa into one. Finally, *S. verrucosus blouchi*, the Javan warty pig of Bawean Island may be one of the most endangered pig taxa. In anticipation of the taxon being elevated to full species, *S. blouchi*, as proposed by Groves and Grubb (2011), a survey is presently ongoing to ascertain the status of the taxon. Such conservation activitities had been planned for a long time, and had partly been implemented (Nijman, 2003; Semiadi and Meijaard, 2004), but the recognition that this could be a distinct species has generated renewed efforts to ensure that it will not go extinct.

Gibbon taxonomy and conservation

A history of taxonomic change in gibbons

The Hylobatidae have experienced not only a significant increase in the number of species, but an increase in the number of genera over time. Most early authors recognised only one genus in the Hylobatidae, that of *Hylobates*, with the occassional addition of *Symphalangus* (e.g. Elliot, 1913; Schultz, 1933). Groves (1968, 1972) tentatively recognised three sub-genera, those of *Hylobates* (which also included the hoolock gibbons), *Nomascus* and *Symphalangus*. This approach was later supported to a large degree by genetic evidence (Hall et al., 1998; Hayashi et al., 1995; Roos and Geissmann, 2001; Takacs et al., 2005). Groves (2001a) added a fourth subgenus, *Bunopithecus*, for the hoolock gibbons, but later changed its name to *Hoolock* (Mootnick and Groves, 2005). These four sub-genera were subsequently raised to generic status (Brandon-Jones et al., 2004; Geissmann, 2002), and are now commonly recognised (e.g. Meyer et al., 2012; Takacs et al., 2005). Groves, therefore played a fundamental role in development of the currently understood systematics of the Hylobatidae at the generic level.

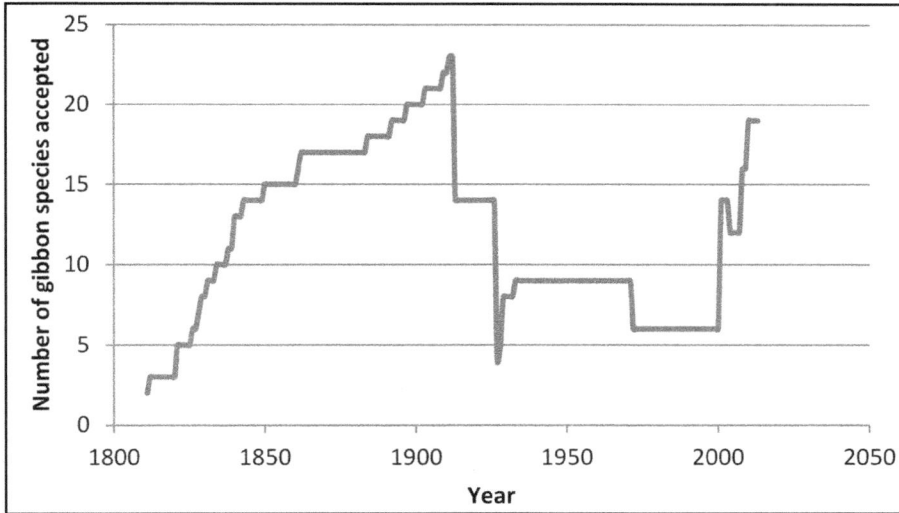

Figure 16.2: Change over time of the number of gibbon species described and generally recognised.

Source: Figure created by authors using authors' own data and data provided in published accounts of species numbers.

Similar to the rapid increase in Suidae species described above, the Hylobatidae experienced a rapid increase in the number of described species between the mid-nineteenth and early twentieth century. From a maximum of 23 species, this was reduced to four by Pocock (1927), revised to seven by Kloss (1929), and subsequently to nine by Schultz (1933). Groves also played a significant role in the species-level taxonomy for the Hylobatidae. His taxonomic review in 1972 reduced the number of species to six (Groves, 1972). In 1967, Groves had already described a second sub-species of the hoolock gibbon, *Hylobates hoolock leuconedys* from east of the Chindwin River, but it was not until much later that this taxon was elevated to generic and specific status as *Hoolock leuconedys* (Mootnick and Groves, 2005). Further taxonomic reviews by Brandon-Jones and others (2004), Mootnick and Fan (2011) and Thinh and others (2010) revealed the distinctness of additional gibbon species, and in the latest review of the gibbons, 19 species of Hylobatidae are recognised (Mittermeier et al., 2013). Sixteen of these are listed on the IUCN Red List and three remain unassessed. Of these 16, four are Critically Endangered, 11 Endangered, and one Vulnerable (IUCN, 2013).

The impact of taxonomic change on gibbon conservation

While some taxa such as *Hylobates klossii*, *Symphalangus syndactylus* and *H. moloch* have undergone no recent taxonomic rearrangements, and as such have not been significantly subjected to changes in priority setting, other taxa have. For example, the raising of the eastern hoolock (*Hoolock leuconedys*) to specific status (Mootnick and Groves, 2005) and discovery of a population in Assam, India reportedly changed conservation priorities in this area, previously believed to only contain *Hoolock hoolock* (J. Das, pers. comm.). The discovery of *Nomascus annamensis* between the distributions of *N. gabriellae* and *N. siki* in 2010 (Thinh et al., 2010) has opened up additional funding opportunities for gibbons located in its area of occurrence (Rawson, pers. obs.).

The Cao Vit gibbon (*Nomascus nasutus*) was until 2004 considered a subspecies of *N. concolor* (Brandon-Jones et al., 2004; Groves, 2001a), and as such the single remaining population of approximately 129 individuals on the Sino–Vietnamese border would have been considered a low priority compared to healthier populations in China, specifically Wuliangshan and Ailaoshan. However, it has now become the second most threatened gibbon taxon after the Hainan gibbon (*Nomascus hainanus*) and considerable conservation funding flows to this population as a result.

To further educate discussion about how taxonomic change has impacted gibbon conservation we sent out a survey to all members (n = 53) of the IUCN Species Survival Commission, Primate Specialist Group, Section on Small Apes and received 24 responses. All 19 Hylobatidae taxa recognised in *Handbook of the mammals of the world, Volume 3: Primates* (Mittermeier et al., 2013) were listed by at least one respondent as a taxon they had worked on. The average time that respondents had been engaged in gibbon conservation was approximately 13 years, providing a sufficiently long period of time and sufficient taxonomic coverage to expect to detect changes in relation to taxonomic inflation.

When asked whether taxonomic inflation had led to the recognition of too many gibbon species, responces were split with 33.3% answering yes, 29.2% answering no and 37.5% expressing no firm opinion. However, when asked whether changes in taxonomy had ever changed their conservation priorities, such as focusing on a newly elevated subspecies, the vast majority said it had not (n = 24, 91.7% answered 'no' while only 8.3% answered 'yes'). The most common rationale cited for not changing priorities was that at the population level, changes in taxonomy do not change conservation interventions required. In other words, the units of conservation commonly of interest to gibbon conservationists are smaller than that of species and occur at the site or population level. Several respondents noted that as gibbons are so rare and under high threat, taxonomic assessments are not significant drivers for conservation interventions and instead they focus on population conservation irrespective of taxonomic affiliation.

Responses to a question concerning whether funding for gibbons had increased, decreased or remained static were split, with 8.3% reporting a decrease, 16.7% an increase, 50% no change and 25% did not know. In fact, changes in funding for gibbon conservation in relation to taxonomic inflation are likely confounded by the increased funding being fed into gibbon conservation in general. Important funding mechanisms such as the USFWS Great Ape Conservation Grant (GACF) (which also includes the Hylobatidae in calls for proposals) and the more recent significant investment by the Arcus Foundation have come online during the most recent expansion in the number of gibbon taxa. Between 2007 and 2011, for example, USFWS GACF provided $25,541,000 for ape conservation, and although the amount specifically allocated for gibbon conservation is unknown, this doubtless represents a significant investment which was not available before the grants inception in 2001. Likewise, the Arcus Foundation has invested at least $3,500,000 in gibbon specific conservation since it began funding gibbons in 2007 (H. Rainer, pers. comm.).

From a captive management perspective, there are implications of taxonomic inflation. One significant issue is that splitting of taxa may result in captive populations of previously taxonomically homogenous individuals being assessed as hybrids. For example, raising *Nomascus siki* and *N. leucogenys* to species level has resulted in a large number of hybrid animals being held in zoos as founder captive populations did not distinguish these taxa at species level (Petersen and Melfi, pers. comm.). The value of hybrid gibbons from a conservation and reintroduction standpoint is much lower than non-hybrids. A further implication of taxonomic inflation in gibbons is that the increased number of taxa means that zoos, because of limited space, are unable to hold significant collections of all species (either independently or collectively). This is resulting in a triage approach limiting which taxa of gibbons are held in regional collections. Quite simply, as the number of taxa increase, the percentage of taxa which can be incorporated into breeding programs decreases. Moreover, the number of wrongly identified gibbons is likely to have increased.

Discussion

We have reviewed the impact of the use of the phylogenetic species concept on the conservation of two groups of Southeast Asian mammals, pigs and gibbons. We find that at least for some taxa there appears to have been a clear benefit of being considered full species, in terms of commitment from national governments to ensure their survival, generate funding for their in situ protection, develop breeding programs, and raise their global profile. This seems obvious for the gibbons for which major funding from the Arcus Foundation and United States Fish and Wildlife Service has been allocated for their protection. This funding has increased with the increasing threat level to the species, which in turn relates to taxonomy. As an example, under the more conservative 1972 taxonomy which largely follows a BSC, the remaining 129 *Nomascus nasutus* gibbons would be unlikely to get the same level of attention as they get now. Similarly, for pigs, renewed funding support was sought for Bawean pig (*Sus (v.) blouchi*) when it became clear that this was likely an endemic species of the island. Similarly, specific programs for Visayan warty pig (*Sus cebifrons*) were only developed once it became clear that the species was distinct. In contrast, we did not encounter examples in which funding per species or attention to their conservation had decreased following their elevation to species level, although, we recognise that our sample of pigs and gibbons is only a small part of the total number of globally threatened taxa, and possibly not representative.

We cannot fully assess the counterfactual, that whether the same kind of conservation support could have been generated if these taxa had remained as subspecies. Total funding for biodiversity conservation is likely to increase over time, with growing global concerns about species extinction and unsustainable use of natural resources (Leiserowitz et al., 2006; Sachs et al., 2009), and a general recognition that more conservation funding is needed to meet biodiversity targets (McCarthy et al., 2012). These processes could take place independent of whichever taxonomic principles or species definitions are followed. It is unclear whether there is a fixed amount of conservation funding available for all species and that adding species would dilute the funding available for each species. Or alternatively, funding levels could actually increase with the number of threatened taxa that require improved conservation management, in which case an increasing number of species would translate into an increasing amount of focus and funding. At least based on our analysis, the latter rather than the former seems the case, and dilution of conservation funding because of taxonomic inflation does not seem to be a major concern.

Another issue of importance, which was revealed through the feedback from the Section on Small Apes of the Primate Specialist Group, is that most people have been working long-term at the site level and do not really feel that taxonomy has much to do with how they approach conservation. They approach conservation from a broader ecological and management point of view and focus efforts on habitat protection. Changing the status of a subspecies to a species is not perceived to be of much consequence to their work. Again, this would indicate that there is limited harm in applying the more stringent PSG, even if this leads to an increase in the number of species. With an increase in the number of species, individual populations and sites generally become more important to that taxon, rather than less important. For example, the recent description of *Nomascus annamensis* (Thinh et al., 2010) significantly reduced the understood range of both *N. siki* and *N. gabriellae* which it falls between, making at least *N. siki* a much higher conservation priority due to a range reduction with fewer sites containing potentially viable populations.

From a conservation standpoint, the discussion about conservation impacts of the use of a BSC rather than PSC ultimately leads to the question of what is the unit of conservation. The IUCN recognises subspecies as a valid unit of conservation, and indeed has assessed 1,838 subspecies of Animalia and Plantae (no Fungi or Protista have been assessed at the subspecies level) compared with 70,289 taxa assessed at the species level (IUCN, 2013). Thus about 2.6% of assessed threatened taxa are subspecies. If the analysis is restricted to South and Southeast Asia, there are 302 subspecies assessed by the IUCN, compared with 16,855 species (i.e. 1.8% of the total). These figures indicate that, although conservation status assessments at the subspecies level exist, such assessments are relatively rare. IUCN status assessments are an important driver of conservation action (Hoffmann et al., 2010; Rodrigues et al., 2006), and considering the relative low uptake of subspecies-level assessments, a precautionary approach would be to err on the side of inflated rather than under-estimated taxonomic diversity.

Finally, there may be good reasons for not underestimating taxonomic diversity. A recent global analysis of avian subspecies showed that 36% of avian subspecies are, in fact, phylogenetically distinct (Phillimore and Owens, 2006). Interestingly, the authors found significant differences in the proportion of subspecies that are phylogenetically distinct, with Nearctic/Palearctic subspecies showing significantly reduced levels of differentiation. Additionally, the authors found differences between island and continental subspecies, with continental subspecies significantly less likely to be genetically distinct. These results indicate that the overall level of congruence between taxonomic subspecies and molecular phylogenetic data is greater than previously thought. Phillimore and Owens suggest that the widespread impression that avian subspecies are not real arises from a predominance of studies focusing on continental subspecies in North America and Eurasia, regions which show unusually low levels of genetic differentiation (Phillimore and Owens, 2006). These findings are reflected by our own phylogenetic and taxonomic studies in Southeast Asia, which indicate ancient patterns of species diversification in a region that has maintained relative climatic stability compared with the higher latitudinal areas that were significantly affected by glacial cycles (de Bruyn et al., 2014; Meijaard, 2004). In addition, the main islands in the Malay Archipelago (Wallace, 1869) and high geological activity and concommitant topographical changes in the region

(Meijaard and Groves, 2006) have resulted in much higher speciation rates than in continental North America and Eurasia. Taking into account that taxonomic research in Southeast Asia is practiced by few scientists (Meijaard and Sheil, 2012), and that many taxonomic studies and updates therefore remain pending, we expect there to be significant undiscovered or overlooked taxonomic diversity in Southeast Asia.

In conclusion, our study suggests that the use of a phylogenetic species concept has been beneficial to the conservation of species in the Southeast Asian region. So-called taxonomic inflation in Southeast Asia is needed to ensure that we identify as many evolutionary distinct species of conservation concern as possible. The efforts by Groves and colleagues to recognise distinct taxa using the PSC therefore appears to be warranted in Southeast Asia, where much taxonomic variation remains hidden and conservation needs are higher than anywhere else (di Marco et al., 2014).

Acknowledgements

Ben Rawson thanks the membership of the IUCN SSC Primate Specialist Group Section on Small Apes membership for their input into discussions on taxonomic inflation and impacts on conservation priorities. We thank Colin Groves for his wonderful mentorship, commitment to good science and conservation, and unfailing willingness to help whenever he can.

References

Bell MA, Travis MP. 2005. Hybridization, transgressive segregation, genetic covariation, and adaptive radiation. *Trends Ecol Evol* 20(7):358–361.

Brandon-Jones D, Eudey AA, Geissmann T, Groves CP, Melnick DJ, Morales J, Shekelle M, Stewart T. 2004. Asian primate classification. *Int J Primatol* 25(1):97–164.

Cracraft J. 1989. Speciation and its ontology: The empirical consequences of alternative species concepts for understanding patterns and processes of differentiation. In: Otte D, and Endler JA, editors. *Speciation and its consequences*. Sunderland, MA: Sinauer Associates. pp. 28–59.

de Bruyn M, Stelbrink B, Morley RJ, Hall R, Carvalho GR, Cannon CH, van den Bergh G, Meijaard E, Boitani L, Maiorano L, Shoup R, von Rintelen T. 2014. Borneo and Indochina are major evolutionary hotspots for Southeast Asian biodiversity. *Systematic Biology* 63:879-901.

Di Marco M, Boitani L, Mallon D, Iacucci A, Visconti P, Meijaard E, Schipper J, Hoffmann M, Rondinini C. 2014. Lessons from the past: A retrospective evaluation of the global decline of carnivores and ungulates. *Conserv Biol* 28(4):1109–1118.

Elliot DG. 1913. *A review of the primates*. American Museum of Natural History Monograph series 1. New York: American Museum of Natural History. pp. 1–317.

Geissmann T. 2002. Taxonomy and evolution of gibbons. *Evol Anth* 11(S1):28–31.

Gippoliti S, and Groves CP. 2013. 'Taxonomic inflation' in the historical context of mammalogy and conservation. *Hystrix* 23(2): 8–11.

Glatston A. 2011. *Regional Visayan warty pig studbook*. Sus cebifrons negrinus. Rotterdam, the Netherlands: Rotterdam Zoo.

Groves CP. 1968. The classification of the gibbons (Primates, Pongidae). *Z für Säugetierkunde* 33:239–246.

Groves CP. 1972. Systematics and phylogeny of gibbons. In: Rumbaugh DM, editor. *Gibbon and Siamang*, Vol. 1. Basel: Karger. pp. 1–89.

Groves CP. 1981. *Ancestors for the pigs*. Canberra: Australian National University Press. p. 96.

Groves CP. 2001a. *Primate taxonomy*. Washington DC: Smithsonian Institution Press. p. 350.

Groves CP. 2001b. Taxonomy of wild pigs of Southeast Asia. *Asian Wild Pig News* 1(1):2–3.

Groves CP. 2013. The nature of species: A rejoinder to Zachos et al. *Mammal Biol – Z für Säugetierkunde* 78(1):7–9.

Groves CP, Grubb P. 2011. *Ungulate taxonomy*. Baltimore, Maryland: Johns Hopkins University Press.

Groves CP, Schaller GB, Amato G, Khounboline K. 1997. Rediscovery of the wild *Sus bucculentus*. *Nature* 386(27 March 1997):335.

Hall LM, Jones DS, Wood BA. 1998. Evolution of the gibbon subgenera inferred from Cytochrome b DNA sequence data. *Mol Phyl Evol* 10(3):281–286.

Hayashi S, Hayasaka K, Takenaka O, Horai S. 1995. Molecular phylogeny of gibbons inferred from mitochondrial DNA sequences: Preliminary report. *J Mol Evol* 41:359–365.

Heude PM. 1888. Étude sur les suilliens. *Mémoires concernant l'histoire naturelle de l'Empire chinois* 2:85–111.

Heude PM. 1892. Étude sur les suilliens, chapitre 4, etc. *Mémoires concernant l'histoire naturelle de l'Empire chinois* 4(3/4):113–211.

Hoffmann M, Hilton-Taylor C, Angulo A, Bohm M, Brooks TM, Butchart SHM, Carpenter KE, Chanson J, Collen B, Cox NA et al. 2010. The impact of conservation on the status of the world's vertebrates. *Science* 330(6010):1503–1509.

Isaac NJB, Mallet J, Macdonald AA. 2004. Taxonomic inflation: Its influence on macroecology and conservation. *Trends Ecol Evol* 19(9):464–469.

International Union for Conservation of Nature (IUCN). 2013. *IUCN Red List of threatened species*. Version 2012.2. www.iucnredlist.org. accessed 10 May 2013.

Kloss CB. 1929. Some remarks on the Gibbons, with the description of a new sub-species. *Proc Zool Soc Lond* 99(1):113–127.

Leiserowitz AA, Kates RW, and Parris TM. 2006. Sustainability values, attitudes, and behaviors: A review of multinational and global trends. *Annu Rev Environ Resour* 31: 413–444.

Lucchini V, Meijaard E, Diong CH, Groves CP, Randi E. 2005. New phylogenetic perspectives among species of Southeast Asian wild pig (*Sus* sp.) based on mtDNA sequences and morphometric data. *J Zool* 266(1):25–35.

McCarthy DP, Donald PF, Scharlemann JPW, Buchanan GM, Balmford A, Green JMH, Bennun LA, Burgess ND, Fishpool LDC, Garnett ST et al. 2012. Financial costs of meeting global biodiversity conservation targets: Current spending and unmet needs. *Science* 338(6109):946–949.

Mallet J. 2007. Hybrid speciation. *Nature* 446(7133):279–283.

Mayr E. 1942. *Systematics and the origin of species from the viewpoint of a zoologist*. New York: Columbia University Press.

Meijaard E. 2004. Solving mammalian riddles. A reconstruction of the Tertiary and Quaternary distribution of mammals and their palaeoenvironments in island Southeast Asia. PhD thesis, Department of Anthropology and Archaeology, The Australian National University.

Meijaard E, d'Huart JP, Oliver WLR. 2011. Family Suidae (Pigs). In: Wilson DE, Mittermeier RA, editors. *Handbook of the mammals of the world*, Vol. 2 *Hoofed mammals*. Barcelona, Spain: Lynx Edicions. pp. 248–291.

Meijaard E, Groves CP. 2002. Upgrading three subspecies of babirusa (*Babyrousa* sp.) to full species level. *Asian Wild Pig News* 2(2):33–39.

Meijaard E, Groves CP. 2006. The geography of mammals and rivers in mainland Southeast Asia. In: Lehman S, and Fleagle J, editors. *Primate biogeography*. New York: Kluwer Academic Publishers. pp. 305–329.

Meijaard E, Groves CP. 2013. New taxonomic proposals for the *Sus scrofa* group in eastern Asia. *Suiform Soundings* 12(1):26–30.

Meijaard E, Sheil D. 2012. The dilemma of green business in tropical forests: How to protect what it cannot identify. *Conserv Lett* 5(5):342–348.

Meyer TJ, McLain AT, Oldenburg JM, Faulk C, Bourgeois MG, Conlin EM, Mootnick AR, de Jong PJ, Roos C, Carbone L et al. 2012. An Alu-based phylogeny of gibbons (Hylobatidae). *Mol Phyl Evol* 29(11):3441–3450.

Mittermeier RA, Rylands AB, Wilson DE, editors. 2013. *Handbook of the Mammals of the World*, Vol. 3 *Primates*. Barcelona, Spain: Lynx Edicions.

Mootnick A, Groves C. 2005. A new generic name for the hoolock gibbon (Hylobatidae). *Int J Primatol* 26(4):971–976.

Mootnick AR, Fan P-F. 2011. A comparative study of crested gibbons (*Nomascus*). *Am J Primatol* 73(2):135–154.

Nijman V. 2003. Notes on the conservation of the Javan Warty Pig *Sus verrucosus blouchi* on the island of Bawean. *Asian Wild Pig News* 3(1):15–16.

Oliver WLR. 1993. *Pigs, peccaries, and hippos. Status survey and conservation action plan*. Gland, Switzerland: IUCN SSC Pigs and Peccaries Specialist group and IUCN SSC Hippos Specialist Group.

Oliver WLR, Cox CR, Groves CP. 1993. The Philippine Warty Pigs (*Sus philippensis* and *S. cebifrons*). In: Oliver WLR, editor. *Pigs, peccaries, and hippos. Status survey and conservation action plan*. Gland, Switzerland: IUCN. pp. 145–155.

Phillimore AB, Owens IPF. 2006. Are subspecies useful in evolutionary and conservation biology? *Proc R Soc Lond B Biol Sci* 273(1590):1049–1053.

Pocock RI. 1927. The gibbons of the genus *Hylobates*. *J Zool Ser A Proc Zool Soc Lond* 97(3):719–741.

Rodrigues ASL, Pilgrim JD, Lamoreux JF, Hoffmann M, Brooks TM. 2006. The value of the IUCN Red List for conservation. *Trends Ecol Evol* 21(2):71–76.

Roos C, Geissmann T. 2001. Molecular phylogeny of the major Hylobatid divisions. *Mol Phyl Evol* 19(3):486–494.

Sachs JD, Baillie JEM, Sutherland WJ, Armsworth PR, Ash N, Beddington J, Blackburn TM, Collen B, Gardiner B, Gaston KJ et al. 2009. Biodiversity conservation and the millennium development goals. *Science* 325(5947):1502–1503.

Schultz AH. 1933. Observations on the growth, classification and evolutionary specialization of gibbons and siamangs. *Hum Biol* 5:212–255, 385–428.

Semiadi G, Meijaard E. 2004. *Survey of the Javan warty pig* (Sus verrucosus) *on Java and Bawean Island, with English summary and detailed survey results in Indonesian.* Bogor, Indonesia: Pusat Penelitian Biologi-LIPI and IUCN SSC Pigs, Peccaries and Hippos Specialist Group.

Takacs Z, Morales JC, Geissmann T, Melnick DJ. 2005. A complete species-level phylogeny of the Hylobatidae based on mitochondrial ND3-ND4 gene sequences. *Mol Phyl Evol* 36(3):456–467.

Thinh VN, Mootnick AR, Thanh VN, Nadler T, Roos C. 2010. A new species of crested gibbon, from the central Annamite mountain range. *Vietnamese J Primatol* 4:1–12.

Wallace AR. 1869. *The Malay Archipelago.* Oxford: Oxford University Press. p. 625.

Zachos FE. 2013. Taxonomy: Species splitting puts conservation at risk. *Nature* 494(7435):35.

Zachos FE, Apollonio M, Bärmann EV, Festa-Bianchet M, Göhlich U, Habel JC, Haring E, Kruckenhauser L, Lovari S, McDevitt AD et al. 2013. Species inflation and taxonomic artefacts: A critical comment on recent trends in mammalian classification. *Mammal Biol – Z für Säugetierkunde* 78(1):1–6.

Zink RM. 2004. The role of subspecies in obscuring avian biological diversity and misleading conservation policy. *Proc R Soc Lond B Biol Sci* 271:561–564.

17. Conserving gorilla diversity

Angela Meder

Introduction

Studying gorillas is fascinating for anybody. Not only are they themselves very special animals, but the environment where they live leaves an everlasting impression on people fortunate enough to visit.

In his book *Extended Family* (Groves, 2008), Colin Groves described how he started to study gorilla skulls in museum collections and how exciting this work was for him. While he examined museum specimens, he remained fascinated by living gorillas and often visited them in zoos. He felt deep empathy for his study animals and did not understand how any researcher could not feel the same. Most impressive for him, however, were the moments when he saw gorillas in the rainforest, even though he did not regard himself as a fieldworker.

When Colin Groves published the results of his PhD thesis on the taxonomy of the genus *Gorilla* (Groves 1967, 1970), he became an internationally renowned gorilla expert. A very important aspect of his work was the effect of ecology on morphological diversity; he found clear differences between populations living in different habitats. Thus, it became important to examine as many specimens as possible that had lived under different conditions to truly understand this variation at the population and species levels. But for many populations, the available sample size was very small, which is something Colin has continually tried to improve.

In 1971, Dian Fossey, who was aware of the importance of his studies, invited Colin to examine some mountain gorilla skulls that she had collected in the Virunga Volcanoes. This was a wonderful chance, not only to examine more specimens, but also to see wild gorillas. Although he had been aware of the threats that existed for gorillas before, his visit with Dian Fossey, who constantly had to fight for their conservation, was an important experience that led him to support conservation activities – something that continues to this day. For me and my gorilla conservation work, this type of lifelong dedication gives constant encouragement.

Gorillas under threat

Mountain gorillas are the best known of all gorilla populations because they have been studied thoroughly for several decades. On the Virunga Volcanoes, the highest mountains where gorillas live, they range higher than 4,000 m into the alpine zone. But most gorillas live in lowland rainforests. Regardless of forest type, all gorillas make their homes in wooded environments. The same wooded environments humans have been cutting down since they developed their own civilisations – for construction, firewood, agriculture and for other purposes. This has increased dramatically more recently with dramatic human population growth. Additionally, the international demand for timber and forest goods has furthered rates of deforestation. Thus, even where human population density is very low, logging companies build roads thereby opening the forest to people who would otherwise not have access, to slowly enter and settle down. Foreign companies as well as Africans exploit natural resources to their maximum for short-term gain while thinking very little about long-term consequences (Oates, 1999). Between the 1990s and 2000s, areas suitable for habitation by African apes decreased very fast; the loss of suitable environment appeared highest for Cross River gorillas (−59%), followed by eastern gorillas (−52%) and western lowland gorillas (−32%) (Junker et al., 2012). In some areas, especially around the mountains of the Albertine Rift, the human pressure on the forest is so high that it results in large areas becoming completely deforested, leaving only small forest islands that are strictly protected. This type of large scale tree felling has been ongoing for many decades and from 1960 to 1996 at least 24% of the eastern gorillas' range was lost (Mehlman, 2007). This is despite scientists such as Emlen and Schaller stating 'above all, the destruction of habitat by forest clearing must be stopped' in a 1960 publication. As a result of this deforestation some gorilla populations survive in small isolated forest islands that are not connected, introducing the new and real threat of surviving in small, isolated and closed populations, which could lead to further speciation.

While the destruction of habitat is a major threat for gorillas throughout their range, there are more, including: hunting for bushmeat and the pet trade; poaching with snares; conflicts between gorillas and people, especially when gorillas raid crops; disease transmission from humans or livestock as well as epidemic diseases (Ebola); war and political instability that may lead to increased hunting, habitat destruction and general lawlessness. All these problems are addressed by many conservation organisations.

In eastern gorillas, two subspecies are distinguished: mountain gorillas and Grauer's gorillas. The mountain gorillas (*Gorilla beringei beringei*) found in the Democratic Republic of Congo, Uganda and Rwanda are critically endangered. They live in two distinct mountain areas that have been isolated from each other and from other gorilla populations for a long time. There are 480 Virunga gorillas and 400 Bwindi gorillas, both populations have been growing continuously over the last two decades, mainly because they are monitored and managed intensely (Gray et al., 2010; Robbins et al., 2011b). Even if the populations are rather small and isolated (and although they face threats like habitat destruction, disease transmission, poaching and political instability), they have very good prospects.

Although the Grauer's gorillas (*Gorilla beringei graueri*) in eastern Democratic Republic of the Congo are the only gorilla subspecies to be listed as merely 'endangered' according to the IUCN Red List (2014.2) rather than 'critically endangered' like the other species, we should still be extremely concerned about them. It is estimated that there are only a few thousand of them remaining, but their numbers have declined dramatically within the last two decades, mainly because of habitat destruction and hunting – particularly as a side-effect of war (Maldonado et al., 2012). The smallest known population is on Mt Tshiaberimu, a part of the Virunga National Park, and contains less than ten individuals. When George Schaller made the first survey at the end of the 1950s, there were other gorilla populations nearby (Emlen and Schaller, 1960), suggesting a dramatic decline has occurred in the past 60 years. Taxonomically, these gorillas are specialised and different from the surrounding species, therefore Colin Groves has always been very interested in them (Groves and Stott, 1979).

Western gorillas also comprise two subspecies: western lowland gorillas and Cross River gorillas. The less numerous subspecies – and in fact the least numerous gorilla subspecies – is the Cross River gorilla (*Gorilla gorilla diehli*) found at the Nigeria/Cameroon border and listed as Critically Endangered by the IUCN Red List (2014.2). There are approximately 250–300 individuals living, however, these are distributed in 12–14 small and fragmented populations (Etiendem et al., 2013). The most pressing threats are the destruction or modification of habitat resulting in further isolation of small populations and hunting for the bushmeat and pet trade. Recently, the genetic work of Bergl and Vigilant (2007) found that there is persistent and recent reproductive contact between many of the smaller populations, which is very good news for the conservation of this subspecies as it suggests genetic isolation may not be as large a threat as previously believed.

Yet, the western lowland gorilla (*Gorilla gorilla gorilla*), the most widely distributed subspecies, is also listed as Critically Endangered (IUCN, 2012). According to the most recent estimate (2008) there are still at least 125,000 individuals in existence (GRASP Scientific Commission, pers. comm. 2009). The most important threats are epidemic diseases (Ebola), destruction of the forest and hunting for the bushmeat and pet trade. As the distribution area of the western lowland gorillas is also affected by deforestation (and bushmeat hunting that often leads to 'empty forests'), there are certainly many small isolated populations that are poorly known. The best known population of this subspecies lives in the Ebo Forest in Cameroon that lies in-between the range of the Cross River gorilla and that of the main western lowland gorilla. This one population is quite small, however, with an estimate of only 25 individuals or less (Morgan, pers. comm. 2012).

The problem with small populations

Traditionally, inbreeding was regarded as the major danger for small populations as limited gene flow results in reduced genetic variability. As early as 1983, Dian Fossey noted certain characteristics that may have been related to inbreeding in Virunga gorillas, but there have not been any signs of inbreeding suppression since then suggesting that this population is not really threatened by inbreeding (Harcourt and Fossey, 1981). Although genetic variation certainly increases individual fitness and population viability, it is possible that populations can survive and recover from reduced population sizes making inbreeding depression less of a concern than it was previously (Strier, 1993).

Despite the ability of some populations to rebound from small numbers, the main reason why it is difficult for very small populations of gorillas to survive is their life history. Both sexes usually look for partners outside of their natal groups. Reaching adulthood, female gorillas generally leave the group they were born in and join a new male. They emigrate only if they encounter another male and they may transfer several times before they settle down in the group in which they will stay (Sicotte, 2001; Stokes et al., 2003). In western gorillas and Grauer's gorillas, male emigration is common, while among mountain gorillas less than 50% of the males emigrate (Stoinski et al., 2009). If they leave, they either become solitary or in some populations may join all-male groups (Watts, 2000; Yamagiwa et al., 2003; Robbins et al., 2004). Solitary males may travel very long distances to find another group to join (Douadi et al., 2007). Thus, if a large area is completely deforested, it becomes less likely a male will be able to cross it, significantly lowering his chance of finding another gorilla. If this happens, and gorillas do not find suitable partners we will see a futher reduction in population size. In addition, if animals cannot cross among populations then no exchange of genes will be possible further isolating populations.

Should we select gorilla populations we want to save?

Some researchers ask the question whether it is necessary at all to save primates (Chapman et al., 2006; Klages, 2010), and although primatologists do not see any reason to question this, it is good to discuss it. The main reasons why primates should be saved are: their role as seed dispersers – they affect ecosystem dynamics; their potential as a flagship species for conservation; as well as ethical reasons. All these reasons apply to gorillas in a special way.

There are many conservation organisations – both big and small, international and local – doing their best to protect gorillas. They all complain that they do not have enough funds to do everything necessary to ensure the survival of the gorilla populations they focus on. Everywhere resources for conservation are limited and, as Martha Robbins and others (2011a) state, 'the channeling of resources toward one species is unavoidably done to the detriment of the conservation for other species. Until sufficient money is made available, conservationists will continue to be faced with the dilemma of devoting more resources to save a few species versus spreading resources too thinly to achieve success with any species.'

To concentrate funds to certain species (or in this case, populations) sounds very reasonable, and indeed the species would be saved, although only a part of it. And that is the exact problem: the species' diversity would be lost. Virunga gorillas are a good example of a population that was saved by large funding investments into conservation efforts – but they are a very specialised population, adapted to extreme heights, which is not typical at all for the eastern gorillas, making findings hard to extrapolate.

I have been working for a small conservation organisation for 22 years now, and our focus has always been to save small gorilla populations at the periphery of their distribution area as they are especially vulnerable. We want to protect them in their natural habitat for what they are – wonderful animals – but also because research can be done and we can learn from them only if we are able to ensure their survival, perhaps not forever, but for as long as possible. This is true for each population, even the smallest ones that contain less than 10 individuals.

Why is the survival of peripheral populations important?

Only if we can save all gorilla populations that still exist, will we save the whole lot and preserve the great variation Colin highlighted in his PhD work. Relict populations at the periphery of the distribution area in particular carry important genes that are lost if these populations are extinguished. In the sense of the biological species concept, this is especially interesting because 'geographically separated populations often exhibit new, fixed character states differentiating them from close relatives, and these are surely "units of evolution" in all meaningful senses' (Groves, 2001).

An especially interesting example is the small gorilla population in Ebo. As it is a relict population in a montane forest area in-between the two western gorilla subspecies, it is very important for our understanding of the evolution and history of this species: 'For quite a number of reasons, therefore, the Ebo Forest gorillas would seem to be a unique and significant population, which should be protected as soon as possible.' (Groves, 2005).

A very well-studied example of the importance of saving the whole diversity of a taxon is the Cross River gorilla. Genetic diversity is not evenly distributed within the Cross River gorilla population; Richard Bergl and others (2008) found three subpopulations that exhibit different levels of genetic variability. They conclude: 'Loss of either peripheral subpopulation would cause the loss of unique alleles, and loss of the Eastern subpopulation would result in a reduction in heterozygosity and a decline in the effective allele number' (Bergl et al., 2008: 855).

What would be lost for research?

The evolution of primates is not yet fully understood with new studies often providing novel insights and directions for study. For reliable results, it is necessary to know as much as possible about living animals, and especially about their environment. Colin Groves noted in 1970 that morphological differences are connected with ecological differences (i.e. the height in which the respective gorilla population lives). 'Thus in phase with the changes in morphological pattern within each of the three subspecies, there are corresponding gradation in altitude, temperature and rainfall' (Groves, 1970).

Since he wrote this, taxonomic methods have changed fundamentally – genetic methods were added to the repertoire making it possible now to analyse relatedness of different populations irrespective of possible effects of phenotypic plasticity. Colin Groves wrote in 2001: 'The coming of "molecular methods" has revolutionised our understanding of evolution more than it has taxonomy.' Genetic analysis allows taxonomists to work with noninvasively collected samples – they can even study animal populations that are not represented in museum collections and that no researcher has ever seen (Thalmann et al., 2006, 2011).

Since genetic methods were developed, they have been used for studies with a wide range of subjects. They have helped us to understand dispersal patterns, mating systems, reproductive strategies, and the influence of kinship on social behavior. Genetic analyses can show whether a population lost genetic diversity and whether such a loss was recent or more ancient (Vigilant and Guschanski, 2009). Today, even whole populations can be studied by sampling all individuals (Gray et al., 2010; Robbins et al., 2011b). Like museum material, this genetic material can serve as the basis for the study of many questions.

To fully understand gorilla evolution, however, it is not sufficient to study material stored in museums and gene banks; the direct observation of the animals in interaction with each other and with their environment provides additional invaluable information. In any case, researchers are interested in the conservation of their study subjects – and many of them actively contribute to their protection. In fact, the presence and interest of researchers can be an important factor for the protection of the gorillas (Tranquilli et al., 2011).

What would be lost for the countries and the people?

Gorillas are charismatic animals that can help to explain the importance of conservation during public awareness campaigns and environmental education. They can draw the attention of local people to the natural resources in their immediate environment. To protect and save these resources is of course not only in the interest of foreign researchers and tourists, but it is important for the future of the countries themselves and the people living there.

For the countries where gorillas live, these animals are not only a flagship species for conservation, but in some cases also an important source of income – particularly where gorilla tourism is practiced. If gorillas become extinct, the authorities may not see any reason to continue the effective protection of the areas where these apes used to live and we would see further loss of forests that are so important to a myriad of wildlife species.

Thus, it is clear that it is not only the gorillas themselves that have to be protected, but also their environment; this cannot be separated because gorillas are not able to survive outside the forest (if they are to survive in a natural environment). An important example of how deforestation can create knock on effects that further put wildlife at risk is the danger of mudslides close to mountains. Close to the Virunga Volcanoes, much of the forest on the mountain slopes was cut during the last few decades, and as a result catastrophic mudslides have become common after heavy rains; potentially destroying whole villages. In addition, deforestation further impacts humans as forested areas often act as water catchments and provide many ecosystem services.

What can we do?

If we want to protect endangered species and preserve them for future generations, surveys are the first step. Such studies have shown in many cases that healthy gorilla populations still survive in remote areas. While a few decades ago a survey always meant a visit to the area, today it is possible to identify potential distribution areas by satellite imaging and remotely sensed data. As Richard Bergl and others (2010) write, by using remotely sensed data 'valuable insights into habitat availability, fragmentation and corridors are possible with relatively small investments of time and money.' These analyses allow survey teams to focus on certain areas in which to conduct transects, during which they map the habitat and identify threats to the forest and the gorillas. Their findings are the basis for planning further research and conservation measures.

To establish protected areas and effectively enforce their conservation is the central gorilla protection activity (Tranquilli et al., 2011). Even local communities have successfully created reserves for gorillas with support of conservation organisations (Bergl et al., 2010). Additional conservation measures can be very different and have to be planned individually for each area. First of all the situation of the gorillas, their habitat and the human population living close to/ in the forest has to be evaluated carefully; this is the basis for the planning of conservation activities.

For some populations, a very special approach may be necessary. In the Virunga Volcanoes, constant monitoring and veterinary management resulted in increased gorilla population growth. According to Robbins and others (2011a), extreme measures were needed to achieve this in addition to conventional conservation efforts. It has to be mentioned, however, that such extreme conservation may influence the natural behavior or life history of a species, potentially disrupting natural selection and introducing new threats. Such measures that are extremely expensive are not possible everywhere of course, but in many areas it is possible to protect gorillas with much smaller financial resources, if the conservation activities are planned carefully and individually, and evaluated constantly.

Many of the gorilla populations that Emlen and Schaller (1960) found in their survey have disappeared – nobody knows when and how it happened exactly, but in general their habitat was destroyed. The same fate threatens several gorilla populations now in many parts of their range. We may not be able to save all the populations – but we should at least try our best to save them for as long as possible. For research, for conservation of the ecosystem (of which gorillas are an important component), for local human populations and for the gorillas themselves. Even taxonomic studies are certainly much more exciting if living populations are the study subjects instead of extinct ones.

On the other hand, researchers should not ignore the threats to their subjects. Colin Groves has a very clear position: 'More and more, the work of taxonomists and other biologists must be put at the service of conservation' (Groves, 2003). And he himself is the best example for a scientist who takes this responsibility seriously.

References

Bergl RA, Vigilant L. 2007. Genetic analysis reveals population structure and recent migration within the highly fragmented range of the Cross River gorilla *(Gorilla gorilla diehli)*. *Molecular Ecol* 16:501–516.

Bergl RA, Bradley BJ, Nsubuga A, Vigilant L. 2008. Effects of habitat fragmentation, population size and gemographic history on genetic diversity: The Cross River gorilla in a comparative context. *Am J Primatol* 70:848–859.

Bergl RA, Warren Y, Nicholas A, Dunn A, Imong I, Sunderland-Groves J, Oates JF. 2010. Remote sensing analysis reveals habitat, dispersal corridors and expanded distribution for the critically endangered Cross River gorilla *Gorilla gorilla diehli*. *Oryx* 46:278–289.

Chapman CA, Lawes MJ, Eeley HAC. 2006. What hope for African primate diversity? *Afr J Ecol* 44:116–133.

Douadi MI, Gatti S, Levrero F, Duhamel G, Bermejo M, Vallet D, Menard N, Petit EJ. 2007. Sex-biased dispersal in western gorilla (*Gorilla gorilla gorilla*). *Molecular Ecol* 16:2247–2259.

Emlen JT, Schaller GB. 1960. Distribution and status of the mountain gorilla, 1959. *Zool Sci Contrib N Y Zool Soc* 45:41-52.

Etiendem DN, Funwi-Gabga N, Tagg N, Hens L, Indah EK. 2013. The Cross River gorillas *(Gorilla gorilla diehli)* at Mawambi Hills, South-West Cameroon: Habitat suitability and vulnerability to anthropogenic disturbance. *Folia Primatol* 84:18–31.

Fossey D. 1983. *Gorillas in the mist*. Boston: Houghton Mifflin Co.

Gray M, Fawcett K, Basabose A, Cranfield M, Vigilant L, Roy J, Uwingeli P, Mburanumwe I, Kagoda E, Robbins MM. 2010. *Virunga Massif Mountain Gorilla census – 2010 summary report*.

Groves CP. 1967. Ecology and taxonomy of the gorilla. *Nature* 213:890–893.

Groves CP. 1970. Population systematics of the gorilla. *J Zool (Lond)* 161:287–300.

Groves CP, Stott KW. 1979. Systematic relationships of gorillas from Kahuzi, Tshiaberimu and Kayonza. *Folia Primatol* 32:161–179.

Groves CP. 2001. *Primate taxonomy*. Washington, DC: Smithsonian Institution Press.

Groves CP. 2003. A history of gorilla taxonomy. In: Taylor AB, Goldsmith ML, editors. *Gorilla biology*. New York: Cambridge University Press. pp. 15–34.

Groves CP. 2005. A note on the affinities of the Ebo Forest gorilla. *Gorilla J* 31:19–21.

Groves CP. 2008. *Extended family*. Arlington, VA: Conservation International.

Harcourt AH, Fossey D. 1981. The Virunga gorillas: Decline of an 'island' population. *Afr J Ecol* 19:83–97.

International Union for Conservation of Nature (IUCN). 2012. *IUCN Red List of Threatened Species*. Version 2012.2. www.iucnredlist.org.

IUCN. 2014. *IUCN Red List of Threatened Species*. Version 2014.2. www.iucnredlist.org.

Junker J, Blake S, Boesch C, Campbell G, du Toit L, Duvall C, Ekobo A, Etoga G, Galat-Luong G, Gamys J, Ganas-Swaray J, Gatti S, Ghiurghi A, Granier N, Hart J, Head J, Herbinger I, Hicks, TC, Huijbregts B, Imong IS, Kuempel N, Lahm S, Lindsell J, Maisels F, McLennan M, Martinez L, Morgan B, Morgan D, Mulindahabi F, Mundry R, N'Goran KP, Normand E, Ntongho A, Okon, DT, Petre C-A, Plumptre A, Rainey H, Regnaut S, Sanz C, Stokes

E, Tondossama A, Tranquilli S, Sunderland-Groves J, Walsh P, Warren Y, Williamson EA, Kuehl HS. 2012. Recent decline in suitable environmental conditions for African great apes. *Divers Distrib* 18:1077–1099.

Klages A. 2010. Triage: conserving primates and competing interests. *Totem: Uni W Ontario J Anth* 18(1):Article 12.

Maldonado O, Aveling C, Cox D, Nixon S, Nishuli R, Merlo D, Pintea L, Williamson EA. 2012. *Grauer's gorillas and chimpanzees in eastern Democratic Republic of Congo (Kahuzi-Biega, Maiko, Tayna and Itombwe landscape): Conservation Action Plan 2012–2022.* Gland: IUCN SSC Primate Specialist Group, Ministry of Environment, Nature Conservation & Tourism, Institut Congolais pour la Conservation de la Nature and the Jane Goodall Institute.

Mehlman PT. 2007. Current status of wild gorilla populations and strategies for their conservation. In: Stoinski TS, Steklis HD, Mehlman PT, editors. *Conservation in the 21st century: Gorillas as a case study.* New York: Springer. pp. 3–54.

Oates JF. 1999. *Myth and reality in the rainforest. How conservation strategies are failing in West Africa.* Berkeley: University of California Press.

Robbins MM, Bermejo M, Cipolletta C, Magliocca F, Parnell RJ, Stokes E. 2004 Social structure and life-history patterns in western gorillas (*Gorilla gorilla gorilla*). *Am J Primatol* 64: 145–159.

Robbins MM, Gray M, Fawcett KA, Nutter FB, Uwingeli P, Mburanumwe I, Kagoda E, Basabose A, Stoinski TS, Cranfield MR, Byamukama J, Spelman LH, Robbins AM. 2011a. Extreme conservation leads to recovery of the Virunga mountain gorillas. *PLoS ONE* 6(6): e19788. doi:10.1371/journal. pone.0019788.

Robbins MM, Roy J, Wright E, Kato R, Kabano P, Basabose A, Tibenda E, Vigilant L, Gray M. 2011b. *Bwindi mountain gorilla census 2011 – summary of results.*

Sicotte P. 2001. Female mate choice in mountain gorillas. In: Robbins MM, Sicotte P, Stewart KJ, editors. *Mountain gorillas.* Cambridge: Cambridge University Press. pp. 59–87.

Stoinski TS, Vecellio V, Ngaboyamahina T, Ndagijimana F, Rosenbaum S, Fawcett KA. 2009. Proximate factors influencing dispersal decisions in male mountain gorillas, *Gorilla beringei beringei. Anim Behav* 77:1155–1164.

Stokes EJ, Parnell RJ, Olejniczak C. 2003. Female dispersal and reproductive success in wild western lowland gorillas. *Behav Ecol Sociobiol* 54:329–339.

Strier KB. 1993. Viability analyses of an isolated population of muriqui monkeys (*Brachyteles arachnoides*): Implications for primate conservation and demography. *Primate Conserv* 14/15:43–52.

Thalmann O, Fischer A, Lankester F, Pääbo S, Vigilant L. 2006. The complex evolutionary history of gorillas: Insights from genomic data. *Mol Biol Evol* 24:146–158.

Thalmann O, Wegmann D, Spitzner M, Arandjelovic M, Guschanski K, Leuenberger C, Bergl RA, Vigilant L. 2011. Historical sampling reveals dramatic demographic changes in western gorilla populations. *BMC Evol Biol* 11:85. doi:10.1186/1471-2148-11-85.

Tranquilli S, Abedi-Lartey M, Amsini F, Arranz L, Asamoah A, Babafemi O, Barakabuye N, Campbell G, Chancellor R, Davenport TRB, Dunn A, Dupain J, Ellis C, Etoga G, Furuichi T, Gatti S, Ghiurghi A, Greengrass E, Hashimoto C, Hart J, Herbinger I, Hicks TC, Holbech LH, Huijbregts B, Imong I, Kumpel N, Maisels F, Marshall P, Nixon S, Normand E, Nziguyimpa L, Nzooh-Dogmo Z, Okon DT, Plumptre A, Rundus A, Sunderland-Groves J, Todd A, Warren Y, Mundry R, Boesch C, Kuehl H. 2011. Lack of conservation effort rapidly increases African great ape extinction risk. *Conserv Lett*. doi: 10.1111/j.1755-263X.2011.00211.x.

Vigilant L, Guschanski K. 2009. Using genetics to understand the dynamics of wild primate populations. *Primates* 50:105–120.

Watts DP. 2000. Causes and consequences of variation in male mountain gorilla life histories and group membership. In: Kappeler PM, editor. *Primate males*. Cambridge: Cambridge University Press. pp. 169–179.

Yamagiwa J, Kahekwa J, Basabose AK. 2003. Intra-specific variation in social organization of gorillas: Implications for their social evolution. *Primates* 44:359–369.

18. The warp and weft: Synthesising our taxonomic tapestry

Marc F Oxenham and Alison M Behie

While this volume has been structured around three key themes, reflecting key research interests of Colin's over the years, what has clearly emerged is the enormous influence Colin has had, at a range of levels and degrees of pervasiveness, on the thinking of many people. Our title, *Taxonomic Tapestries*, evokes Colin's legacy in a number of ways with Kelley and Sussman's (2007) paper on academic genealogies being particularly relevant in this context. Kelley and Sussman explore a series of ancestor (academic supervisor) and descent (PhD student) lineages in order to document, for the most part, the academic intellectual history of primatology. Indeed, we can all engage in such genealogical detective work either to explore our own intellectual roots, or simply to locate famous ancestors. Both of us, for instance, can directly (in a unilineal fashion) trace our roots back to Hooton and his descendent (student) Washburn, while our respective ancestries diverge quite markedly after that. Colin only needs to go back a single generation to John Napier to locate his illustrious ancestor and perhaps chief intellectual influence, during his formative years, at least. Such formative periods are relevant to our tapestry of influences as we are all shaped to a greater of lesser degree by our subsequent histories and networks of colleagues and acquaintances. The contributions to this volume are clearly illustrative of this point.

In this final chapter we wish to briefly explore a range of additional ideas and questions that have emerged from contributor's approaches to the three meta-themes (the warp if you like) encompassed by this volume: behaviour and morphology; evolution; and conservation. While you, as readers, may have constructed your own framework upon which to engage with the various contributions presented here, we, the editors, have been particularly struck by the following (various strands forming the weft if you like): (1) ways of looking at the universe (incorporating ideas on taxonomy, speciation and evolution); (2) how one actually does taxonomy; (3) the causes of speciation; and (4) the value (and pitfalls) of the phylogenetic species concept (PSC). The main warp and emergent weft go to form a complex tapestry of tightly woven ideas and themes that illustrate the breadth of research interests and influence of Colin Groves.

Ways of looking at the universe: Species, taxonomy and evolution

A rather disparate collection of approaches to this theme can arguably be said to characterise the entire set of contributions to this volume, and this is despite the quite specific set of themes authors were asked to address. Far from being a problem, the rich quarry of ideas and positions has been served. We have restricted our discussion of contributions to this emergent theme by selecting those that have a historical basis. Ulrich Welsh provides us with a biography of the German polymath Adolf Remane, who published some 300 papers, including a significant number on primate evolution and dental morphology. He presents the idea that Remane's interests resonate with those of Colin and in exploring some of the influences on research and methodological approaches of Remane, a range of prominent historical figures and themes were explored. The main point being that none of us work in an intellectual vacuum and many of the meta-ideas we cherish may in fact represent the repackaging of theories and approaches that have been circulating for centuries or are currently circulating elsewhere by scientists from a different, possibly unconnected, academic lineage.

Juliet Clutton-Brock provides a very useful perspective on how biologists have viewed their universe by reviewing aspects of the intellectual history of the naming of things, or more specifically, the naming of living organisms. Given the number of chapters dealing with the concept of species, and other taxonomic categories, in one way or another, this chapter provides a very solid context for these discussions (see below for Clutton-Brock's contribution to 'doing taxonomy'). The theme of the recycling of meta-ideas implicit in Juliet's work, is an issue explored in some detail, while in a highly specific sense, in Oxenham's chapter on Lamarck.

As noted, in some instances, ideas ascribed to one prominent scholar can be seen in the work of others that were somewhat less fortunate in terms of the success of their career and subsequent academic traction they received: Oxenham deals with the pre-Darwinian evolutionist Lamarck. Much of his chapter discusses what the author sees as Lamarck's actual views on the processes and mechanisms of evolutionary change, an important aspect of which prefigures some views usually ascribed to Darwin. Moreover, Lamarck's understanding of the nature of species is dealt with in some detail as well, as it was fundamental to his particular evolutionary scheme.

Historical precedents aside, Natasha Fijn provides a very important insight into how we conceptualise difference and similarity. Fijn looks at wild and domestic horses in Mongolia from the perspective of both scientific and cultural taxonomies. Colin is connected in as much as he has worked on horse taxonomy in the past (and published on it), and was the author's doctoral supervisor or academic ancestor. Indeed, Fijn states Colin is the world expert 'on the morphology and taxonomy of *Equus przewalskii*, or the takhi'. In trying to see past the domestic/wild dualism, she poses some interesting questions regarding the role of animal behaviour (and human–animal behavioural interactions) in taxonomic or classificatory determinations (see below for Fijn's contribution to 'doing taxonomy').

Perhaps the least intuitively obvious exploration of how taxonomies inform our understanding of biological and cultural processes comes from the pen of archaeologists Peter Hiscock and Chris Clarkson. They review models seeking to explain the processes involved in the production of what are often seen as various types of lithic artefacts in Mousterian (Neanderthal) assemblages. They assess two of these models in the context of a specific assemblage and argue that tool types, for this assemblage at least, are not simply a function of varying levels of flake retouch (curation) over time, but that different forms are arrested stages in a continuum. They suggest the initial lithic blank form plays a significant role in finished flake form or type. They go on to argue that different models may explain particular artefact assemblages and no one approach fits them all. They caution that simply seeing flake end stage morphology as a function of retouch intensity may lead one to misinterpret past (hominid) behaviour to a significant degree. There is no easy pathway to Neanderthal cognitive abilities and behaviour through lithic artefact form. There is an implicit link here to the nature of things, or in the context of this book, the nature of species. The transformational processes Hiscock and Clarkson discuss with reference to final forms (types of lithic artefact) have analogues in the biological world. Their argument that blank form predicates final form will stir reflections by biologists on canalisation and evolution.

How does one actually do taxonomy?

While not a main theme in this volume, one contribution deals with the nuts and bolts of taxonomic methodology and principles in a contribution that details the process of naming a new species of Sulawesi rat after Colin (*Lenomys grovesi*). And what better tribute to Colin?

A further two chapters deal with the operational aspects of taxonomy a little less directly. For instance, Juliet Clutton-Brock directly tackles the issue of defining, or correctly naming, domestic species. This particular approach is quite compatible with Colin's own definition of the Phylogenetic Species Concept. And, yes, we agree the dingoes should have their own name and *Canis dingo* is more than appropriate! Natasha Fijn's chapter (mentioned above under a different thematic heading), read in conjunction with Juliet Clutton-Brock's discussion of domestic species, provides an interesting discourse on the way we see, think about and categorise domestic animals and these two approaches are all the more interesting as they wrote their chapters independent of the content and approach of each other.

Finally, we wish to discuss Kees Rookmaaker's contribution here, even though it is perhaps more at home under the theme (see below) looking at the pros and cons of the PSC. Kees provides us with a history of rhino systematics which forms a solid basis to an understanding of the development of systematics in general. The paucity of recognised rhino species in Linneaus's time, to a slight increase in the time of Cuvier, sees an explosion in recognised forms when otherwise small differences (e.g. horn shape and temperament) were considered sufficient taxonomic signifiers. By the end of the nineteenth century a more considered approach to rhino taxonomy prevailed, a sort of compromise between extreme splitting and lumping as it were. The author then looks at Colin's rather substantive contributions to the taxonomy of rhinos, much in the form of his subscription to the PSC. Kees makes the same link back to the issue of conservation that other chapters have made, with the somewhat dour prediction of the possibility that taxonomy may end up serving the role of a way to study past biodiversity rather than to serve one of its key roles in helping protect current biodiversity. This idea is also alluded to in Angela Meder's chapter (discussed under the theme below) where she addresses the fact that 'while taxonomy is more exciting to study for living rather than extinct animals' the rapid decline of gorillas may not give us that luxury.

Some causes of speciation

A number of contributions explored, either implicitly or explicitly, mechanisms leading to speciation events, or contributing to morphological change or elevated levels of variation. For instance, Peter Andrews and Richard Johnson in reviewing the fossil evidence for gibbon ancestors find it somewhat lacking. They provide a thoroughly useful discussion of problems with actually identifying last common ancestors, which by definition cannot have homologous characters defining their descendent new species. With respect to the speciation theme Peter and Richard explore three scenarios, or potentially (future) testable hypotheses, for the origins of the hylobatids, but stress that however they came into being it was relatively recent (within the last few million years) and likely in the context of some severe environmental stressor.

This idea of the role of environmental stressors is also pursued by one of us, (AB) along with a number of colleagues, where the possible role of severe weather conditions as a variable in speciation events or, minimally, the biogeographic distribution of closely related species of primates (in this instance) is explored. It was found that group size was a key variable in adaptation to severe environments (i.e. frequent severe storms that destroy significant portions of the regular food supply). This, coupled with establishment of energy conservation behaviours by species faced with environmental stochasticity and the resultant stress (e.g. a hurricane wipes out the forest), provides a selective advantage in regions that are at high risk of extreme weather on a somewhat regular basis. Clearly, ecological niches can be characterised by extreme weather, as much as by any other variables.

Angela Meder's chapter focuses on the conservation of gorillas and the fact that due to habitat disturbance and hunting one of the major threats current gorilla populations deal with is small population size. She describes how many of these populations (as small as 10 animals in some instances) are isolated from each other, posing the threat that they may easily die off. However, another risk from this scenario is that they will potentially become further speciated from related species. While Angela notes that recent studies of the Cross River gorilla have shown populations that were once believed to be isolated are in fact exchanging genetic material, it is important to consider that these small, isolated populations could further change gorilla taxonomy.

While not a speciation event per se, David Bulbeck (a former graduate student of Colin's and later colleague) used human tooth size and shape analyses to explore the population history of Indo-Malaysia from the Late Pleistocene through to the modern period. He found a generalised trend in reducing tooth size, but no specific evidence for a reduction in tooth size from the Neolithic into the late Holocene. He favours a migration model, rather than an environmentally based one, whereby newcomers, perhaps originating in Sulawesi, have influenced the dental size and shape history of the region from the late Holocene to modern times.

Another former graduate student, and current colleague of Colin's, Debbie Argue, has also looked at speciation events, this time among a range of Early and Middle Pleistocene hominins. Argue has carried out a cladistic analysis of the Ceprano, Bodo, Daka and Kabwe fossils and presents data in support of placing these specimens into a common species: *H. rhodesiensis*. Colin had previously seen Bodo and Kabwe, at least, as distinct to *H. erectus* and preferred to see them as *H. sapiens heidelbergensis*. However, Debbie agrees with Colin's earlier work that these specimens, which she now sees as belonging to the same species, are quite distinct to *H. erectus*. She argues that the key evolutionary trait shared by members of this new species is a marked expansion in the cranial vault. Whether Colin would agree with these findings or not, he would be the first to applaud the approach which is, after all, in line with his version of the phylogenetic species concept. Standing on the shoulders of giants is perhaps something Debbie is quite proud of in this instance.

Before leaving the emergent theme of speciation, a further past graduate student of Colin's, and now colleague, Michael Westaway (and his associates) stress the importance of Colin's interpretation of the PSC and the resultant definition of more, rather than fewer, species. Minimally, splitting facilitates hypothesis testing as to the validity of such 'split' species, while the opposing approach (lumping) does not. Colin's influence in this regard can be seen in the expansion in the number of recognised hominin species in Pleistocene Southeast Asia, an early 'pre-erectine' species of which may be ancestral to *Homo floresiensis*. In terms of anatomically modern human variation in the region, the authors see microevolutionary change as key, perhaps in the context of environmental and climatic influences and not ruling out the possibility of more prosaic explanations; sexual dimorphism for instance. They see no role for previously held views of genetic inputs from *Homo erectus*, or similar such species. Notwithstanding, the new genetic data indicating Neandertal and Denosovan influences seems quite exciting.

The value (and pitfalls) of the phylogenetic species concept (PSC)

A number of contributors have dealt more or less explicitly with the PSC (with Kees Rookmaaker's chapter, while relevant in this regard, discussed under the theme of 'doing taxonomy' above). The nature of the species concept biologists ascribe to would seem to have important positive and negative ramifications for a range of sub-disciplines, not the least being conservation and parasitic disease control. Robert Attenborough, a long-time colleague of Colin's, explores Colin's version of the Phylogenetic Species Concept, which had led him to split numerous taxonomic units into smaller ones by identifying more species than other operational species concepts might allow. Robert develops an argument for the value of this approach, particularly where seeing more species may be of some practical benefit. He applies this idea to his review of the taxonomy of mosquitoes that have the capacity to infect humans with the malarial parasite and stresses the need to fine tune our taxonomy of these organisms if we are ever to make advances in interrupting the malarial cycle.

John Oates and Nelson Ting draw a detailed and well-argued direct link between the realities of institutionally funded conservation efforts and taxonomy. While they are quick to stress that taxonomy needs to be carried out independently of the direct needs of conservation priorities (in order to preserve the scientific rigour and independence of systematics), they stress that:

> Species are the common primary 'currency' used in biodiversity conservation planning. Regions and ecosystems are often prioritised for conservation action based on measures of species richness and endemism and species judged to be in danger of extinction are usually given special attention … [but] unstable classifications can have implications for conservation priorities and policies.

While the authors try and avoid taking sides in terms of subscription to one major approach (BSC vs PSC) over another, they tend to side with taxonomies that are a result of PSC approaches, in as much as the operational practicality of the PSC appears more objective, even if it does tend to lead to a proliferation of taxonomic units rather than a reduction. Notwithstanding, even subscribing to the PSC has its own problems. For instance, the authors note that with respect to red colobus monkey classification, even those who agree on subscribing to the PSC argue over which traits are important in classifying/differentiating this group. At the end of the day, the authors advocate consistency and consensus (and soon!) with regard to colobus monkey systematics, before a lack of such inadvertently leads to more extinction events (see Kees Rookmaaker's similar lament with respect to rhino taxonomy).

Eric Meijaard and Ben Rawson, both previous students of Colin, note the significant and far reaching impact Colin has had on both pig and gibbon taxonomy. Indeed, they argue that his work in this area has made a positive contribution to the conservation of both groups. In general, the authors address the issue of the effect on conservation efforts of greater or lesser taxonomic differentiation. For some Southeast Asian pigs, at least, better identification of endangered species (recognised using the PSC) has been a boon in terms of conservation success. The same can be argued for several species (recognised by Colin) of gibbon. However, interestingly, a survey of those involved in gibbon conservation efforts suggested that taxonomic classifications, per se, were not a driver, or detractor, with respect to conservation as conservation efforts tend to focus on endangered populations and not species or sub-species. One important point, in the context of the current heated discourse over taxonomic instability, is that the authors do not see a trend whereby conservation funds are diluted in the face of taxonomic inflation.

The main purpose of this edited volume was to pay tribute to the incredibly productive career of Colin Groves in the context of a complex tapestry of bio-evolutionary themes linked by the common glue of taxonomy. We believe we have been successful in this regard and as editors, are honoured to have the privilege of packaging this tribute to a remarkable scholar: Colin Groves.

References

Kelley E, Sussman RW. 2007. Academic genealogy on the history of American Field Primatology. *Am J Phys Anthrop* 132:406-425.

Index